Inherited Neuromuscular Diseases

Volume 652

**ADVANCES IN EXPERIMENTAL MEDICINE AND BIOLOGY**



---

Recent Volumes in this Series

Volume 641
CELLULAR OSCILLATORY MECHANISMS
Edited by Miguel Maroto and Nick Monk

Volume 642
THE SARCOMERE AND SKELETAL MUSCLE DISEASE
Edited by Nigel G. Laing

Volume 643
TAURINE 7
Edited by Junichi Azuma

Volume 644
TROPOMYOSIN
Edited by Peter Gunning

Volume 645
OXYGEN TRANSPORT TO TISSUE XXX
Edited by Per Liss, Peter Hansell, Duane F. Bruley, and David K. Harrison

Volume 646
EARLY NUTRITION PROGRAMMING AND HEALTH OUTCOMES IN LATER LIFE
Edited by Berthold Koletzko, Tamás Desci, Denes Molnár, and Anne De la Hunty

Volume 647
THERAPEUTIC TARGETS OF THE TNF SUPERFAMILY
Edited by Iqbal Grewal

Volume 648
ARTERIAL CHEMORECEPTORS
Edited by Constancio Gonzalez, Colin A. Nurse, and Chris Peers

---

Carmen Espinós · Vicente Felipo · Francesc Palau
Editors

# Inherited Neuromuscular Diseases

## Translation from Pathomechanisms to Therapies

Springer

*Editors*
Dr. Carmen Espinós
CIBERER-BIOBANK
CIBER de Enfermedades Raras
Avda. Campanar, 21
46009 Valencia
Spain
cespinos@ibv.csic.es

Dr. Vicente Felipo
Centro de Investigación
Príncipe Felipe
Lab. Neurobiología
Av. Autopista del Saler, 16
46012 Valencia
Spain
vfelipo@cipf.es

Dr. Francesc Palau
Instituto de Biomedicina, CSIC
CIBER de Enfermedades Raras
C/Jaume Roig, 11
46010 Valencia
Spain
fpalau@ibv.csic.es

ISSN 0065-2598
ISBN 978-90-481-2812-9 e-ISBN 978-90-481-2813-6
DOI 10.1007/978-90-481-2813-6
Springer Dordrecht Heidelberg London New York

Library of Congress Control Number: 2009936205



*Cover design*: Boekhorst Design b.v.

Printed on acid-free paper

Springer is part of Springer Science+Business Media (www.springer.com)

# Preface

This volume contains the text of the presentations delivered at the International Symposium on Rare Diseases "Inherited Neuromuscular Diseases: Translation from Pathomechanisms to Therapies", held in Valencia, Spain, from November 16 to 18, 2008. The symposium represents a part of the continuous efforts on diffusion of science to the society of the Cátedra Santiago Grisolía and the Fundación Ciudad de las Artes y las Ciencias –Comunitat Valenciana. More than 200 international scientists from different countries of Europe, the USA, and Australia attended the meeting. The venue was the Auditorium of the Science Museum Príncipe Felipe.

Ten years ago Alan Emery wrote in the preface of *Neuromuscular Disorders: Clinical and Molecular Genetics* the following comments: "It has been estimated that more than one person in every 3,000 has a serious disabling inherited neuromuscular disorder. The suffering caused by these disorders is considerable, but, until the last decade or so, virtually nothing was known of their pathogenesis. Any rationale approach to treatment was therefore out of the question. However, matters are now changing rapidly. The genes for many of these disorders have been localised and characterised and their gene products identified and studied. The detection of preclinical disease, the identification of heterozygous carriers and prenatal diagnosis are all becoming possible, and, hopefully, effective treatments may no be too far distant." Ten years later, more genes associated with neuromuscular disorders have been reported, confirming the wide genetic heterogeneity of most of diseases of the peripheral nervous system. Thinking genetically has become more important and more compelling. It allows the unequivocal diagnosis of neonatal, pediatric and adult diseases whose etiology has a genetic basis, thus providing a more accurate prediction of natural history and prognosis, and reproductive planning for the family, not only offering genetic counselling and prenatal diagnosis but also preimplantational genetic diagnosis. Moreover, for a number of them molecular and cell pathogenesis is suggesting new molecular targets and, more relevant, novel therapeutic approaches are currently developing to manage and treat these disorders, including new drugs and gene and cellular therapies. We thought that was the moment to share with clinicians, biomedical researchers, colleagues, medical specialists in training and students which is the current status of the knowledge and treatment of neuromuscular disorders.

Thus, the main aim of the Symposium was to join outstanding physicians and scientists from over the world who develop their activities in different areas of the neuromuscular disorders, from pathogenic mechanisms to therapies. Discussion was oriented to talk about the state-of-the-art of inherited neuromuscular diseases as a whole, including muscular dystrophies, mitochondrial disorders, peripheral neuropathies, spinal muscular atrophy, motoneurone disease, ataxias and spastic paraplegias. Participants, both speakers and attendees, had the opportunity to confront and compare pathogenic mechanisms and molecular targets for the different diseases, as a forum for discussion of the rational basis of the new therapeutic approaches.

Current days are exciting for the field of neuromuscular diseases. Knowledge of the epidemiology and natural history of the diseases is improving the perception of these disorders by the general and specialized physicians, the general population and political authorities. In the last fifteen years molecular genetics findings have improve our understanding of the genes and molecules involved in muscle and nerve diseases, and also have provide new tools for diagnosis. These tools allow the clinicians and geneticists to offer new ways for disease prevention such genetic counselling based on specific tests and prenatal and preimplantational genetic diagnoses. More recently new therapeutic approaches are coming into the field and are of major interest for researchers. Cell therapies and gene-addressed therapies such exon-skipping in Duchenne muscular dystrophy are an encouraging example. Treatment of mitochondrial disorders is not easy in many instances but in some cases it is possible as it happens in the recent group of coenzyme $Q_{10}$ deficiencies associated with neurological syndromes for which replacement therapies may be curative or have a good palliative response. Clinical trials are ongoing for Charcot-Marie-Tooth neuropathies and Friedreich ataxia, changing the panorama we had just few years ago.

This book provides, therefore, an update on the knowledge of certain crucial aspects regarding to mechanisms and treatment of disorders of the peripheral nervous system.

We would like to express our gratitude to all participants for their written contributions and the very fruitful discussion.

We also thank the CIBER de Enfermedades Raras (CIBERER), the Instituto de Biomedicina de Valencia, CSIC, the Cátedra Santiago Grisolía and the Fundación Ciudad de las Artes y las Ciencias – Comunitat Valenciana, which provided the personnel and the facilities to organize the Symposium. The Symposium was partially funded by grants from the Ministry of Science and Innovation, Conselleria de Educación de la Generalitat Valenciana, Fundación Premios Rey Jaime I, the CIBERER, Genzyme, S.L., Lilly, S.A and Merck Sharp & Dohme de España, S.A., and TREAT-NMD.

Valencia, Spain — Carmen Espinós
Valencia, Spain — Vicente Felipo
Valencia, Spain — Francesc Palau

# Contents

# Contributors

**J. Arenas** Mitochondrial Diseases Laboratory, Research Center, "12 de Octubre" University Hospital and U723, CIBER de Enfermedades Raras (CIBERER), Instituto de Salud Carlos III, Madrid, Spain

**R. Artuch** Biochemistry Department, Hospital Sant Joan de Déu and CIBER de Enfermedades Raras (CIBERER), ISCIII, Barcelona, Spain

**R. Attali** Department of Human Genetics, Hadassah University Hospital, PO Box 91120, Jerusalem, Israel

**J. Berciano** Service of Neurology, Clinical Neurophysiology and Radiology, University Hospital "Marqués de Valdecilla", University of Cantabria, CIBERNED and IFIMAV, Santander, Spain, jaberciano@humv.es

**A. Blázquez** Mitochondrial Diseases Laboratory, Research Center, "12 de Octubre" University Hospital and U723, CIBER de Enfermedades Raras (CIBERER), Instituto de Salud Carlos III, Madrid, Spain

**M. Brewer** University of Sydney at the ANZAC, Research Institute, Concord Hospital, Sydney NSW 2139, Australia

**K. Bushby** Institute of Human Genetics, International Centre for Life, Newcastle upon Tyne, UK, kate.bushby@ncl.ac.ul

**M. Carreras** CIBER de Enfermedades Raras (CIBERER), ISCIII, and Departamento de Bioquímica, Biología, Molecular y Celular, Universidad de Zaragoza, Zaragoza, Spain

**O. Combarros** Services of Neurology, Clinical Neurophysiology and Radiology, University Hospital "Marqués de Valdecilla", University of Cantabria, CIBERNED and IFIMAV, Santander, Spain

**S. Dimauro** Department of Neurology, Columbia University Medical Center, New York, 10032 USA

**C. Espinós** Laboratory of Genetics and Molecular Medicine, Instituto de Biomedicina de Valencia CSIC and CIBER de Enfermedades Raras (CIBERER), Valencia, Spain

**A. Estela** Laboratory of Genetics and Molecular Medicine, Instituto de Biomedicina de Valencia, CSIC and CIBER de Enfermedades Raras (CIBERER), Valencia, Spain

**E. Gallardo** Services of Neurology, Clinical Neurophysiology and Radiology, University Hospital "Marqués de Valdecilla", University of Cantabria, CIBERNED and IFIMAV, Santander, Spain

**J. Garbern** Department of Neurology, Wayne State University School of Medicine, Detroit, Michigan, USA

**A. García** Services of Neurology, Clinical Neurophysiology and Radiology, University Hospital "Marqués de Valdecilla", University of Cantabria, CIBERNED and IFIMAV, Santander, Spain

**A. Gómez-Durán** CIBER de Enfermedades Raras (CIBERER), ISCIII and Departamento de Bioquímica, Biología Molecular y Celular, Universidad de Zaragoza, Zaragoza, Spain

**P. González-Cabo** Laboratory of Genetics and Molecular Medicine, Instituto de Biomedicina de Valencia, CSIC and CIBER de Enfermedades Raras (CIBERER), Valencia, Spain

**I. Gurt** Department of Human Genetics, Hadassah University Hospital, PO Box 91120, Jerusalem, Israel

**M. D. Herrero-Martín** CIBER de Enfermedades Raras (CIBERER), ISCIII and Departamento de Bioquímica, Biología Molecular y Celular, Universidad de Zaragoza, Zaragoza, Spain

**M. Hirano** Department of Neurology, Columbia University Medical Center, New York, USA

**S. Jackson** Department of Neurology, Technical University, Dresden, Germany

**M. Kennerson** University of Sydney at the ANZAC Research Institute, Concord Hospital, Sydney NSW 2139, Australia

**J. V. Llorens** Departament Genètica, Facultat de Biologia, Universitat de València, Burjassot, Valencia, Spain

**E. López-Gallardo** CIBER de Enfermedades Raras (CIBERER), ISCIII and Departamento de Bioquímica, Biología Molecular y Celular, Universidad de Zaragoza, Zaragoza, Spain

**M. J. López-Pérez** CIBER de Enfermedades Raras (CIBERER), ISCIII and Departamento de Bioquímica, Biología Molecular y Celular, Universidad de Zaragoza, Zaragoza, Spain

**C. Marchesi** IRCCS Foundation, "C. Besta" Neurological Institute, Via Celoria 11, 20133 Milan, Italy

**M. A. Martín** Mitochondrial Diseases Laboratory, Research Center, "12 de Octubre" University Hospital and U723, CIBER de Enfermedades Raras (CIBERER), Instituto de Salud Carlos III, Madrid, Spain

**I. Martínez-Romero** CIBER de Enfermedades Raras (CIBERER), ISCIII and Departamento de Bioquímica, Biología Molecular y Celular, Universidad de Zaragoza, Zaragoza, Spain

**J. Melki** Molecular Neurogenetics Laboratory, Institut National de la Santé et de la Recherche Médicale (INSERM), U-798, CP 5724, 91057, Evry, France and Department of Human Genetics, Hadassah University Hospital, PO Box 91120, Jerusalem, Israel, judith.melki@inserm.fr

**G. Meyer Zu Hörste** Department of Neurology, Heinrich-Heine-University, Düsseldorf, Germany

**M. D. Moltó** Department Genètica, Facultat de Biologia, Universitat de València, and CIBER de Enfermedades Mentales (CIBERSAM), Burjassot, Valencia, Spain

**J. Montoya** CIBER de Enfermedades Raras (CIBERER), ISCIII Instituto Aragonés de Ciencias de la Salud and Departamento de Bioquímica, Biología Molecular y Celular, Universidad de Zaragoza, Zaragoza, Spain, jmontoya@unizar.es

**M. Morán** Mitochondrial Diseases Laboratory, Research Center, "12 de Octubre" University Hospital and U723, CIBER de Enfermedades Raras (CIBERER), Instituto de Salud Carlos III, Madrid, Spain

**C. Navarro** Department of Pathology and Neuropathology, University Hospital of Vigo (Meixoeiro), Pontevedra, Spain, carmen.navarro.fernandez.balbuena@sergas.es

**P. Navas** Centro Andaluz de Biología del Desarrollo, Universidad Pablo de Olavide-CSIC and CIBER de Enfermedades Raras (CIBERER), ISCIII, Sevilla, Spain

**G. Nicholson** University of Sydney at the ANZAC Research Institute, Concord Hospital, Sydney NSW 2139, Australia, garthn@med.usyd.edu.au

**D. Pacheu** CIBER de Enfermedades Raras (CIBERER), ISCIII and Departamento de Bioquímica, Biología Molecular y Celular, Universidad de Zaragoza, Zaragoza, Spain

**F. Palau** Laboratory of Genetics and Molecular Medicine, Instituto de Biomedicina de Valencia, CSIC and CIBER de Enfermedades Raras (CIBERER), Valencia, Spain

**D. Pareyson** IRCCS Foundation, "C. Besta" Neurological Institute, Via Celoria 11, 20133 Milan, Italy, dpareys@istituto-besta.it

**S. I. Pascual Pascual** Servicio de Neuropediatría, Hospital Universitario La Paz, Universidad Autónoma de Madrid, Madrid, Spain

**D. Pla-Martin** Laboratory of Genetics and Molecular Medicine, Instituto de Biomedicina de Valencia, CSIC and CIBER de Enfermedades Raras (CIBERER), Valencia, Spain

**C. Ramón** Services of Neurology, Clinical Neurophysiology and Radiology, University Hospital "Marqués de Valdecilla", University of Cantabria, CIBERNED and IFIMAV, Santander, Spain

**E. Ruiz-Pesini** CIBER de Enfermedades Raras (CIBERER), ISCIII, Fundación ARAID and Departamento de Bioquímica, Biología Molecular y Celular, Universidad de Zaragoza, Zaragoza, Spain

**L. Salviati** Clinical Genetics Unit, Department of Pediatrics, University of Padova, Padova, Italy

**B. San Millán** Department of Pathology and Neuropathology, University Hospital of Vigo (Meixoeiro), Pontevedra, Spain

**M. Sánchez-Piris** Laboratory of Genetics and Molecular Medicine, Instituto de Biomedicina de Valencia, CSIC and CIBER de Enfermedades Raras (CIBERER), Valencia, Spain

**T. Sejerson** Astrid Lindgrens Barnsjukhus, Karolinska University Hospital, Stockholm, Sweden

**M. W. Sereda** Department of Clinical Neurophysiology, University of Göttingen and Department of Neurogenetics, Max-Planck Institute of Experimental Medicine, Göttingen, Germany

**M. Shy** Department of Neurology, Wayne State University School of Medicine, Detroit, Michigan, USA

**A. Spinazzola** Unit of Molecular Neurogenetics, Foundation IRCCS Neurological Institute "C. Besta", Milano, Italy

**S. Teijeira** Department of Pathology and Neuropathology, University Hospital of Vigo (Meixoeiro), Pontevedra, Spain

**E. Tizzano** Genetics and Research Institute, Hospital Sant Pau and CIBER de Enfermedades Raras (CIBERER), Barcelona, Spain

**B. Udd** Neuromuscular Centre, Tampere University Hospital and University of Tampere, and Folkhalsan Institute of Genetics, University of Helsinki, Finland

**C. Ugalde** Mitochondrial Diseases Laboratory, Research Center, "12 de Octubre" University Hospital and U723, CIBER de Enfermedades Raras (CIBERER), Instituto de Salud Carlos III, Madrid, Spain

**J. Vitte** Molecular Neurogenetics Laboratory, Institut National de la Santé et de la Recherche Médicale (INSERM), U-798, CP 5724, 91057 Evry, France

**N. Warwar** Department of Human Genetics, Hadassah University Hospital, PO Box 91120, Jerusalem, Israel

**M. Zeviani** Unit of Molecular Neurogenetics, Foundation IRCCS Neurological Institute "C. Besta", Milano, Italy, zeviani@istituto-besta.it

# Chapter 1
# Pathology and Diagnosis of Muscular Dystrophies

**Carmen Navarro, Susana Teijeira, and Beatriz San Millán**

**Abstract** Limb girdle muscular dystrophies are a heterogeneous group of muscle disorders predominantly affecting the pelvic and shoulder girdles. They have been classified into autosomal recessive and dominant forms, the former being by far the commonest. Genetic analyses have led to the description of 14 recessive forms and 7 dominant. In four of the dominant forms the chromosomal loci have been localised but the gene and the protein have not been yet identified. The genes and proteins in these subgroups are localised over a wide range across the muscle fiber and at the extracellular matrix.

Immunohistochemistry and Western blotting analyses of the proteins involved in the various forms of muscular dystrophies, have permitted a refined pathological approach, necessary to conduct genetic studies and to offer an appropriate genetic counseling. The application of molecular medicine in genetic muscular dystrophies also brings great expectations to the therapeutic management of these patients.

**Keywords:** Dystrophin · Extracellular matrix · Immunohistochemistry · Limb Girdle Muscular Dystrophies · Molecular genetics

## 1.1 Introduction

Over recent decades, major scientific advances have led to a better understanding of the cellular and molecular physiopathology of muscular dystrophies, thus providing the basis for better diagnostic and therapeutic capabilities.

Muscular dystrophies are defined as inherited muscle disorders characterised by muscle wasting and weakness of variable distribution, age of onset, rate of disease progression and severity. They are clinically heterogeneous with few clinical phenotypes being well characterised decades ago. Duchenne muscular dystrophy (DMD) [11, 12] and facioscapulohumeral muscular dystrophy (FSH) [15, 33] were described in detail before the end of the 19th century, followed by

C. Navarro (✉)
Department of Pathology and Neuropathology, University Hospital of Vigo (Meixoeiro), Pontevedra, Spain
e-mail: carmen.navarro.fernandez.balbuena@sergas.es

C. Espinós et al. (eds.), *Inherited Neuromuscular Diseases*, Advances in Experimental Medicine and Biology 652, DOI 10.1007/978-90-481-2813-6_1,

oculopharyngeal muscular dystrophy (OPMD) [45] and distal myopathies (DM) [23]. The allelic form of DMD, Becker muscular dystrophy (BMD) was described in the mid-nineteens [1].

The term "limb girdle muscular dystrophy" (LGMD) was coined by Walton and Nattraas in 1954 [47], to describe muscle primary diseases with less distinctive phenotype, predominantly affecting the shoulder and pelvic girdles. Invariably, a large number of muscle disorders fulfilled these wide and non-specific criteria, and the LGMD group, before the "genetic era," included multiple diseases that needed further characterisation.

The extraordinary discovery in 1987, of the muscle cytoskeleton protein "dystrophin" [28], a few months after the identification of the dystrophin gene by positional cloning [18, 22, 30–32], provided a new and almost endless way for the identification of many other genes and proteins and therefore, accuracy in diagnosis, identification of carriers, genetic counseling and prenatal diagnosis.

Concerning LGMD, a new classification was proposed in 1995 at a workshop held at the European Neuromuscular Centre (ENMC) [8], and revised in successive workshops. Limb Girdle Muscular Dystrophies (LGMD) were classified into two broad groups according to the type of inheritance: type 1 included autosomal dominant diseases, and type 2 those of autosomal recessive inheritance. They were further subdivided into subtypes, with a letter designation in chronological order of gene location [2, 7].

The recessive forms are by far the commonest, accounting for 90% of all LGMD. Fourteen AR- and three AD-LGMD genes have been cloned and their protein products identified. In four other AD-LGMD the chromosomal loci have been localized, but the gene and the protein have not yet been identified [25].

The genes and proteins causing diseases in these two groups are localized over a wide range across the muscle fiber, from the sarcolemma to the nuclear envelope and the extracellular matrix (Table 1.1). Thus, a rational way to understand the pathogenesis of muscular dystrophies, includes the knowledge of gene mutations and their secondary protein deficiency.

Dystrophin, the gene product deficient in Duchenne Muscular Dystrophy (DMD), is a component of the cytoskeleton which links the intracellular actin network to the sarcolemma. Dystrophin associates with several glycoproteins to form the "dystrophin associated protein complex" (DAPC), and the entire structure acts as a physical link between actin cytoskeleton and the extracellular matrix, assuring the integrity of the sarcolemma during contraction and relaxation [10, 13, 16, 17, 48].

Various integrants of the DAPC, four transmembrane proteins, α-, β-, γ- and δ-sarcoglycans (SG), are associated with the autosomal recessive forms of muscular dystrophy called LGMD2D, 2E, 2C and 2F respectively. The β-, γ- and δ-SG co-purify, with β and δ forming a tight link, whereas α SG may be spatially separated. The four LGMD related to SG deficiency are clinically variable regarding age of onset and severity [3, 4]. Alpha-sarcoglycanopathy is the less severe, probably due to its looser relation with the other SG, whereas β- and γ-sarcoplycanopathies (LGMD2E and 2C) present a clinical phenotype comparable to that of DMD. Several authors have reported clinical variability within families and siblings [37].

**Table 1.1** Distribution of muscle proteins accross the muscle fiber, gene location and disease nomenclature

| | |
|---|---|
| **Sarcolemma** | **Sarcoplasma** |
| Dystrophin (Xp21) DMD/DMB | Calpain-3 (15q15) LGMD2A |
| α-SG (17q12) LGMD2D | TRIM32 (9q33) LGMD2H |
| β-SG (4q12) LGMD2E | * **Related to glycosylation** |
| γ-SG (13q12) LGMD2C | Fukutin (9q31) Fukuyama CMD/LGMD2L |
| δ-SG (5q33) LGMD2F | FKRP (19q13) LGMD2I/CMD1C |
| Dysferlin (2p13) LGMD2B/ Miyoshi | POMGnT1 (1p34) Muscle-Eye-Brain CMD/LGMD2M |
| Caveolin-3 (3p25) LGMD1C/RMD/HyperCK | POMT1 (9q34) Walker-Warburg CMD/LGMD2K |
| | POMT2 (14q34) Walker-Warburg CMD/LGMD2N |
| | LARGE (22q12) CMD1D |
| **Extracellular matrix** | **Nucleus** |
| Laminin-alfa-2 (6q22) Merosin Negative CMD | Emerin (Xq28) Emery-Dreifuss MD |
| Collagen VI alpha (21q22)Ullrich/Bethlem CMD | Lamin A/C (1q21) LGMD1B |
| **Sarcomeric** | Emery-Dreifuss MD (AD) |
| Titin (2q32) LGMD2J/TMD | Emery-Dreifuss MD (AR) |
| Myotilin (5q31) LGMD1A | |
| Telethonin (17q12) LGMD2G | |

Other proteins located at the sarcolemma are dysferlin and caveolin-3. Dysferlin is a protein that immunolocalises to the muscle fiber periphery but is not part of the dystrophin-associated proteins complex. It is a member of the ferlin family, with an important role in plasma membrane repair after injury. Mutations in the dysferlin gene produce at least two distinct clinical phenotypes, Miyoshi distal myopathy and LGMD2B [35]. The absence of dysferlin causes early alterations of the plasma membrane in non-necrotic fibers and the presence of rows of small vesicles close to the plasmalemma. Muscle fibers displayed abnormal papillary projections. These lesions, although basically located at the plasma membrane level, clearly differ from those seen in DMD and sarcoglycanopathies.

Caveolin-3, which is known to interact with dysferlin, is associated with the dystrophin glycoprotein complex, but is not a core part of it. It has functions in organizing the T-tubule system during myogenesis as well as regulating numerous signaling pathways. Mutations in caveolin-3 have been demonstrated in a number of clinical conditions and it is referred as LGMD1C. Clinical phenotypes include limb girdle muscle dystrophy, rippling muscle disease, distal myopathy, hypertrophic cardiomyopathy and isolated hyperCKemia [20, 38].

Mutations in *CAPN3*, which encodes the muscle-specific cytosolic protease calpain-3, results in LGMD2A [43]. This autosomal recessive muscular dystrophy was described in the original population of the island of La Reunion, in whom the gene was localized to chromosome 15. Mutations in *CANP3* were subsequently identified in families from such diverse geographical origins as America, Brazil,

Turkey and the Basque Country, in Spain. In America, LGMD2A was described in the Amish community of Indiana.

Two nuclear envelope proteins, emerin and lamin A/C are involved in the X-linked and dominant form of Emery Dreifuss MD respectively [5, 14, 40]. The nuclear lamina provides structural integrity for the inner nuclear membranes and nuclear pore complexes. Mutations in the lamin A/C gene (*LMNA*) are increasingly recognized in patients with muscular dystrophy and in a wide range of clinical phenotypes. Sarcomeric proteins are related with LGMD2J, 2G and 1A among others.

There is a spectrum of disorders caused by mutations in genes encoding several proteins of the extracellular matrix (ECM). Approximately 50% of the patients presenting a congenital form of muscular dystrophy have a deficiency of laminin-α2, due to mutations in the *LAMA2* gene. It is known as "merosin-negative congenital muscular dystrophy", often associated with central nervous system abnormalities. Laminin-α2 expression is absent in muscle and other tissues. Patients with splice site or missense mutations which do not entirely abolish the production of protein, present a variable reduction of laminin-α2 and a milder clinical course [27, 39]. Collagen VIα is another constituent of the ECM. Mutations in *COL6A* have been shown to cause a mild late-onset myopathy with contractures (Bethlem myopathy) and a severe congenital muscular dystrophy (Ullrich congenital muscular dystrophy) [9, 26, 29, 36].

An increasing number of congenital muscular dystrophies have been linked to genes that regulate the glycosylation of α-dystroglycan, which establishes the transmembrane link between laminin-2 and dystrophin. Mutations in the fukutin-related protein gene (*FKRP*) cause a spectrum of disorders ranging from a severe congenital dystrophy, MDC1C, to a relatively mild dystrophy, LGMD2I. Other CMD are the Fukuyama type, Muscle-Eye-Brain CMD and Walker-Warburg CMD [6, 21, 44].

## 1.2 Diagnosis of Muscular Dystrophies

### *1.2.1 Protein Studies*

Until the discovery of dystrophin and other muscle proteins, pathological diagnosis of muscle biopsies was restricted to traditional and histochemical techniques, which provide crucial information regarding the severity and the stage of the disease, the presence of inflammatory processes or mitochondrial abnormalities, and help to characterize some congenital myopathies such as nemaline or central core myopathies. Histochemistry is of great value in the diagnosis of some metabolic disorders such as glycogenoses or lipid metabolism disorders.

Nowadays, the use of monoclonal antibodies against muscle proteins has added an extraordinary tool in pathology for the diagnosis of the different types of muscular dystrophies. In the majority of cases, immunohistochemistry (IHC) has to

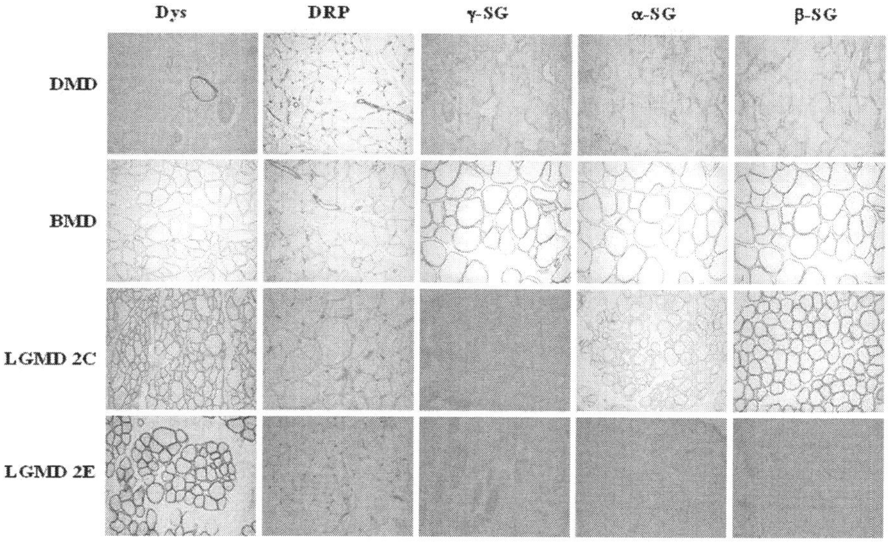


**Fig. 1.1** Immunohistochemical analysis of dystrophin (carboxyl terminal), dystrophin-related protein, γ-sarcoglycan, α-sarcoglycan and β-sarcoglycan in cases of DMD, BMD, LGMD2C and LGMD2E. Note negative labelling for dystrophin in DMD and a revertant positive fiber. Increased expression of DRP and variable sarcolemmal staining for sarcoglycans in DMD and BMD. Dystrophin staining is irregular in BMD and normal in LGMD2C and LGMD2E. Gamma-sarcoglycan labelling is negative in LGMD2C. In a case of LGMD2E with M100K mutation the three sarcoglycans are negative

be performed simultaneously for a wide range of antibodies, in order to allow a correct diagnosis. In certain cases, immunoblotting is complementary. Interpretation of results helps to understand the pathogenesis of the disease [34].

In many muscle disorders, loss of the primary causing-disease protein results in the concomitant reduction of other interacting proteins. This has been well documented in DMD, in which the almost complete absence of dystrophin leads to a considerable reduction of the sarcoglycan proteins, as members of the dystrophin-associated protein complex. In DMD the loss of dystrophin is almost complete, except for the so-called revertant fibers that can appear in about 5% of all the fibers (Fig. 1.1).

In Becker MD changes are more subtle, and dystrophin IHC alone may lead to an uncertain diagnosis. Both, IHC for dystrophin-related protein (DRP), also called utrophin, and Western blotting are crucial to correctly interpret the results [46]. Dystrophin-related protein is a membrane protein with similar characteristics to dystrophin. It appears at the muscle membrane in the foetal period when dystrophin is not yet present, and disappears as soon as dystrophin develops later, in about the 26th week of intrauterine life. In normal adult muscle, DRP is only present around vessels, nerve endings and the neuromuscular junction, but not in the sarcolemma, except in regenerative fibers. When dystrophin is pathologically absent or decreased as in DMD and BMD, utrophin is upregulated and expressed around muscle fibers with negative labeling for dystrophin (Fig. 1.1). This is of great help in the interpretation of muscle biopsies from symptomatic women with Xp21 mutations. The typical mosaic pattern of symptomatic carriers, with patchy loss of dystrophin can be reproduced as an artifact in muscle biopsies when the sample has not been well preserved before cryostat sections are taken. Utrophin immunolabeling is normal in cases of artifact, and positive in the dystrophin-negative fibers in carrier women.

In BMD, the dystrophin band seen by Western blotting is of reduced molecular weight in accordance to the number of deleted exons, and has a diminished intensity related to the reduction in the amount of protein. Exceptionally, molecular weight can be increased as happens when the mutation in Xp21 is a duplication. This occurs in about 5% of all cases of BMD.

In LGMD related to SG gene mutations, giving rise to LGMD2D, 2E, 2C and 2F, the expected absence of the protein product is accompanied with a variable decrease of the three other sarcoglycans from the sarcolemma, which is demonstrated both by IHC and Western blotting. In the case of β-sarcoglycanopathy (LGMD2E), a dramatic loss of all four proteins of the SG complex is generally observed (Fig. 1.1) [42].

Immunoblotting had been considered the ideal technique for the diagnosis of LGMD2A or calpain-3 gene defects. However, interpretation of results is difficult, and normal Western blotting in genetically proven cases has been reported. This would account for the presence of a non-functional protein. LGMD2A should be better diagnosed through the direct analysis and sequencing of the calpain-3 gene [7, 24, 25, 34].

The immunohistochemical approach for the diagnosis of muscular dystrophies provides new strategies in diseases with nuclear envelope protein defects [40]. Mutations in the emerin and lamin A/C genes are responsible for the X-linked and autosomal dominant forms of Emery-Dreifuss muscular dystrophy, respectively. Immunohistochemistry is an extraordinary diagnostic tool in X-linked cases, emerin being negative in any cell nucleus, including white blood cell nuclei. Under the clinical suspicion of mutations in the emerin gene, diagnosis can be easily performed in peripheral blood samples or other types of cells, without the need for a muscle biopsy. In our laboratory, we generally obtain cells from the oral mucosa with excellent results. In affected males all nuclei are negative, whereas a variable proportion (around 50%) of nuclei are immunolabeled in the female carriers, ranging about 50% (Fig. 1.2).

In other situations, when monoclonal specific antibodies are not available, secondary changes in the immunolabeling of other proteins may be the clue to detect the underlying disorder.

As commented earlier, mutations in *FKRP*, a gene that regulates the glycosylation of α-dystroplycan, cause a spectrum of disorders ranging from a severe congenital dystrophy, MDC1C, to a relatively mild dystrophy, LGMD2I [6, 39]. Protein analysis of LGMD2I muscle biopsies demonstrates a reduction in α-dystroglycan immunostaining (Fig. 1.3) and a reduction in its molecular weight by

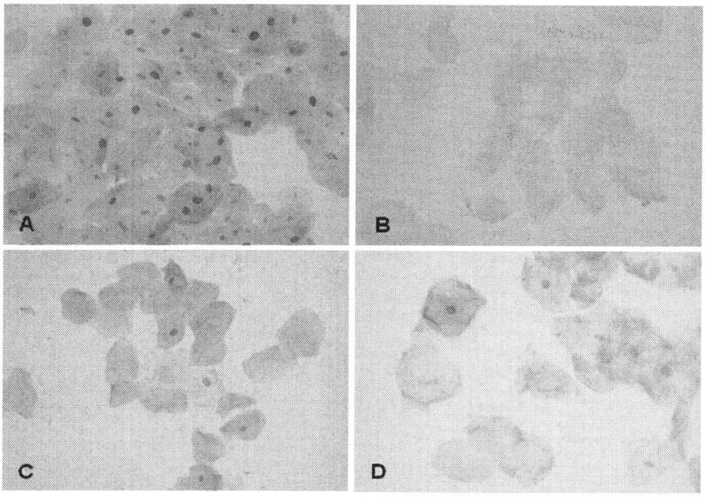


**Fig. 1.2** Immunostaining for emerin of oral mucosa cytology. (**A**) all epithelial cell nuclei are positive in a control. (**B**) all nuclei are negative in an Emery Dreifuss case. (**C**) and (**D**) immunostaining for emerin in two female carriers, with negative and positive nuclei

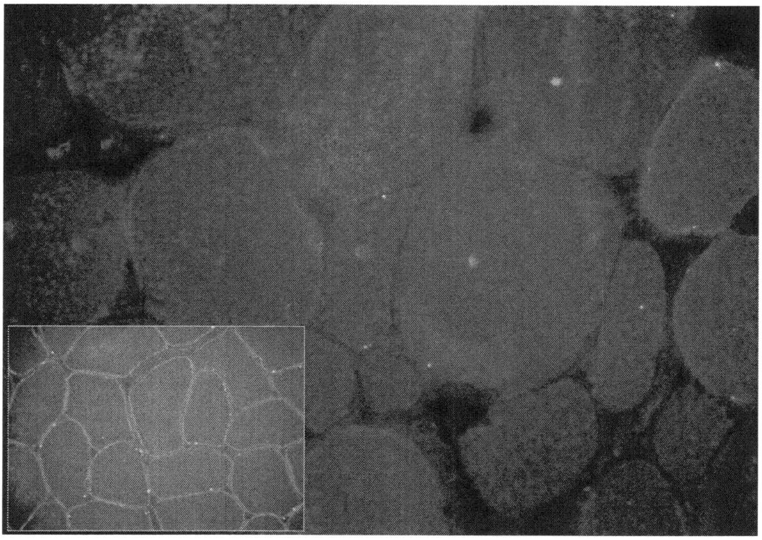

**Fig. 1.3** Negative immunostaining for alpha-dystroglycan in a case of LGMD1C, with a homozygous mutation L276I in the *FKRP* gene. Inset: control for alpha-dystroglycan

Western blotting. Secondary protein abnormalities are frequent in this group, the most common being a reduction in laminin- α2 immunolabeling. Other genes that regulate α-dystroglycan glycosylation are *LARGE*, *POMT1*, *POMT2*, *POMGnT1*, *fukutin* and *FCMD*.

### *1.2.2 Molecular Genetics*

Molecular genetic analysis is the golden standard of diagnosis, and is necessary to detect carriers in the families of affected patients and to offer appropriate genetic counseling.

Precise IHC diagnosis is the first step to guide the mutational search by genetic analysis. In some forms of muscle dystrophies, genetic analysis can be easily performed if there is a known common mutation. This is the case of LGMD2I with the common mutation C826A in the *FKRP* gene, in the Northern Europe population. Interestingly, the second LGMD gene among the Manitoba Hutterite population is known to be *FKRP* with all patients being homozygous for the same C826A mutation seen in European patients [19].

Some ethnic groups have a common ancestral mutation for certain diseases, as happens in the Gypsy population with LGMD2C and the founder mutation C283Y in the $\gamma$-sarcoglycan gene [41]. In European and America populations the commonest mutation for this muscular dystrophy is del521T. Genetic search in other MD could be laborious and expensive, due to the length of the gene, and the absence of hot spots, as in LGMD2A or 2B (calpain and dysferlin genes).

In conclusion, the wide field of muscular dystrophies is increasingly complex at the level of genetics and physiopathology. The ability to detect the genetic abnormality has transformed the diagnostic approach to these disorders. Reduction of cases by genetic counseling and prenatal diagnosis is now possible in most muscular dystrophies.

Moreover, the application of molecular medicine in genetic muscular dystrophies brings great expectations to the therapeutic management of these patients and their families.

## References

1. Becker PE, Kiener F. Eine neue X-chromosomale muskeldystrophie. Arch Psychiat-Z Neurol 1955; 193: 427–448.
2. Beckmann JS, Brown RH, Muntoni F, Urtizberea A, Bonnemann C, Bushby KM. 66th/67th ENMC sponsored international workshop: The limb-girdle muscular dystrophies, 26-28 March 1999, Naarden, The Netherlands. Neuromuscul Disord 1999; 9: 436–445.
3. Ben Hamida M, Fardeau M, Attia N. Severe childhood muscular dystrophy affecting both sexes and frequent in Tunisia. Muscle Nerve 1983; 6: 469–480.
4. Ben Hamida M, Marrakchi D. Dystrophie musculaire progressive de type Duchenne en Tunisie. A propos de 13 familles et 31 cas d'une forme en apparence récessive autosomique. J Génét Hum 1980; 28: 1–9.
5. Bonne G, Mercuri E, Muchir A, Urtizberea A, Bécane HM, Recan D, Merlini L, Wehnert M, Boor R, Reuner U, Vorgerd M, Wicklein EM, Eymard B, Duboc D, Penisson-Besnier I, Cuisset JM, Ferrer X, Desguerre I, Lacombe D, Bushby K, Pollitt C, Toniolo D, Fardeau M, Schwartz K, Muntoni F. Clinical and molecular genetic spectrum of autosomal dominant Emery-Dreifuss muscular dystrophy due to mutations of the lamin A/C gene. Ann Neurol 2000; 48: 170–180.
6. Brockington M, Blake DJ, Prandini P, Brown SC, Torelli S, Benson MA, Ponting CP, Estournet B, Romero NB, Mercuri E, Voit T, Sewry CA, Guicheney P, Muntoni F. Mutations

in the fukutin-related protein gene (FKRP) cause a form of congenital muscular dystrophy with secondary laminin alpha2 deficiency and abnormal glycosylation of alpha-dystroglycan. Am J Hum Genet 2001; 69: 1198–1209.
7. Bushby KMD. Making sense of the limb-girdle muscular dystrophies. Brain 1999; 122: 1403–1420.
8. Bushby KMD, Beckmann JS. The limb-girdle muscular dystrophies – Proposal for a new nomenclature. 30th and 31st ENMC International Workshops, Naarden, The Netherlands, held 6-8 January 1995. Neuromuscul Disord 1995; 5: 337–343.
9. Camacho Vanegas O, Bertini E, Zhang RZ, Petrini S, Minosse C, Sabatelli P, Giusti B, Chu ML, Pepe G. Ullrich scleroatonic muscular dystrophy is caused by recessive mutations in collagen type VI. Proc Natl Acad Sci U S A 2001; 98: 7516–7521.
10. Campbell KP, Kahl SD. Association of dystrophin and an integral membrane glycoprotein. Nature 1989; 338: 259–262.
11. Duchenne de Boulogne GBA. De l'électrisation localisée et son application à la pathologie et à la thérapeutique. 2a ed. Paris, JB Baillière & Fils, 1861.
12. Duchenne de Boulogne GBA. Recherches sur la paralysie musculaire pseudo-hypertrophique, ou paralysie myo-sclérosique. Arch Gen Med 1868; 11: 5–25, 179–209, 305–321, 421–443, 552–588.
13. Ehmsen J, Poon E, Davies K. The dystrophin-associated protein complex. J Cell Sci 2002; 115: 2801–2803.
14. Emery AEH, Dreifuss FE. Unusual type of benign X-linked muscular dystrophy. J Neurol Neurosurg Psychiat 1966; 29: 338–342.
15. Erb W. Muskelbefund bei der juvenilen form der Dystrophia muscularis progressiva. Neurol Zbl 1986; 13: 286–295.
16. Ervasti JM, Campbell KP. Membrane organization of the dystrophin-glycoprotein complex. Cell 1991; 66: 1121–1131.
17. Ervasti JM, Ohlendieck K, Kahl SD, et al. Deficiency of a glycoprotein component of the dystrophin complex in dystrophic muscle. Nature 1990; 345: 315–319.
18. Francke U, Ochs HD, de Martinville B, et al. Minor Xp21 chromosome deletion in a male associated with expression of Duchenne muscular dystrophy, chronic granulomatous disease, retinitis pigmentosa and McLeod syndrome. Am J Hum Genet 1985; 37: 250–267.
19. Frosk P, Del Bigio MR, Wrogemann K, Greenberg CR. Hutterite brothers both affected with two forms of limb girdle muscular dystrophy: LGMD2H and LGMD2I. Eur J Hum Genet 2005; 13: 978–982.
20. Fulizio L, Nascimbeni AC, Fanin M, Piluso G, Politano L, Nigro V, Angelini C. Molecular and muscle pathology in a series of caveolinopathy patients. Hum Mutat 2005; 25: 82–89.
21. Godfrey C, Clement E, Mein R, Brockington M, Smith J, Talim B, Straub V, Robb S, Quinlivan R, Feng L, Jimenez-Mallebrera C, Mercuri E, Manzur AY, Kinali M, Torelli S, Brown SC, Sewry CA, Bushby K, Topaloglu H, North K, Abbs S, Muntoni F. Refining genotype phenotype correlations in muscular dystrophies with defective glycosylation of dystroglycan. Brain 2007; 130: 2725–2735.
22. Goodfellow PN, Davies KE, Ropers HH. Report of the committee on the genetic constitution of the X and Y chromosomes. Human Gene Mapping 8. Cytogenet Cell Genet 1985; 40: 296–352.
23. Gowers WR. A lecture on myopathy and a distal form. Br Med J 1902; 2: 89.
24. Guglieri M, Magri F, D'Angelo MG, Prelle A, Morandi L, Rodolico C, Cagliani R, Mora M, Fortunato F, Bordoni A, Del Bo R, Ghezzi S, Pagliarani S, Lucchiari S, Salani S, Zecca C, Lamperti C, Ronchi D, Aguennouz M, Ciscato P, Di Blasi C, Ruggieri A, Moroni I, Turconi A, Toscano A, Moggio M, Bresolin N, Comi GP. Clinical, molecular, and protein correlations in a large sample of genetically diagnosed Italian limb girdle muscular dystrophy patients. Hum Mutat 2008; 29: 258–266.
25. Guglieri M, Straub V, Bushby K, Lochmüller H. Limb-girdle muscular dystrophies. Curr Opin Neurol 2008; 21: 576–584.

26. Hicks D, Lampe AK, Barresi R, Charlton R, Fiorillo C, Bonnemann CG, Hudson J, Sutton R, Lochmüller H, Straub V, Bushby K. A refined diagnostic algorithm for Bethlem myopathy. Neurology 2008; 70: 1192–1199.
27. Hillaire D, Leclerc A, Fauré S, Topaloglu H, Chiannilkulchaï N, Guicheney P, Grinas L, Legos P, Philpot J, Evangelista T, et al. Localization of merosin-negative congenital muscular dystrophy to chromosome 6q2 by homozygosity mapping. Hum Mol Genet 1994; 3: 1657–1661.
28. Hoffman EP, Brown RH, Kunkel LM. Dystrophin: the protein product of the Duchenne muscular dystrophy locus. Cell 1987; 51: 919–928.
29. Jöbsis GJ, Keizers H, Vreijling JP, de Visser M, Speer MC, Wolterman RA, Baas F, Bolhuis PA. Type VI collagen mutations in Bethlem myopathy, an autosomal dominant myopathy with contractures. Nat Genet 1996; 14: 113–115.
30. Koenig M, Hoffman EP, Bertelson CJ, et al. Complete cloning of the Duchenne muscular dystrophy (DMD) cDNA and preliminary genomic organization of the DMD gene in normal and affected individuals. Cell 1987; 50: 509–517.
31. Kunkel LM and co-authors. Analysis of deletions in DNA from patients with Becker and Duchenne muscular dystrophy. Nature 1986; 322: 73–77.
32. Kunkel LM, Monaco AP, Middlesworth W, et al. Specific cloning of DNA fragments absent from the DNA of a male patient with a X-chromosome deletion. Proc Natl Acad Sci USA 1985; 82: 4778–4782.
33. Landouzy L, Dejerine J. De la myopathie atrophique progressive (myopathie héréditaire, débutant dans l'enfance par la face, sans altération du système nerveux). CR Acad Sci 1884; 98: 53–55.
34. Laval SH, Bushby KMD. Limb-girdle muscular dystrophies – from genetics to molecular pathology. Neuropathol Appl Neurobiol 2004; 30: 91–105.
35. Liu J, Aoki M, Illa I, Wu C, Fardeau M, Angelini C, Serrano C, Urtizberea JA, Hentati F, Hamida MB, Bohlega S, Culper EJ, Amato AA, Bossie K, Oeltjen J, Bejaoui K, McKenna-Yasek D, Hosler BA, Schurr E, Arahata K, de Jong PJ, Brown RH Jr. Dysferlin, a novel skeletal muscle gene, is mutated in Miyoshi myopathy and limb girdle muscular dystrophy. Nat Genet 1998; 20: 31–36.
36. Mercuri E, Yuva Y, Brown SC, Brockington M, Kinali M, Jungbluth H, Feng L, Sewry CA, Muntoni F. Collagen VI involvement in Ullrich syndrome: a clinical, genetic, and immunohistochemical study. Neurology 2002; 58: 1354–1359.
37. Merlini L, Kaplan J-C, Navarro C, Barois A, Bonneau D, Brasa J, Echenne B, Gallano P, Jarre L, Jeanpierre M, Kalaydjieva L, Leturcq F, Levi-Gomes A, Toutain A, Tournev I, Urtizberea A, Vallat J-M, Voit T, Warter J-M. Homogeneous phenotype of the gypsy limb-girdle MD with the gamma-sarcoglycan C283Y mutation. Neurology 2000; 54: 1075–1079.
38. Minetti C, Sotgia F, Bruno C, Scartezzini P, Broda P, Bado M, Masetti E, Mazzocco M, Egeo A, Donati MA, Volonte D, Galbiati F, Cordone G, Bricarelli FD, Lisanti MP, Zara F. Mutations in the caveolin-3 gene cause autosomal dominant limb-girdle muscular dystrophy. Nat Genet 1998; 18: 365–368.
39. Muntoni F, Valero de Bernabe B, Bittner R, Blake D, van Bokhoven H, Brockington M, Brown S, Bushby K, Campbell KP, Fiszman M, Gruenewald S, Merlini L, Quijano-Roy S, Romero N, Sabatelli P, Sewry CA, Straub V, Talim B, Topaloglu H, Voit T, Yurchenco PD, Urtizberea JA, Wewer UM, Guicheney P. 114th ENMC International Workshop on Congenital Muscular Dystrophy (CMD) 17-19 January 2003, Naarden, The Netherlands: (8th Workshop of the International Consortium on CMD; 3rd Workshop of the MYO-CLUSTER project GENRE). Neuromuscul Disord 2003; 13: 579–588.
40. Nagano A, Koga R, Ogawa M, Kurano Y, Kawada J, Okada R, Hayashi YK, Tsukahara T, Arahata K. Emerin deficiency at the nuclear membrane in patients with Emery–Dreifuss muscular dystrophy. Nat Genet 1996; 12: 254–259.
41. Piccolo F, Jeanpierre M, Leturcq F, Dodé C, Azibi K, Toutain A, Merlini L, Jarre L, Navarro C, Krishnamoorthy R, Tomé FM, Urtizberea JA, Beckmann JS, Campbell KP, Kaplan JC.

A founder mutation in the gamma-sarcoglycan gene of gypsies possibly predating their migration out of India. Hum Mol Genet 1996; 5: 2019–2022.
42. Rivas E, Teijeira S, Dos Santos MR, Porrit I, Lecturcq F, Fernández JM, Navarro C. Beta-sarcoglycanophaty (LGMD2E) in a Spanish family. Acta Myologica*** 2004; XXIII: 159–162.
43. Sáenz A, Leturcq F, Cobo AM, Poza JJ, Ferrer X, Otaegui D, Camaño P, Urtasun M, Vílchez J, Gutiérrez-Rivas E, Emparanza J, Merlini L, Paisán C, Goicoechea M, Blázquez L, Eymard B, Lochmuller H, Walter M, Bonnemann C, Figarella-Branger D, Kaplan JC, Urtizberea JA, Martí-Massó JF, López de Munain A. LGMD2A: genotype-phenotype correlations based on a large mutational survey on the calpain 3 gene. Brain 2005; 128: 732–742.
44. Sveen ML, Schwartz M, Vissing J. High prevalence and phenotype-genotype correlations of limb girdle muscular dystrophy type 2I in Denmark. Ann Neurol 2006; 59: 808–815.
45. Taylor, E. W. Progressive vagus-glossopharyngeal paralysis with ptosis: contribution to group of family diseases. J Nerv Ment Dis 1915; 42: 129–139.
46. Teijeira S, Teijeiro A, Fernández R, Navarro C. Subsarcolemmal expression of utrophin in neuromuscular disorders: an immunohistochemical study of 80 cases. Acta Neuropathol 1998; 96: 481–486.
47. Walton JN, Nattrass FJ. On the classification, natural history and treatment of the myopathies. Brain 1954; 77: 169–231.
48. Yoshida M, Ozawa E. Glycoprotein complex anchoring dystrophin to sarcolemma. J Biochem 1990; 108: 748–752.

# Chapter 2
# Standards of Care for Duchenne Muscular Dystrophy: Brief Treat-NMD Recommendations

**Thomas Sejerson and Kate Bushby on Behalf of the TREAT-NMD EU Network of Excellence**

**Abstract** Care for patients with Duchenne muscular dystrophy (DMD) is poorly standardised. There are many interventions in different systems which are known to improve outcomes in DMD but these are not uniformly applied. This leads to inequality in access to treatment, as well as problems for planning controlled trials of future therapeutics. A worldwide effort is underway to generate care guidelines for DMD, which involves the Centre for Disease Control in the USA and the TREAT-NMD network of excellence for neuromuscular diseases in Europe. In advance of the full consensus document, TREAT-NMD has worked on the generation of brief standards of care for DMD, which are presented here and are available via the TREAT-NMD website (www.treat-nmd.eu). Guidelines are presented for diagnostics, neurological follow up, gastrointestinal and nutritional issues, respiratory and cardiac care as well as orthopaedics, rehabilitation, psychosocial interventions and oral care.

**Keywords:** Duchenne muscular dystrophy · Management · Corticosteroids · Cardiomyopathy · Respiratory support · Rehabilitation · Oral care

## 2.1 Introduction

The present document is a brief TREAT-NMD summary of standards of care (SOC) recommendations for diagnosis and management of Duchenne muscular dystrophy (DMD).

A first draft, based where possible on already available and published guidelines (see Key references), was revised following discussions by expert groups for each of the following areas: Diagnosis, Neurology, GI-Nutrition, Respiratory Care, Cardiac Care, Orthopaedics, Psychosocial, Rehabilitation, and Oral Care.

---

K. Bushby (✉)
Institute of Human Genetics, International Centre for Life, Central Parkway, Newcastle upon Tyne NE1 3BZ, UK
e-mail: kate.bushby@ncl.ac.ul

C. Espinós et al. (eds.), *Inherited Neuromuscular Diseases*, Advances in Experimental Medicine and Biology 652, DOI 10.1007/978-90-481-2813-6_2,

The aim of these brief SOC recommendations for DMD is to achieve the rapid dissemination of existing knowledge in this area while awaiting the more detailed recommendations presently being drawn up by the US Centre for Disease Control in collaboration with TREAT-NMD. The present guidelines are to be considered as expert opinion, and are not based on systematic review processes, although an evidence-based approach has guided the work. The feasibility to follow the recommendations will vary considerably between different countries and regions. However, also in regions where it is not presently possible to follow the guidelines, they may act as goals to be aimed for.

The present document, the CDC recommendations, and key references related to SOC for DMD will be available in the SOC section of the TREAT-NMD website (www.treat-nmd.eu/soc).

## 2.2 Diagnosis of DMD

Clinical examination: must include seeing the child try to run, jump, climb stairs and get up from the floor. Common presenting symptoms include abnormal gait with frequent falls, difficulties in rising from the floor, tip-toe walking, and pseudo-hypertrophy of calves. Examination may reveal decreased or lost muscle reflexes, and commonly a positive Gower sign, i.e. requirement to make use of arms to push to erect position from lying by moving hands up the thighs. Many signs of proximal muscle weakness will be detected much more easily in the corridor than the consulting room.

Serum creatine kinase (CK): Massive elevation of the serum CK (at least 10–20 × normal and often much more) is non-specific but always present. The finding of a high CK level should prompt urgent specialist referral for confirmation of the diagnosis. Clinician should be made aware of the association of non-hepatic elevation of AST and ALT in DMD. Unexpected elevation of these enzymes should raise the suspicion of high CK.

Genetic testing should be performed. A deletion of the dystrophin gene will be found in around 70% of cases, a duplication in around 6% and the remaining cases will have a point mutation. Readily available genetic tests for DMD are not always exhaustive and a negative result on initial testing does not exclude the disease. It is very important to understand the tests offered by a particular laboratory and their limitations – further specialist input may be necessary. A laboratory diagnosis should be possible in >95%.

Muscle biopsy: General signs of muscular dystrophy will be seen, including muscle fibre degeneration, muscle regeneration, and increased content of connective tissue and fat. Dystrophin analysis on a muscle biopsy specimen will always be abnormal and offers a route to confirm the diagnosis, complementary to genetic testing. Dystrophin analysis needs to be followed by molecular genetic testing in order to be able to offer genetic counselling to other family members.

An integral part of the diagnostic process is determining the carrier status of the mother by molecular genetic testing. Even if the condition has arisen as a result of a new mutation, there is an average 10% risk of recurrence due to germline mosaicism. Carrier status should also be determined in sisters and aunts (mother's side) in reproductive age if the mother carries the mutation.

Support: around the time of diagnosis it is useful to provide contact with a named member of support staff and imperative to offer details of parent/patient support groups such as the national muscular dystrophy charities and the Parent Project.

## 2.3 Neurology

The use of corticosteroids in DMD:

Timing: experience suggests that the best improvement in performance will be seen with the introduction of medication at or before the point at which the physical performance of the child plateaus (as assessed by sequential functional testing); this is most typically seen around the age of 4–6. Less functional gain may be seen if initiation of steroids is delayed until close to the loss of ambulation.

Regimes: the most common daily dosage regimes are 0.75 mg/kg/day prednisone/prednisolone and 0.9 mg/kg/day deflazacort. They are likely to be equally effective, but have slightly different side-effect profiles. Deflazacort may produce less weight gain but has a higher risk of asymptomatic cataracts. Other regimes suggested to reduce the incidence of steroid-associated side-effects include alternate day dosing, lower dose daily regimes and intermittent regimes (e.g. 10 days on/10 days off; high dose on weekends). It is important to note that none of these regimes have been tested against the daily dosing schedules so that their relative efficacy in the long term is not known.

Tests before starting steroids: immunity to chicken pox (and in high risk populations, tuberculosis) should be ensured ahead of starting steroids. Boosters of vaccinations should be up to date, and consider giving the 6 year booster early if needed.

Efficacy: monitoring for efficacy should include tests of muscle function and strength (e.g. timed function tests, Hammersmith motor ability score, MRC muscle strength score), FVC and parent and child perception of the value of the treatment.

Side effects: monitoring and prophylaxis of the predictable side-effects of steroid use should go hand in hand (http://enmc.org/workshop/?id=21&mid=88). Major side effects to consider are behavioural changes, failure to gain height, excessive weight gain, osteoporosis, impaired glucose tolerance, immune/adrenal suppression, dyspepsia/peptic ulceration, cataract, and skin changes. It is for this reason important to monitor weight, height, blood pressure, urinary dipstix (glucose), cushingoid features, mood/behaviour/personality/GI skin changes, red reflex of eyes, bone fractures, and recurrent infections.

Many side effects can be dealt with without dose reduction or withdrawal of the corticosteroid. Monitoring for weight gain should be accompanied by dietary advice and support prior to starting steroids, behavioural changes should be supported by psychological input and advice on behaviour management, advice on bone health should be provided alongside monitoring of fracture frequency. Concomitant treatment with non-steroidal anti-inflammatory agents should be avoided. Abdominal pain/peptic ulcerations can be treated with antacids.

Dose reduction: Despite prophylactic measures above certain events may necessitate dose reduction. These include behavioural changes disrupting family/school life, weight gain 25% or 3 centile increase from baseline, failure to gain height or skin changes (e.g. acne, striae, hirsuitism) unacceptable to child/family, fasting blood glucose >110 mg/dl (>6.1 mmol/l) or blood glucose 2 h after meal >140 mg/dl (7.8 mmol/l), unusually high frequency of infections/unusual organism, persistent GI symptoms (abdominal pain, heartburn, GI bleeding) despite treatment with antacids.

Withdrawal of corticosteroids: Severe side effects must lead to discontinuation of corticosteroid medication. Events that may necessitate this include severe behaviour changes disrupting family/school life, weight gain/failure to gain height or skin changes unacceptable to child/family despite dose reduction, diabetes mellitus defined as fasting blood glucose >126 mg/dl (7.0 mmol/l) or blood glucose 2 h after meal >200 mg/dl (>11.1 mmol/l), or confirmed hypertension (systolic blood pressure increased 15–30 mmHg over 97th centile or diastolic blood pressure increased 10–30 mmHg over 97th centile for height), unusually high frequency of infections/unusual organism on lowered dose corticosteroids, or GI symptoms not satisfactorily controlled by antacids and lowered dosage of corticosteroids.

Tapering: In case corticosteroid medication is decided to be withdrawn, then it should not be stopped suddenly. Suggested tapering of drug dosage is to take ½ the regular corticosteroid dose the first week, ¼ the dose during the second week, 1/8 the dose during the third week and thereafter stop corticosteroid medication.

How long to continue steroid treatment: continuation of steroids beyond the loss of ambulation is common practise in some centres for the possible protective effect on spinal alignment, respiratory and cardiac function. There is, as yet, no evidence for any beneficial effect of starting steroids after the loss of ambulation in steroid naive patients, but some patients may notice an improvement in function and forced vital capacity.

Patient information resources: are available from the European Neuro Muscular centre (ENMC) (www.enmc.org/workshop/?id=21&mid=88), as well as via the national muscular dystrophy charities.

## 2.4 GI-Nutrition

Adequate dietary advice should be offered from a young age, with focus on healthy eating habits that the whole family could benefit from, with specific focus on weight control, adequate Calcium and Vitamin D intake and controlled sodium intake. Particular emphasis should be made on appetite control around time the corticosteroids are started.

The weight in boys without nutritional problems should be measured 1–2 times a year. If there is active concern about overweight or underweight the weight should be measured more frequently. Situations where changes in weight are to be expected should also initiate weight monitoring (e.g. loss of walking ability, before major surgery).

A child's ideal weight is determined by its height and is influenced by loss of lean body mass (such as in Duchenne Muscular Dystrophy). Tracking a child's weight and height on a centile chart gives an indication of excessive weight gain. Body Mass Index (BMI) body weight divided by the square of height (kg/m, centile adjusted for age and sex) is a more reliable measure of adiposity and can also be tracked on a chart. Clinical judgment, taking into account all issues such as emotional, psychosocial and familial aspects will influence the dietary advice.

To prevent excessive weight gain a dietician should be involved at diagnosis, at initiation of steroids and at loss of ambulation. A dietician should also be involved if there is a tendency to underweight.

In case of overweight it is preferable to aim for a weight loss of 0.5 kg per month, or stabilisation of weight in cases when a long term normalisation is preferred.

Problems with under-nutrition are most likely after the young person is using a wheelchair (approximately 12–13 years of age) and may be multifactorial. The first step is to evaluate intake and to optimize if necessary the existing diet with energy and protein. The next step with more severe malnutrition is enteral nutrition during night time.

Nutritional status should always be examined before major surgery and in particular under-nutrition be addressed before surgery. Particularly overweight patients may have respiratory dysfunction during sleep so may require additional sleep evaluation of oxygen saturation prior to surgery.

Especially in steroid treated boys, if the diet is not adequate in supply of calcium and vitamin D, these should be added separately to reach a recommended intake of calcium (4–8 years: 800 mg/d; 9–18 years: 1300 mg/d) and vitamin D (400 IU).

In later stages of the disease there can be difficulty swallowing and where this leads to aspiration and/or under-nutrition, discussion of feeding by tube or percutaneous endoscopic gastrostomy (PEG) is indicated.

## 2.5 Respiratory Care

Respiratory surveillance: serial measurement of forced vital capacity (FVC: absolute values and as predicted for height, arm span, or ulna length) provides an easy way to document the progression of respiratory muscle weakness. Once clinical signs of nocturnal hypoventilation develop or FVC drops to 1.25 l or <40% predicted value, then serial measurement of overnight oxymetry allows the recognition of the development of nocturnal respiratory failure. This can be done easily at home through the use of small portable machines. Symptoms should also be sought for at every clinic attendance.

Surveillance of cough effectiveness: Serial measurement of peak cough flow will enable monitoring of cough effectiveness. Methods of augmenting cough such as assisted coughing, volume recruitment techniques, cough assist machine should be considered when the PCF is below 270 l/min in non-ambulant boys and introduced before PCF is less than 160 l/min.

Prophylaxis of chest infections: once a patient's FVC begins to drop, boys are susceptible to chest infections and should be offered flu, pertussis, and pneumonococcal vaccination.

Management of chest infections: when coughing is ineffective antibiotics should be provided promptly. Chest physiotherapy such as postural drainage and assisted coughing should be taught when coughing is ineffective and may need to be supplemented with cough assist machine or other volume recruitment techniques such as glossopharyngeal breathing.

Management of nocturnal hypoventilation: Ask for symptoms of nocturnal hypoventilation at every visit. Symptomatic nocturnal hypoventilation is an indication for elective non-invasive nocturnal ventilation (NIV). NIV should also be considered if nocturnal oxycapnography/polisomnography show low SaO2 or elevated pCO2.

Extending the ventilation to the diurnal period should be considered if the patient presents elevated pCO2 lowered or SaO2 while awake. More comfort for the patient is achieved with intermittent ventilation with positive pressure, using a mouthpiece.

Structured education of the ventilator user and the attendants and regular follow-up should be an integral part of the treatment, and there should be surveillance of NIV complications, e.g. air leaks, gastric distension, mucosal dryness, facial bone deformation.

Anaesthetic techniques must be tailored to minimise intra and post-operative respiratory and cardiovascular depression and may require invasive monitoring and access to intensive care. Depolarising muscle relaxants should be avoided because of the risk of hyperkalaemia.

## 2.6 Cardiac Care

Surveillance: cardiac investigation (echocardiogram and ECG) is indicated at diagnosis, every 2 years thereafter to age 10 and then annually, or more often, if abnormalities detected. Cardiac investigation should be done at any age prior to general anaesthesia. Cardiac MRI may be useful in patients with limited echocardiographic acoustic window.

Abnormalities of cardiac rhythm should be promptly investigated and treated. Periodic Holter monitoring should be considered for patients with demonstrated cardiac dysfunction.

Prophylaxis: ACE-inhibitors should be started at a sub-clinical deterioration of cardiac function as detected by echocardiography. One long term study suggests that even earlier start of ACE-inhibitor medication prevents later deterioration. Early prophylactic treatment with ACE-inhibitors at a pre-clinical stage is therefore recommended by several centres from 5 to 10 years of age, although there is still no general consensus on this.

Treatment: Dependent on type and stage of cardiomyopathy. Dilated cardiomyopathy is the most common form. ACE-inhibition and beta-blocker both at once or

ACE inhibitors first followed, when indicated, by beta-blockers should be initiated in the presence of progressive abnormalities, with the addition of diuretics and other with onset of heart failure. Anticoagulation therapy should be considered in patients with severe cardiac dysfunction to prevent systemic thromboembolic events.

Arrhythmogenic stage: Ventricular arrhythmias occur in the late stage of dystrophinopathic cardiomyopathy. Isolated premature ventricular beats require no treatment, but it is important to monitor the cardiac status carefully. If major ventricular arrhythmias occur, prolonged treatment with amiodarone (400–1000 mg per day) is suggested while carefully monitoring thyroid function.

Carriers should have cardiac investigation (echocardiogram and ECG) every 5 years or more frequently if abnormalities are found.

## 2.7 Orthopaedics

Splinting: in ambulant children night splints should be provided when there is loss of dorsiflexion at the ankle when there is loss of normal grade of dorsiflexion and before the foot can only achieve plantagrade. Daytime AFOs are not recommended before loss of ambulation.

In non-ambulant children: sitting AFOs are recommended as painful contractures will develop that also impact negatively on posture. Some children will require tenotomies but AFOs are still needed after surgery.

KAFOs can be considered to delay contracture development and prolong ambulation. Standing frames or swivel walkers can delay contracture development in non-ambulant children.

DMD spine deformity presents an onset in the early teens if not long term treated with corticosteroids. A fusion surgery can be recommended as soon as there is clear progression and the Cobb angle passes 25–30°.

## 2.8 Psychosocial

Every family should be offered a home visit at diagnosis to help come to terms with the emotional and practical problems that the knowledge of the diagnosis brings. For example, feelings of loss, guilt, anger, talking to affected child and siblings about the disease. Issues about access for home, school, leisure and barriers to independence.

Social (information, advocacy and advice) and psychological support should be offered at times of changing needs and crises. For example, considering more children, moving/adapting house, loss of ambulation, surgery, cardiac & respiratory problems, starting university/employment, end of life.

Psychological support should be offered to affected children and their families during times of emotional/behavioral problems.

Learning difficulties/autism spectrum disorders should be identified early and recommendations given to parents and educators about managing these difficulties.

## 2.9 Rehabilitation

Annual neurological, respiratory and cardiological assessments should ideally be co-ordinated via a centralised DMD rehabilitation unit.

From the time of diagnosis the boys must be assessed once to twice a year by therapists (physiotherapists and occupational therapists) with special experience in neuromuscular disorders. The interval between the assessments depends on the boy's age, the progression of the disease and his functional ability.

The aim of the assessments is to set up a plan for interventions for the child to optimize his physical, social and intellectual abilities. The plan should ensure that professionals and parents are ahead of the events and prepared in advance on the next stage of disease. Assessments of physical abilities, repeated with fixed intervals, are needed to determine the rate of progression of the disease.

Major aims of the physiotherapist and occupational therapist are to encourage activity and promote function. This includes interventions to delay or reduce complications due to the deterioration of muscle strength, and to give guidelines regarding activities, possibilities, adaptations and adjustments enabling the boys/men to live a socially active life together with family and friends.

Annual home visit/assessments performed by an interdisciplinary team from a special DMD rehabilitation unit is recommended to support the family and the local team of physiotherapists, occupational therapists, social worker and teacher.

Exercise: resisted exercises should not be prescribed as there is no evidence that they are useful but there are concerns that they may accelerate muscle damage. Moderate levels of active exercise particularly in the hydrotherapy pool are recommended. Children taking steroids may acquire additional motor skills such as riding a bike and this encourages independent play and interaction with peers.

Wheelchairs: should be supplied to improve mobility and independence. Tilt-in-space/recline electric wheelchairs with supportive seating should be supplied early to avoid postural contractures and poor sitting posture.

Annual, centralised courses for young and adults with DMD and their families are recommended, arranged by the NMD-association in co-operation with the centralised rehabilitation unit.

## 2.10 Oral Care

Boys with DMD should see a dentist with extended experience and detailed knowledge of the disease, preferably at a centralised or specialist clinic. The dentist's mission should be to strive for high-quality treatment, oral health and wellbeing and to function as a resource for the families and the boy's own dentist in his home community. This dentist should be aware of the specific differences in dental and skeletal development in boys with DMD and collaborate with a well-informed and experienced orthodontist.

Oral and dental care is to be based on prophylactic measures with a view to maintaining good oral and dental hygiene.

Individually adapted assistive devices and technical aids for oral hygiene are of particular importance when the muscular strength of the patient's hands, arms and neck begins to decrease.

**Acknowledgements** TREAT-NMD is funded by the EU via FP6 contract number 036825. We are grateful for the participation of a number of experts in the generation of these care standards. These include:

Giovanni Nigro, Italy; Elizabeth Wraige, UK; Heinz Jungbluth, UK; John Bourke, UK; Thomas Sejersen, Sweden; Virginia Spehrs-Ciaffi, Switzerland; Gunna Sjoberg; Caroline Sewry, UK; Carina Wallgren-Pettersson, Finland; Francesco Muntoni, UK; H Ginjaar, Netherlands; Annie Aloysius, UK; Nathalie Goemans, Belgium; Trak Davies, UK; Adnan Manzur, UK; Birgit Steffensen, Denmark; Chiara Straathof, Netherlands; Corrado Angelini, Italy; Doug Biggar, Canada; Kate Bushby, UK; Rudolf Korinthenberg, Germany; Wendy King, USA; Elizabeth Vroom, Netherlands; Olé Nőrregård, Denmark; Benjamin Alman, Canada; Garaldine Edge, UK; Helena Saraste, Sweden; Corrado Angelini, Italy; Jos Hendriksen, Netherlands; Martin Chianani, UK; Anna-Karin Kroksmark, Sweden; Elaine Scott, UK; Jes Rahbek, Denmark; Michelle Eagle, UK; Anita Simonds, UK; Jonathan Finder, USA.

## Further Reading

1. American Thoracic Society Consensus Conference (Finder JD, Chair). Respiratory care of the patient with Duchenne muscular dystrophy. Am J Crit Care Med 2004; 170:456–465.
2. Bushby K, et al. Report on the 124th ENMC International Workshop. Treatment of Duchenne muscular dystrophy; defining the gold standards of management in the use of corticosteroids. 2–4 April 2004, Naarden, The Netherlands. Neuromuscul Disord 2004; 4:526–534.
3. Bushby K, et al. The multidisciplinary management of Duchenne muscular dystrophy. Curr Paediat 2005; 15:292–300.
4. Duboc D, et al. Perindopril preventive treatment on mortality in Duchenne muscular dystrophy: 10 years' follow-up. Am Heart J 2007; 154:596–602.
5. Section on Cardiology and Cardiac Surgery. Cardiovascular health supervision for individuals affected by Duchenne or Becker muscular dystrophy. Pediatrics 2005; 116:1569–1573.
6. Quinlivan R, et al.. Report of a Muscular Dystrophy Campaign Funded Workshop Birmingham, UK, January 16th 2004. Osteoporosis in Duchenne muscular dystrophy; its prevalence, treatment and prevention. Neuromuscul Disord 2005; 15:72–79.

# Chapter 3
# Genetics and Pathogenesis of Distal Muscular Dystrophies

**Bjarne Udd**

**Abstract** Distal myopathies are distal muscular dystrophies because they are genetic disorders with progressive loss of muscle tissue. The true distal dystrophies not only show a distal onset; they also remain more distal than proximal throughout the course of the disease. Currently almost 20 different entities of distal muscular dystrophies have been genetically determined, compared to just five entities delineated on clinical grounds in the 1980s. Half of the genes underlying these disorders have been associated with distal phenotypes only, whereas the other genes can manifest also with other than distal phenotypes such as proximal, scapuloperoneal or generalized phenotypes. Interestingly, most of the genes causing distal muscular dystrophies code for protein components of the sarcomere, in contrast to the proximal dystrophies in which most of the genes cause defects in sarcolemmal proteins. The reason for why some gene defects predominantly affect distal muscles is not well understood. The fact that the majority of these defects are due to structural and functional components of the sarcomere is intriguing but so far it does not provide further clues for understanding or for therapeutic approaches. The highly selective involvement of muscles in many of the distal dystrophies is even less well understood.

**Keywords:** Distal myopathy · Distal muscular dystrophy · Molecular genetics · Molecular pathogenesis

## 3.1 Introduction

Many distal dystrophies have been clarified regarding the genetic underlying cause due to advances in molecular genetics and immunohistochemical techniques. Specific chromosomal localization is now known for all major disorders, and in

B. Udd (✉)
Neuromuscular Centre, Tampere University Hospital and University of Tampere, and Folkhalsan Institute of Genetics, University of Helsinki, Finland
e-mail: bjarne.udd@netikka.fi

C. Espinós et al. (eds.), *Inherited Neuromuscular Diseases*, Advances in Experimental Medicine and Biology 652, DOI 10.1007/978-90-481-2813-6_3,

some also early understanding of potential molecular pathophysiology is available. The defect gene has, among others, been published for Miyoshi myopathy [1], distal myopathy with rimmed vacuoles (DRMV, hereditary inclusion body myopathy h-IBM, Nonaka myopathy) [2], tibial muscular dystrophy (TMD, Udd myopathy) [3] and early onset distal myopathy (MPD1, Laing myopathy) [4]. A new avenue was opened the last few years when the genes responsible for the pathologically defined category of myofibrillar myopathy were shown to frequently display a distal phenotype [5]. This was first known for desminopathy [6] and alphaB-crystallin [7], and later shown also for myotilinopathy [8] and zaspopathy as with the Markesbery-Griggs late onset distal myopathy [9]. For some of these genes, mutations in the same gene can cause other phenotypes, such as limb girdle muscular dystrophy LGMD2B in dysferlinopathy, LGMD1A in myotilinopathy or scapuloperoneal myopathy in desminopathy. Nonaka myopathy and quadriceps sparing-vacuolar myopathy (h-IBM) are caused by mutations in GNE (UDP-N-acetylglucosamine 2-epimerase/N-acetylmannosamine kinase) and consequently considered the same disease [10]. Thus, future classifications of distal myopathies may be based on the gene protein product or gene mutation rather than on the distribution of weakness. However, the current classification considering also age at onset, pattern of muscle involvement, and mode of inheritance and pathology is clinically useful and facilitates further diagnostic procedures (Table 3.1).

## 3.2 Late Adult Onset Distal Dystrophies

### *3.2.1 Welander Myopathy*

#### 3.2.1.1 Clinical Features

An autosomal dominant disorder so far found only in families of Scandinavian descent. Onset is usually in the fifth–sixth decade with weakness in the distal upper extremities, typically finger and wrist extensors. A later feature is marked atrophy of thenar and intrinsic hand muscles after several years of disease progression. The weakness and atrophy can be moderately asymmetric. Finger and wrist flexor weakness follows and gradually extends to the lower extremities, with weakness in toe and ankle extensors. Proximal weakness is very uncommon. Deep tendon reflexes are preserved except for the ankle reflexes that are lost later in the disease. Patients may complain about cold fingers but gross sensory examination is usually normal. The progression is slow and most patients continue full activities and have normal life span. Cardiomyopathy does not occur. Rare homozygous patients have more severe disease with early age at onset, proximal muscle weakness, and lose their walking ability.

#### 3.2.1.2 Laboratory Findings

Serum creatine kinase (CK) level is normal or mildly elevated. Electromyography (EMG) shows small motor unit potentials with early recruitment. Fibrillations and

**Table 3.1** Classification of distal dystrophies

| *Type* | *Description* | *Heredity* | *Locus/Gene* | *Onset age (yr)* | *Early symptoms* | *CK* | *Pathology* |
|---|---|---|---|---|---|---|---|
| Welander myopathy | Welander 1951 | AD | 2p13 | >40 | hands, finger extensors | 1–3× | Dystrophic, rimmed vacuoles |
| Tibial muscular dystrophy (Udd myopathy) | Udd 1993 | AD | 2q31/titin | >25 | anterior lower leg | 1–4× | Dystrophic, rimmed vacuoles in tibial anterior muscle |
| Nonaka DMRV | Nonaka 1981 | AR | 9p1-q1/GNE | 5–40 | anterior lower leg | 3–4× | Dystrophic, prominent rimmed vacuoles |
| Miyoshi myopathy | Miyoshi 1985 | AR | 2p13/Dysferlin | 15–40 | posterior lower leg, calf | 20–150× | Dystrophic, dysferlin defect |
| Laing myopathy (MPDI) | Laing 1995 | AD | 14q/myosin MYH7 | 3–25 | anterior lower leg | 1–3× | Mild to moderate myopathic/dystrophic |
| Distal nebulin myopathy | Wallgren-Pettersson 2007 | AR | 2q21/Nebulin | 1–20 | Anterior lower leg | 1–3× | Myopathic, group atrophy, No rods on light microscopy |
| **Distal myofibrillar myopathies** | | | | | | | |
| Distal desminopathy | Milhorat and Wolf 1943 | AD | 2q35/Desmin | variable | distal and proximal weakness, cardiomyopathy | 1–4× | Myopathic/dystrophic, rimmed vacuoles, desmin aggregates |
| Desmin-related myopathy | Vicart 1998 | AD/sporad | 11q/αB-crystallin | variable | distal leg and hands, cardiomyopathy | – | Myopathic/dystrophic, rimmed vacuoles,desmin aggregated |

**Table 3.1** (continued)

| *Type* | *Description* | *Heredity* | *Locus/Gene* | *Onset age (yr)* | *Early symptoms* | *CK* | *Pathology* |
|---|---|---|---|---|---|---|---|
| Distal myotilinopathy | Penisson-Besnier 1998 | AD | 5q31/myotilin | 50–60 | Posterior more than anterior distal leg | 1–3× | Dystrophic, rimmed and non-rimmed vacuoles, desmin-myotilin aggregates |
| Distal Zaspopathy | Markesbery and Griggs 1974 | AD | 10q22/ZASP | 40–50 | Clinically anterior but posterior lower leg on muscle imaging | 1–4× | Dystrophic, rimmed and non-rimmed vacuoles, desmin-myotilin aggregates |
| Single families | | | | | | | |
| Adult onset distal myopathy (MPD3) | Mahjneh 2003 | AD | 8p-q and 12q | 30 | Hand thenar and hypothenar | 1–4× | Dystrophic, rimmed vacuoles, eosinophilic bodies |
| Distal myopathy with vocal cord and pharyngeal weakness (MPD2) | Feit 1998 | AD | 5q31/MATR3 | 35–60 | anterior lower leg, dysphonia | 1–8× | Rimmed vacuoles |
| Distal myopathy with pes cavus and areflexia | Servidei 1999 | AD | 19p13 | 15–50 | lower leg, dysphonia, dysphagia | 2–6× | Dystrophic, rimmed vacuoles |
| Juvenile-adult onset distal myopathy | Williams 2005 | AD | All known loci excluded | 16–30 | Hand grip and posterior lower leg | 1–2× | Myopathic |

complex repetitive discharges are frequent in weak muscles. Nerve conduction velocities are normal, but mildly increased vibration and temperature thresholds by quantitative sensory testing may suggest some small-fiber neuropathy. Muscle biopsy in affected muscle shows rimmed vacuolated dystrophic changes: fiber size variability, increase in connective and fat tissue, central nuclei, and split fibers and 15- to 18-nm cytoplasmic and nuclear filaments. By muscle imaging considerable involvement of posterior calf muscles besides fatty degenerative changes in the anterior compartment can be detected, usually not causing any clinical plantar flexion weakness.

#### 3.2.1.3 Molecular Genetics

Welander myopathy was linked to chromosome 2p13 already 10 years ago [11]. The locus is close to but outside of the *dysferlin* gene. All known genes and transcripts in the linked region of interest have been sequenced without mutations found (Udd, personal communication).

### *3.2.2 Tibial Muscular Dystrophy (TMD, UDD Myopathy)*

#### 3.2.2.1 Clinical Features

First symptoms of weakness in ankle dorsiflexion are usually observed after age 35. Visible atrophy of anterior ompartment lower leg muscles comes later. Symptoms and signs may be asymmetric for years and progression is slow. Mild to moderate proximal weakness in lower extremities may occur in the eight decade but usually patients remain ambulant. A typical clinical clue is the sparing of short toe extensors (extensor digitorum brevis) and hand muscles are only exceptionally affected. Some phenotypic heterogeneity is observed in 9% of the patients [12]. In Finland TMD is the most common muscle disease with an estimated prevalence of 20/100,000. TMD in patients with Finnish ancestry has been identified in Sweden, Norway, Germany and Canada. TMD families without Finnish ancestry have been identified in France, Belgium, Spain and Italy.

#### 3.2.2.2 Laboratory Findings

EMG findings include low-amplitude, short-duration motor unit potentials, and in tibial anterior muscle fibrillation potentials and high-frequency and complex repetitive discharges. Fatty degeneration in clinically weak anterior anterior compartment muscles is easily demonstrated by imaging. Later lesions are observed in hamstring muscles and focal lesions may occur in soleus and medial gastrocnemius. Serum CK is normal or mildly elevated.

In the muscle biopsy variation of fiber size, thin atrophic fibers, central nuclei, internal structural changes and endomysial fibrosis is observed in mildly affected muscle. In the tibial anterior also frequently rimmed vacuoles and rare necrotic fibers occur, later ending up with total fatty replacement in the end stage muscle. Both fiber types are involved. Some rimmed vacuoles are acid phosphatase positive while others are ubiquitin positive and they are usually not lined by sarcolemmal membrane proteins. Congo red stains and immunohistochemistry for beta-amyloid and amyloid precursor protein are negative. On electron microscopy the sarcomere structure is largely preserved even in the homozygote LGMD2J mutant patients.

#### 3.2.2.3 Molecular Genetics

Mutations in the giant sarcomeric protein titin underlie TMD disease. All known mutations causing this phenotype are located in the c-terminus of the protein which physically resides in the M-line of the sarcomere. Finnish patients carry one common founder mutation (FINmaj), a complex 11 bp insertion-deletion mutation changing four amino acids without frameshift [3]. Point mutations were found in the same last exon of titin in unrelated French and Belgian TMD families [13, 14]. Truncating mutations in the last and second to last exons have been identified in Spanish and other French families [14b], and recently the first Italian TMD family with a point mutation in the last exon was determined (Udd unpublished data 2008). Sequencing the last three *titin* exons is the diagnostic method of choice in new families.

#### 3.2.2.4 Molecular Pathogenesis

Interestingly the TMD mutations in the last two domains of c-terminal titin cause a completely different LGMD2J phenotype when inherited homozygously [12]. Titin is the third most abundant protein (after myosin and actin) in muscle and constitutes the third filament system of the sarcomere. Titin binds calpain3 in the second last domain and in the N2A-line of I-band titin. Homozygous mutant muscle shows secondary calpain3 deficiency [15]. In primary calpain3 defect, LGMD2A, perturbations of IκBα/NF-κB pathway and apoptotic myonuclei have been observed [16]. Similar changes of apoptotic myonuclei were also found TMD/LGMD2J suggesting in part similar molecular pathways. C-terminal antibodies do not label mutant titin, indicating that the c-terminus may be totally cleaved off by the mutation. New ligands of C-terminal titin are obscurin and obscurin-like protein which are disrupted by the mutations. There are also motifs for signaling and a catalytic kinase domain with a signaling complex in the M-line part of the protein [3].

### *3.2.3 Zaspopathy (Markesbery-Griggs Distal Myopathy)*

#### 3.2.3.1 Clinical Features

This disease has been mainly reported and in families of Central European descent. Ankle weakness usually begins after age 40 years with ankle weakness; finger and wrist extensors may be affected later as the disease progresses towards proximal weakness late in life. Usually the progression is slow but walking ability may be lost after 15 or 20 years of disease duration. Cardiomyopathy with heart block requiring a pacemaker has been described, but facial, bulbar, and respiratory muscles are not affected [17].

#### 3.2.3.2 Laboratory Findings

EMG shows small amplitude motor-unit potentials with early recruitment in affected muscles. CK levels are normal or up to three-fourfold elevated. Earliest muscle changes can be found in posterior compartment of lower leg muscles by MRI imaging, even before subjective symptoms. Later involvement of all lower leg muscles and moderate involvement in proximal leg muscles is obtained. Rimmed vacuoles together with non-rimmed vacuoles and dark structures in trichrome stain compatible with myofibrillar myopathy are the main findings on routine muscle biopsy. Abnormal cytoplasmic aggregation with ectopic dystrophin expression is shown by immunohistochemical stains for desmin, myotilin, and alphaB-crystallin [9].

#### 3.2.3.3 Molecular Genetics

Two recurrent mutations in ZASP (Z-disk alternatively spliced PDZ-domain containing protein, also termed *LDB3* gene) are usually associated with this type of distal dystrophy [18]. In the Markesbery-Griggs family the causative A165V mutation is one of these ancient European founder mutations [9]. The other recurring European ZASP mutation, A147T, causes an identical phenotype.

#### 3.2.3.4 Molecular Pathomechanism

See below 2.5.

### *3.2.4 Distal Myotilinopathy*

#### 3.2.4.1 Clinical Features

Mutations in myotilin were initially described in two families with dominant limb-girdle phenotype (LGMD1A). In later studies myotilin mutations are more frequently identified in patients with late onset distal dystrophy [8]. Symptoms usually begin with loss of ankle dorsiflexion between age 50 and 60 followed by plantar flexion weakness [19]. In other patients weakness and atrophy of calf muscles may be the first sign followed by pains and cramps [20, 21]. At later stages involvement of upper limbs or proximal leg muscles may be moderate or even severe. In patients with distal dystrophy, dysphonia or respiratory defect is infrequent.

#### 3.2.4.2 Laboratory Findings

On EMG, myopathic changes with fibrillations and complex repetitive discharges are obtained in affected muscle groups. By MRI imaging extensive fatty degenerative changes in calf muscles and milder proximal leg muscle involvement is the usual finding. CK levels vary from normal to less than 2-fold. On muscle biopsy frequent large non-rimmed vacuoles, focal cytoplasmic HE-basophilic and trichrome dark material in both fiber types consistent with myofibrillar myopathy is observed. Other features included occasional rimmed vacuoles and fiber splitting. The pathologically defined spheroid body myopathy also proved to be a myotilinopathy [22].Studies on ultrastructure shows autophagic vacuoles and large zones of myofibrillar disorganisation including also 15–18 nm tubular filaments close to the vacuoles. Widening of the dark Z-disk occurred as an early change. Immunohistochemistry confirms irregularly aggregated and mislocated myotilin protein.

#### 3.2.4.3 Molecular Genetics

So far all myotilin mutations are dominant and located within the serine rich second domain of the protein. Frequently the mutations involve a serine or threonine residue. In our series of late onset distal myotilinopathy patients the mutations S60F and S60C are recurring events.

### *3.2.5 Molecular Pathomechanisms in Distal Dystrophies with Myofibrillar Myopathy*

These are characterized pathologically by disintegrated myofibrillar structures and abnormal accumulation of proteins. Two major structural abnormalities are foci of myofibrillar destruction, and lesions composed of compacted and degraded myofibrillar elements (cytoplasmic or spheroid bodies). These lesions include desmin, dystrophin, neural cell adhesion molecule, gelsolin, and β-amyloid precursor protein but do not contain myosin, actin, or α-actinin. Myotilin reactivity in the aggregations has proved to be a very sensitive marker. Primary *desminopathies* very frequently have a pronounced distal phenotype and typical accumulations but there are exceptions [6, 23]. The classic "neurogenic" scapulo-peroneal syndrome of Kaeser, proved to be a desmin mutated disease [24], but desmin accumulation was very weak.

The pathomechanism of protein aggregation is apparently a dominant negative effect of the mutant allele. Overexpression of mutant desmin protein (L385P) in cultured cells formed intracytoplasmic aggregates and the majority of these cells died within 72 h by apoptosis, suggesting a direct toxic effect of the aggregation proned mutant protein [25].

Another myofibrillar myopathy protein, αB-crystallin [7], is a molecular chaperone and believed to interact with desmin in the assembly of intermediary filaments. Recently mutated FilaminC mutation was identified in one myofibrillar myopathy family [26]. About half of the myofibrillar myopathies are explained by these five genes involved in myofibrillar myopathies.

## 3.3 Early Adult-Onset Distal Myopathies

### *3.3.1 Distal Myopathy with Rimmed Vacuoles (DRMV, Nonaka Myopathy)*

#### 3.3.1.1 Clinical Features

The autosomal recessive disease was first described in Japan and has later been reported in many other populations. Usually the onset is in the second or third decade, but both younger and older onset may occur. Weakness of ankle dorsiflexors and toe extensors, causing foot drop and a steppage gait are the first symptoms. Later patients regularly develop proximal weakness, however with relative sparing of quadriceps muscles. The progression can vary from slow to lost ambulation 10–15 years after disease onset. Cranial muscles are not involved. Cardiac arrhythmia with pacemaker implant has been reported but may be coincidental.

#### 3.3.1.2 Laboratory Findings

Small amplitude and brief motor-unit potentials and fibrillation potentials are recorded on EMG. CK level is increased threefold to fourfold. Rimmed vacuolar pathology is the most prominent finding on muscle biopsy together with later dystrophic changes. On ultrastructure 15- to 18-nm filamentous inclusions in the nucleus and cytoplasm in addition to the autophagic vacuoles are observed.

#### 3.3.1.3 Molecular Genetics

The responsible gene was first identified in patients with quadriceps sparing myopathy in Middle East (hereditary inclusion body myopathy, h-IBM) [2], and subsequently confirmed in Nonaka DRMV families [10, 27, 28]. Practically all families from Middle Eastern countries are homozygous for missense mutation (2186T→C, M712T) in the kinase domain of UDP-N-acetylglucosamine 2 epimerase/N-acetyl mannosamine kinase (GNE) [2]. In Japanese patients one mutation is a more frequent founder mutation V572L [27], and families of different ethnic origins (European, North American, and Caribbean) were heterozygous for distinct missense mutations in the kinase and epimerase domains of the GNE [2].

#### 3.3.1.4 Molecular Pathogenesis

GNE is a bifunctional enzyme that catalyzes the first two rate limiting steps in the biosynthesis of N-acetylneuraminic acid or sialic acid. Dominant mutations in the epimerase domain of the gene cause sialuria. None of the h-IBM patients had elevated sialic acid levels. GNE is exclusively shared by vertebrates and bacteria.

There is no GNE ortholog in *Drosophila melanogaster, Caenorhabditis elegans,* or yeast. The two enzymatic activities of GNE are carried out by separate proteins in bacteria. GNE is a ubiquitous molecule, encoded by a single gene. Sialic acid modification of glycoproteins and glycolipids expressed at the cell surface is crucial for their function in many biologic processes, including cell adhesion and signal transduction. Hyposialylation of proteins in affected muscles has been proposed in Nonaka myopathy [29], and successive treatment of a mouse model was recently reported [30].

## *3.3.2 Miyoshi Distal Myopathy*

### 3.3.2.1 Clinical Features

Symptoms in this recessive disorder begin between ages 15 and 25 years with weakness and atrophy of the calf muscles. Patients complain that they cannot walk on their toes or climb stairs and aching discomfort in the calves is common. The anterior compartment muscles are spared initially but later become affected. Proximal muscles in legs and arms are always involved later, and the two phenotypes in dysferlinopathy, Miyoshi myopathy and LGMD2B, will merge into one. About one third of patients are confined to a wheelchair within 10 years of symptom onset. Patients with anterior onset and fairly rapid progression have also been reported [30]. Patients were first reported from Japan and has later been reported in many ethnic populations.

### 3.3.2.2 Laboratory Findings

Serum CK levels range from 20 to 150 times the normal upper limit, and elevated CK levels are often detected before other symptoms or signs. EMG is myopathic with small amplitude motor-unit potentials. The severely affected gastrocnemius muscle biopsy may show "end-stage" disease with widespread fibrosis, fatty replacement, and loss of muscle fibers. The diagnosis can now be confirmed by immunohistochemical technique. In patients with Miyoshi myopathy dysferlin is absent in the plasma membrane, whereas scattered granular staining in the cytoplasm or nuclear membrane may be observed.

### 3.3.2.3 Molecular Genetics

Identical mutations in the *dysferlin* gene on chromosome 2p13 can cause both distal dystrophy and proximal onset LGMD2B [1]. The patterns of muscle involvement may be distinct at onset even if the phenotypes are indistinguishable at a late stage. Families with typical phenotype and families with adult and later onset of symptoms have been identified without *dysferlin* mutations, and in some with exclusion of linkage to the locus 2p [32, 33].

#### 3.3.2.4 Molecular Pathomechanisms

Dysferlin is expressed in many tissues, including heart, skeletal muscle, kidney, stomach, liver, spleen, lung, uterus and, to a lesser extent, brain and spinal cord [34]. Despite its localization to the plasma membrane dysferlin does not interact with dystrophin associated glycoprotein complex. Immunoprecipitation studies revealed that dysferlin interacts with caveolin-3 [35]. Mutations in caveolin-3 normally cause LGMD1C but can also manifest with distal dystrophy [36]. Possible caveolin-3 binding domains have been identified in dysferlin. Structural abnormalities, including subsarcolemmal vacuoles and papillary projections, have been observed [37]. Most studies, however, implicate malfunctions in the membrane repair mechanisms [38].

## 3.4 Early Onset Distal Myopathies

### *3.4.1 Laing Distal Myopathy (MPD1)*

#### 3.4.1.1 Clinical Features

First symptoms in this autosomal dominant disease appear from early childhood to early adulthood with weakness of ankle dorsiflexion, neck flexors and a hanging big toe [39–41]. Later, patients have weakness of finger extensors and shoulder muscles. The progression is very slow and most patients retain independent walking.

#### 3.4.1.2 Laboratory Findings

EMG is myopathic with reduced motor unit potentials. CK levels are normal or mildly elevated. Muscle biopsy findings may be compatible with congenital fiber type disproportion and have been misinterpreted as neurogenic. Rimmed vacuoles are usually not present. Using myosin heavy chain (MyHC) immunohistochemistry fiber types are abnormally distributed in the affected tibial anterior muscle, showing very atrophic and hybrid type1 fibers expressing also fast MyHC [42].

#### 3.4.1.3 Molecular Genetics

The disease responsible gene is *MYH7* encoding slow beta myosin heavy chain protein which is the main myosin isoform in type 1 slow muscle fibers and in the heart. Interestingly, heart is usually not part of the phenotype in Laing distal myopathy. All mutations causing this phenotype are located in the tail region of the MYH7 dimer, whereas mutations in other parts of the protein may cause cardiomyopathy or hyaline body myopathy [4]. Many mutations are recurring hotspot mutations and many are *de novo*.

## 3.4.2 *Distal Nebulin Myopathy*

#### 3.4.2.1 Clinical Features

In patients with a TMD-like phenotype but with early onset and recessive inheritance we have identified nebulin mutations. Extensor muscles of the lower legs, feet and later hands are predominantly affected. The progression is very mild and patients do not have major disability in adult life.

#### 3.4.2.2 Laboratory Findings

EMG shows myopathic or mixed findings. CK is normal or mildly elevated. Similarly to TMD, selective fatty degeneration in the anterior tibial muscles is shown by muscle imaging. Biopsy of a distal muscle may show scattered and grouped atrophic fibers mimicking neurogenic changes. Light microscopy did not show nemaline rods but small rod-like structures may or may not be observed on electron microscopy.

#### 3.4.2.3 Molecular Genetics

Previously recessive mutations in nebulin are well known in severe and classic nemaline myopathy [43]. The reason for the phenotype differences is that nemaline myopathy patients have more disruptive mutations while patients with this milder distal phenotype have two missense mutations in the large gene [44].

# 3.5 Distal Dystrophies with Other Molecular Cause

## 3.5.1 *Distal Myopathy with Vocal Cord and Pharyngeal Weakness (VCPDM, MPD2)*

A separate autosomal dominant distal phenotype with onset between age 35 and 60 of distal upper and lower limb weakness together with vocal cord and pharyngeal weakness was described by Feit et al. [45]. CK levels ranged from normal to eightfold increase. Muscle biopsy showed rimmed vacuolated myopathy. The disease was linked to 5q overlapping the locus for *LGMD1A,* but mutations in myotilin were excluded and mutations in a nuclear gene *matrin3* were recently reported [45b].

## 3.5.2 *Dynaminopathy*

Dominant mutations in dynamin *DNM2* frequently have the clinical presentation with a distal phenotype involving more the posterior calf muscles than the anterior compartment. Onset can range from childhood to late adulthood and progression is usually slow. The morphological hallmark is that of centronuclear myopathy without rimmed vacuolations.

### *3.5.3 VCP Mutated Distal Myopathy*

Autosomal dominant diseases caused by mutations in VCP are known to have multiorgan involvement including rimmed vacuolated muscle pathology, variable Paget disease and late frontotemporal dementia. The clinical muscle phenotype was first described as a proximal one, but later families more often show a scapulo-peroneal presentation [46]. However, mutations in VCP can in some families cause a very pronounced distal phenotype without scapular or proximal involvement (Udd, personal communication).

### *3.5.4 Distal Caveolinopathy*

While mutations in caveolin3 are best known to cause LGMD1C or rippling muscle disease, some mutations may cause a progressive distal dystrophy with atrophies first observed in the hands [36]. The reason for the different phenotypic outcomes with caveolin-3 mutations is not further delineated but the different signaling and regulatory functions of the sarcolemmal caveolae are obviously affected differentially depending on the mutational defects in the protein.

### *3.5.5 19p13 Linked Distal Phenotype*

Italian autosomal dominant families with distal leg anterior and posterior weakness, atrophies, areflexia, pes cavus starting between the second and 6th decades of life and progressing to upper limbs and proximal muscles have been reported [47, 48]. Dysphagia and dysphonia were early signs and muscle biopsy showed rimmed vacuoles with dystrophic changes.

### *3.5.6 Australian Distal Myopathy*

A large Australian family with a dominant distal myopathy showing posterior-lateral calf muscle involvement and sparing of anterior lower leg muscles is a separate entity based on molecular genetic studies exclusion of 12 different distal myopathy loci. CK was only mildly elevated and dysferlin staining was normal [49].

### *3.5.7 Finnish Adult Onset Distal Dystrophy (MPD3)*

A new dominant distal dystrophy has been described in a Finnish family [50]. Weakness of the intrinsic hand muscles or asymmetric weakness in the anterior lower leg muscles started after age 30. Muscle morphology showed frequent rimmed vacuoles with prominent eosinophilic inclusions. Other distal myopathies were excluded by linkage and significant LOD scores >3 were shown for two separate loci on 8p22-q11 and 12q13-q22 [51].

## 3.6 Conclusion

Distal dystrophies were initially reported in Scandinavian and Japanese populations but are now recognized in all populations. Advances in molecular genetics has markedly moved the field forward and increased awareness of these diseases. Dysferlinopathy and caveolinopathy can be diagnosed by immunohistochemistry and/or Western blotting. Characterization of genes underlying the heterogenous group of myofibrillar myopathies and early onset distal myopathies has facilitated molecular diagnosis in many families and sporadic patients. With exact determination of protein mutations and defects the first ideas on the molecular pathophysiology and disease mechanisms have emerged. At the next stage these may define targets for genetic treatment strategies. Since most distal dystrophies are dominant disorders downregulation of the mutant allele by siRNA techniques would probably be one theoretical option.

## References


1. Liu J, Aoki M, Illa I, et al. Dysferlin, a novel skeletal muscle gene, is mutated in Miyoshi myopathy and limb girdle muscular dystrophy. Nat Genet 1998;20:31–36.
2. Eisenberg I, Avidan N, Potikha T, et al. The UDP-N-acetylglucosamine 2-epimerase/N-acetylmannosamine kinase gene is mutated in recessive hereditary inclusion body myopathy. Nat Genet 2001;29:83–87.
3. Hackman P, Vihola A, Haravuori H, et al. Tibial muscular dystrophy is a titinopathy caused by mutations in TTN, the gene encoding the giant skeletal-muscle protein titin. Am J Hum Genet 2002;71:492–500.
4. Meredith C, Herrmann R, Parry C, et al. Mutations in the slow skeletal muscle fiber myosin heavy chain gene (MYH7) cause Laing early-onset distal myopathy (MPD1). Am J Hum Genet. 2004;75:703–708.
5. Selcen D, Ohno K, Engel AG. Myofibrillar myopathy: clinical, morphological and genetic studies in 63 patients. Brain 2004;127:439–451.
6. Sjöberg G, Saavedra-Matiz C, Rosen D, et al. A missense mutation in the desmin rod domain is associated with autosomal dominant distal myopathy, and exerts a dominant negative effect on filament formation. Hum Mol Genet 1999;8:2191–2198.
7. Vicart P, Caron A, Guicheney P, et al. A missense mutation in the alphaB-crystallin chaperone gene causes a desmin-related myopathy. Nat Genet 1998;20:92–5.
8. Selcen D, Engel AG. Mutations in myotilin cause myofibrillar myopathy. Neurology 2004;62:1363–1371.
9. Griggs R, Vihola A, Hackman P, et al. Zaspopathy in a large classic late onset distal myopathy family. Brain 2007;130:1477–1484.
10. Nishino I, Noguchi S, Murayama K, et al. Distal myopathy with rimmed vacuoles is allelic to hereditary inclusion body myopathy. Neurology 2002;59:1689–1693.
11. Ahlberg G, Tell D, Borg K, et al. Genetic linkage of Welander distal myopathy to chromosome 2p13. Ann Neurol 1999;46:399–404.
12. Udd B, Vihola A, Sarparanta J, et al. Titinopathies and extension of the M-line mutation phenotype beyond distal myopathy and LGMD2J. Neurology 2005;64:636–642.
13. de Seze J, Udd B, Haravuori H, et al. The first European tibial muscular dystrophy family outside the Finnish population. Neurology 1998;51:1746–1748.
14. Van den Bergh P, Bouquiaux O, Verellen C, Marchand S, Richard I, Hackman P, Udd B. Tibial muscular dystrophy in a Belgian family. Annal Neurol 2003;54:248–251.

14b. Hackman P, Marchand S, Sarparanta J, Vihola A, Pénisson-Besnier I, Eymard B, Pardal-Fernández J, Hammouda E, Richard I, Illa I and Udd B. Truncating mutations in C-terminal titin may cause more severe tibial muscular dystrophy (TMD). Neuromusc Disord 2008;18:922–8.
15. Haravuori H, Vihola A, Straub V, et al. Secondary calpain3 deficiency in 2q linked muscular dystrophy – titin is the candidate gene. Neurology 2001;56:869–877.
16. Richard I, Broux O, Allamand V, et al. Mutations in the proteolytic enzyme calpain3 cause limb-girdle muscular dystrophy type 2A. Cell 1995;81:27–40.
17. Markesbery WR, Griggs RC, Leach RP, et al. Late onset hereditary distal –myopathy. Neurology 1974;23:127.
18. Selcen D, Engel AG. Mutations in ZASP define a novel form of muscular dystrophy in humans. Ann Neurol 2005;57:269–276.
19. Olive M, Goldfarb LG, Shatunov A, et al. Myotilinopathy: refining the clinical and myopathological phenotype. Brain 2005;128:2315–2326.
20. Penisson-Besnier I, Dumez C, Chateau D, et al. Autosomal dominant late adult onset distal leg myopathy. Neuromuscul Disord 1998;8:459–466.
21. Penisson-Besnier I, Talvinen K, Dumez C, et al. Myotilinopathy in a family with late onset myopathy. Neuromuscul Disord 2006;16:427–431.
22. Foroud T, Pankratz N, Batchman AP, et al. A mutation in myotilin causes spheroid body myopathy. Neurology 2005;65:1936–1940.
23. Milhorat AT, Wolff HG. Studies in diseases of muscle: XIII. Progressive muscular dystrophy of atrophic distal type: Report on a family: Report of autopsy. Arch Neurol Psychiatry 1943;49:655.
24. Walter M, Reichlich P, Hübner A, et al. Identification of a desmin gene mutation in scapuloperoneal syndrome type Kaeser. Neuromusc Disord 2006;16:708–709.
25. Sugawara M, Kato K, Komatsu M, et al. A novel de novo mutation in the desmin gene causes desmin myopathy with toxic aggregates. Neurology 2000;55:986–990.
26. Vorgerd M, van der Ven PF, Bruchertseifer V, et al. A mutation in the dimerization domain of filamin c causes a novel type of autosomal dominant myofibrillar myopathy. Am J Hum Genet 2005;77:297–304.
27. Tomimitsu H, Ishikawa K, Shimizu J, et al. Distal myopathy with rimmed vacuoles: novel mutations in the GNE gene. Neurology 2002;59:451–454.
28. Kayashima T, Matsuo H, Satoh A, et al. Nonaka myopathy is caused by mutations in the UDP-N-acetylglucosamine-2-epimerase/N-acetylmannosamine kinase gene (GNE). J Hum Genet 2002; 47:77–79.
29. Nishino I, Malicdan MC, Murayama K, et al. Molecular pathomechanism of distal myopathy with rimmed vacuoles. Acta Myol 2005; 24:80–83.
30. Malicdan MC, Noguchi S, Hayashi YK, et al. Prophylactic treatment with sialic acid metabolites precludes the development of the myopathic phenotype in the DMRV-hIBM mouse model. Nat Med 2009;15:690–5.
31. Illa I, Serrano-Munuera C, Gallardo E, et al. Distal anterior compartment myopathy: A dysferlin mutation causing a new muscular dystrophy phenotype. Ann Neurol 2001; 49: 130–134.
32. Linssen W, de Visser M, Notermans N, et al. Genetic heterogeneity in Miyoshi type distal muscular dystrophy. Neuromusc Disord 1998;8:317–320.
33. Jaiswal JK, Marlow G, Summerill G, et al. Patients with a non-dysferlin Miyoshi myopathy have a novel membrane repair defect. Traffic 2006 [Epub ahead of print].
34. Anderson L, Davison K, Moss J, et al. Dysferlin is a plasma membrane protein and is expressed early in human development. Hum Mol Genet 1999; 8:855–861.
35. Matsuda C, Hayashi Y, Ogawa M, et al. The sarcolemmal proteins dysferlin and caveolin-3 interact in skeletal muscle. Hum Mol Genet 2001; 10:1761–1766.
36. Tateyama M, Aoki M, Nishino I, et al. Mutation in the caveolin-3 gene causes a peculiar form of distal myopathy. Neurology 2002; 58:323–325.

37. Selcen D, Stilling G, Engel A. The earliest pathologic alterations in dysferlinopathy. Neurology 2001; 56:1472–1481.
38. Bansal D, Campbell KP. Dysferlin and the plasma membrane repair in muscular dystrophy. Trends Cell Biol 2004; 14:206–213.
39. Laing N, Laing B, Meredith C, et al. Autosomal dominant distal myopathy: Linkage to chromosome 14. Am J Hum Genet 1995; 56:422–427.
40. Voit T, Kutz P, Leube B, et al. Autosomal dominant distal myopathy: Further evidence of a chromosome 14 locus. Neuromuscul Disord 2001;11:11–19.
41. Zimprich F, Djamshidian A, Hainfellner J, et al. An autosomal dominant early adult onset distal muscular dystrophy. Muscle Nerve 2000;23:1876–1879.
42. Lamont P, Udd B, Mastaglia F, et al. Laing early-onset distal myopathy – slow myosin defect with variable abnormalities on muscle biopsy. J Neurol Neurosurg Psychiatry 2006; 77: 208–215.
43. Wallgren-Pettersson C, Pelin K, Nowak KJ, et al. Genotype-phenotype correlations in nemaline myopathy caused by mutations in the genes for nebulin and skeletal muscle alpha-actin. Neuromuscul Disord 2004; 14:461–470.
44. Wallgren-Pettersson C, Lehtokari V-L, Kalimo H, et al. Distal myopathy caused by homozygous missense mutations in the nebulin gene. Brain 2007; 130:1465–1476.
45. Feit H, Silbergleit A, Schneider L, et al. Vocal cord and pharyngeal weakness with autosomal dominant distal myopathy: clinical description and gene localization to 5q31. Am J Hum Genet 1998; 63:1732–1742.
45b. Senderek J, Garvey SM, Krieger M, et al. Autosomal-dominant distal myopathy associated with a recurrent missense mutation in the gene encoding the nuclear matrix protein, matrin 3. Am J Hum Genet 2009;84:511–8.
46. Kimonis VE, Mehta SG, Fulchiero EC, et al. Clinical studies in familial VCP myopathy associated with Paget disease of bone and frontotemporal dementia. Am J Med Genet A 2008 15;146:745–757.
47. Servidei S, Capon F, Spinazzola M, et al. A distinctive autosomal dominant vacuolar neuromyopathy linked to 19p13. Neurology 1999; 53:830–837.
48. Di Blasi C, Moghadaszadeh B, Ciano C, et al. Abnormal lysosomal and ubiquitin-proteasome pathways in 19p13.3 distal myopathy. Ann Neurol 2004;56:133–138.
49. Williams DR, Reardon K, Roberts L, et al. A new dominant distal myopathy affecting posterior leg and anterior upper limb muscles. Neurology 2005; 64:1245–54.
50. Mahjneh I, Haravuori H, Paetau A, et al. A distinct phenotype of distal myopathy in a large Finnish family. Neurology 2003; 61:87–92.
51. Haravuori H, Siitonen A, Mahjneh I, et al. Linkage to two separate loci in a family with a novel distal myopathy phenotype (MPD3). Neuromusc Disord 2004; 14:183–187.

# Chapter 4
# Phenotype Variations in Early Onset Pompe Disease: Diagnosis and Treatment Results with Myozyme®

**Samuel Ignacio Pascual Pascual**

**Abstract** Pompe disease is a rare autosomal recessive lysosomal storage disease caused by deficiency of acid-a-glucosidase (GAA). This deficiency results in glycogen accumulation in the lysosomes, leading to lysosomal swelling, cellular damage and organ dysfunction. Patient age at the onset of Pompe disease symptoms and the rate of deterioration can vary considerably.

In early onset patients (the classical infantile form) this glycogen accumulation leads to death usually before the age of 1 year. Some patients with early onset don't develop cardiomiopathy and their progression is slower (atypical infantile form). The late-onset form (juvenile and adult forms) have more slow and variable course. The gene is localized in 17q25. More than 200 different mutations have already been described. Diagnosis has been classically made by mean of muscular biopsy. Nowadays is more convenient the screening of GAA in dried blood sample followed by GAA assessment in lymphocytes or fibroblasts or by the genetic study of mutations.

Besides non specific multiprofesional management, since 2006 there is a specific enzyme replacement therapy (ERT), Myozyme®), which compensates for the missing enzyme by i.v. administration of recombinant produced enzyme.

In classic Pompe disease the reported results improve significantly the survival, the motor development and the cardiac function. The sooner ERT starts, the better are the results.

In late onset Pompe disease ERT has also demonstrated significant improvement in muscular function and quality of life.

**Keywords:** Pompe disease · GSD-II · Glycogen storage disease · Classic infantile Pompe disease · Late-onset Pompe disease · Acid-a-glucosidase · GAA · rhGAA · Myozyme ® · Enzymatic replacement therapy · ERT

---

S.I. Pascual Pascual (✉)
Servicio de Neuropediatría. Hospital Universitario La Paz. Profesor Asociado de Pediatría. Universidad Autónoma de Madrid, Madrid, Spain
e-mail: ipascualp.hulp@salud.madrid.org

C. Espinós et al. (eds.), *Inherited Neuromuscular Diseases*, Advances in Experimental Medicine and Biology 652, DOI 10.1007/978-90-481-2813-6_4,

## 4.1 Introduction

Pompe disease is a rare autosomal recessive lysosomal storage disease caused by deficiency of acid-a-glucosidase (GAA). It has been named also Glycogen Storage Disease type II (GSD-II). This deficiency results in glycogen accumulation in the lysosomes, leading to lysosomal swelling, cellular damage and organ dysfunction. It was described by Pompe in 1932. In 1963 was recognized as a lysosomal disorder after Duve et al. discovered this cellular organelle (1955), when Hers et al. discovered the metabolic defect of acid-alfaglucosidase (GAA).

Although in first decades Pompe disease was used to name only the early onset classic form with a very fast evolution, nowadays this is the name given to all the clinical forms of GSD-II.

## 4.2 Incidence

The global incidence is estimated in 1 out of 40.000 [9]. There are an ethnic or geographic variations, being more frequent among Afro-Americans and in the south of China. The early infantile form has an incidence of 1/138.000 whereas the late onset is more frequent, 1/57.000. In non Caucasian people the global incidence is 1/100.000 [7].

## 4.3 Symtomatology

It has been classified in different phenotypes according to their severity. The earlier the onset, the more severe, faster and worse the evolution is, because the deficit of GAA is more severe. Actually, there is not a clear division into phenotypes but a continuous spectrum from the complete lack of GAA that produces severe signs in early infancy to the slightest disease that begins in adult age and produces only a slight and steady respiratory distress (diaphragm weakness) or limbs weakness.

### *4.3.1 Infantile-Onset Pompe Disease*

This means that their onset is before 1 year of age. It is subdivided in two forms:

*A.* Classic infantile disease, that begins in the first months of age and progress rapidly to the death before 2 years old. Most of the cases have this classic form.
*B.* Non-classic infantile or atypical infantile form, whose evolution is slower.

### *4.3.2 Late-Onset Disease, Whose Symptoms Appear After the First Year of Age*

It can be sub-classified in (1) juvenile or (2) adult form.

In cases of slow course it is difficult to date the onset.

The onset of the symptomatology is related with the grade of enzymatic deficit but also with other factors, either modifiers genes or non-genetic. When the enzymatic activity is below 1% the muscular dysfunction is so severe than give a classic infantile form, while if it is above 10% muscle degenerates more slowly and will give a juvenile or adult form.

Myopathy is presented in all the cases, and it is progressive. Some muscles are more severely involved, as diaphragm and proximal lower limbs and trunk. Classic infantile Pompe disease also gives hypertrophic cardiomyopathy and cardiac conduction defects.

Children with classic Pompe disease have disorders of feeding, sucking or swallowing since birth or the first months of age, resulting in low weight gain and respiratory infections and failure because of diaphragm involvement. Cervical flexor and extensor muscles have severe weakness. Tendon reflexes are absent. More than 50% of patients show macroglosia and hepatomegaly. CK and transaminases are increased. Glycogen is accumulated in striated and smooth muscle, in liver and in other tissues. There is a neuronal accumulation of glycogen, mostly in motor neurons of anterior horn of spinal cord and brainstem. As a consequence of it the course is rapidly progressive and the death occurs at an average age of 7 months. Less than 25% of cases live more than 1 year [10, 16].

Some cases, the atypical infantile cases, show early symptoms but they don't have cardiomyopathy, and survive 2 years.

The diagnosis of the classical infantile type is easy. There are other diseases with severe hypotonia and weakness, like spinal muscular atrophy, congenital muscular dystrophies, congenital myopathies, peroxisomal or mitochondrial diseases, but the coexistence of cardiomyopathy with cardiomegaly at the beginning must have to suspect a classic Pompe disease.

Opposite to other myopathies which increase CK more than trasaminases, in Pompe disease the ratio CK/transaminases is usually 2–5. EMG is usually myopathic, and repetitive discharges are often present, as result of impairment of the function of muscular fibres membranes.

The late-onset Pompe disease can develop the symptoms at any time after 1 year of age, often it is impossible to date the onset. It gives no cardiomyopathy. A slowly progressive limb-girdle myopathy, with hypotonia and muscular atrophy is the rule, and sometimes myalgias are present. The differential diagnosis must be made with other congenital myopaties or muscular dystrophies. There is not usually any particular sign, and this is the cause of a delayed diagnosis of more than 5 years in average, especially in cases of an almost stable course when a muscular biopsy can be not proposed or accepted by the patients. New methods of diagnosis that avoid the invasive biopsy facilitate the diagnosis when suspected. We must think in Pompe disease in cases with a proximal myopathy with severe involvement of cervical flexor or extensors and of diaphragm. In lower limbs there is usually more involvement of ischiotibials than of quadriceps [14]. Periodic exams, and especially taking pictures or video, helps us to realize that a slow progression is occurring. The percentiles of weight are falling during childhood. CK, generally raised, can also be normal.

## 4.4 Evolution

Pompe disease is progressive [10, 17]. In classic infantile disease the onset is at 2 months of live, the need for invasive ventilatory support comes at 5.9 months, and death at 8.7 months of age in average. Only 25% of cases live beyond 1 year and 7.1% beyond 2 years [10].

The late onset Pompe disease is also progressive, but the rate of deterioration can vary considerably.

## 4.5 Diagnosis

The diagnosis is made with a muscular biopsy and corroborated with the assessment of GAA.

Muscular biopsy has been the usual method so far. It shows a vacuolar myopathy with glycogen deposits in lysosomes or freely distributed in cytoplasm, with destructive changes of myofibrils.

Enzymatic assessment can be made in fibroblasts, lymphocytes or muscle. When studying blood cells, it must be done in lymphocytes instead of in leucocytes because the latter have neutral maltases with some grade of activity in acid pH that can give false normal results. Fibroblast GAA assessment is less aggressive than muscle biopsy, but the procedure takes months.

The screening of GAA deficit in dried blood spots with fluorometric techniques [4] is possible and easily done, and gives a presumptive diagnosis in few days. It has to be confirmed with fibroblast GAA assessment or genetic testing.

The neonatal screening of GAA in dry blood has been proposed. At present it is important because there is enzymatic replacement therapy with results in correlation with the precocity of the replacement. A trial done in Taiwan in 2005–2206 in 70.000 newborns has detected three cases, two with an early type and one with a juvenile type [12].

The assessment of an increase of glucose tetrasacharide in urine, Glc4, a substance resulting of the action of amylases on blood glycogen is also a good screening test for Pompe disease and a good indicative of the response to enzymatic replacement therapy [2, 19].

## 4.6 Genetics

The gene is localized in 17q25. There are more than 200 different mutation already described (www.pompecenter.nl). The mutations of the 22 first cases studied in Spain have been reported [5]. There is usually a relationship between mutations and severity (genotype and phenotype), some mutations are severe and others milder, but it doesn't always happen [13].

Pompe disease is recessive. Parents are heterozygote for mutations and the risk for others children of the same couple is 25%. Prenatal diagnosis can be made. Taking the estimated incidence of Pompe disease of 1/100.000 in Caucasian people, the incidence of asymptomatic carriers is 1/160. That gives to a carrier, for example, to a brother of a patient, a risk more than 100 times above that of general population of having a child with Pompe disease. Thus, the genetic diagnosis of relatives is generally requested.

The genetic testing of the patient and parents is currently compulsory in order to: 1) start the ERT and 2) carrier and prenatal diagnosis.

## 4.7 Treatment

A multiprofessional management is needed because of the multiorganic involvement. Paediatricians, neuropaediatricians or neurologist, cardiologist, pneumologist, nutrition and rehabilitation doctors, physiotherapist, logotherapist, and others must be part of the team. There are two types of treatment: non specific and specific enzymatic treatment.

### *4.7.1 Non Specific or Symptomatic Treatment*

Respiratory support, non invasive or invasive, is needed when the disease progresses.

The anaesthesic procedures have more risk because of the muscular weakness. The cardiomyopathy requires a precise treatment depending on the stage of the disease. The use of inotropic, diuretic, hypotensive, betabockers drugs must be tailored to the patient. The loss of weight or lack of gain of it is a factor that aggravates the disease. Caloric and proteic supplements improve the general situation. Infant with classic Pompe disease need nasogastric tube or gastrostomy. There are some reported cases that led good evolution with a hyperproteic and low-carbohydrate diet [15].

Like in other neuromuscular diseases, physiotherapy is necessary. In this case it means not only limbs but mainly respiratory physiotherapy. Orthopaedic supports like orthoses, wheelchairs, adapted elements at school or at home improve the quality of live.

### *4.7.2 Enzimatic Replacement Therapy (ERT)*

GAA has been synthesized with recombinant technology (rhGAA), at first from rabbit milk and later from Chinese hamster ovaries, thanks to the discovery of the mechanisms of cellular release and capture (cellular receptor of mannose-6-phosphate). At present ERT is possible in several lysosomal disorders as Gaucher, Fabry, MPS-1, Niemann-Pick and Pompe diseases.

rhGAA (Myozyme®) was approved as ERT treatment for Pompe disease by EMEA and FDA in 2006.

The treatment consists of periodic intravenous infusion of GAA, usually every other week. Most part of Myozyme® goes to the liver and in lesser proportion to cardiac and striated muscles. The enzyme can be taken by most cells [18], but cannot cross the hematoencephalic barrier. Fortunately there is not important neuronal involvement in Pompe disease, but it may be an important concern at long term evolution in the severe classic type of Pompe disease, because neuronal glycogen deposits can be developed over the years.

Cardiac muscle improves more than skeletal muscle after rhGAA. It may be due to the fact that the former has more mannose-6-phosphate receptors. The response is better and earlier in other tissues like endotelio and peripheral nerves than in muscles [18].

The response to ERT is different among patients, as results of several factors like the precocity, the severity, the type of muscular fibers (type II fibers are less sensitive [6]) or the development of antibodies against GAA.

In classic Pompe disease the reported results improve significantly the survival and the cardiac function, thought because of the severe multiorganic involvement the improvement not always imply significant advance in body function and quality of life.

The sooner ERT starts, the better are the results. That is why early diagnosis is crucial. In 1999 was demonstrated the in vivo efficacy of ERT in the mouse-KO model of Pompe disease, reducing the glycogen storage [3]. Soon after, pilot studies in humans with classic Pompe disease treated with rhGAA purified from rabbit milk [16] or Chinese hamster ovary [1] showed that survival was prolonged beyond a year in all cases. There are so far more than 270 cases in the world treated with GAA. The data, from treated and non treated cases, can be recruited by doctors, with the patients consent, in an international registry (www.pomperegistry.com). The objective is to know as exactly as possible the natural evolution of the disease and the way the treatments modify if.

There are two core studies of the effect of Myozyme® in classic Pompe disease, starting in 2003, with a total of 39 cases. They differ on the age of onset of the treatment. The first study treated 18 infants younger than 6 months and the aim was to demonstrate if it will improve the survival free of assisted ventilation compared to the cohort of non treated patients. The 2nd study assessed the effect of the treatment when started from 6 to 26 months of age. There were 21 infants, and the primary goal was to compare the survival compared with the historical cohort. Secondary end points of both studies were to analyze the effects on cardiac ventricular mass, growth and motor development of the children.

The first study demonstrated increase of survival. All the 18 cases treated were alive at 18 months, and 15 out of 18 did not need invasive ventilation. In 15 cases the mass ventricular index was improved. Growth and weight was normal in those 15. The development improved in 13 out of 18. The cases treated with double doses (40 mg/k instead of 20 mg/k every 2 weeks) didn't get better results and, on the contrary, they suffered more adverse effects [8].

The second study, with 21 cases whose treatment started after 6 months of age, showed slightly lesser efficacy. The conclusion was that early treatment is essential.

A recently reported open multicentre study [11] with part of the patients of the former study, consisting of 8 patients with classic Pompe disease with less than 1% of GAA shows that 6/8 remained alive after 1 year of treatment with Myozyme®, and 5 of them free of ventilatory support. 5 got new motor milestones, even 3 of them got independent gait. Four of the cases died after the initial phase of the study. The average age of death was 21 months, much higher than the natural evolution.

Summarising, rHGAA treatment improves the survival free of ventilator, the cardiomyopathy, growth and motor function as compared with that of patients without treatment [9].

In late onset severe Pompe disease, the rhGAA ERT for patients ventilator and wheelchair dependent, has reduced the ventilator use in 8/17, has improved the muscular strength in 14/18, after treatment lasting from 8.3 months to 6.3 years [12].

The adverse effects of the treatment are mainly related with allergic reaction to infusion, and consisted in rash, fever, malaise, cough, tachycardia, that arise during infusion, more often in the firsts infusion. They are mild or moderate in intensity and disappear reducing the speed of infusion and treating with antihistamines or steroids.

Patients treated with double dose (40 mg/k every 2 weeks) are in higher risk of adverse reactions to the infusion. As no better results were obtained at double dose, the best dose is considered 20 mg/k every 2 weeks, for classic and for late-onset Pompe disease. Severe anaphylactic reactions have rarely reported. The deaths of the treated patients were considered related with the severity of the illness, not with the treatment itsef. It is compulsory to monitor the long-term evolution of the patients (Pompe registry) in order to know if some long-term adverse effect is presented.

Limitations of ERT are the high costs and the high accumulated doses required, and the development of antibodies against rhGAA, because it is recognized as a foreign body [12]. Almost all the patients produce antibodies against Myozyme®, mostly in the firsts weeks of treatment, and they tend to remain or to diminish along time, but they don`t inhibit the clinical response. The ERT continues to be useful [8].

Many patients have residual natural GAA, but inactive. They are called CRIM positive ("cross-reacting immune material"). Patients with no activity of GAA are considered CRIM-negative and tend to produce more antibodies and to get less clinical benefits [1], the same thing that happens to the mouse-KO model without GAA [16].

## 4.8 Conclusion

Pompe disease is a genetic muscular progressive disorder with a broad spectrum of severity that covers from a neonatal onset and fatal early evolution to a very mild adult myopathy that can be undiagnosed for several years.

Nowadays ERT is posible. It has demonstrated to improve survival and muscular function. Vigilance to long term evolution and long term effects of the drug is needed, but ERT allows the patients and doctors to look at the future in a more optimistic way, with the hope to gain the time necessary to develop other genetic treatments.

## References


1. Amalfitano A, Bengur AR, Morse RP, et al. Recombinant human acid alpha-glucosidase enzyme therapy for infantile glycogen storage disease type II: results of a phase I/II clinical trial. Genet Med 2001; 3:132–138.
2. An Y, Young SP, Kishnani PS, et al. Glucose tetrasaccharide as a biomarker for monitoring the therapeutic response to enzyme replacement therapy for Pompe disease. Mol Genet Metab 2005; 85:247–254.
3. Bijvoet AG, Van Hirtum H, Kroos MA, et al. Human acid alphaglucosidase from rabbit milk has therapeutic effect in mice with glycogen storage disease type II. Hum Mol Genet 1999; 8:2145–2153.
4. Chamoles NA, Niizawa G, Blanco M, et al. Glycogen storage disease type II: enzymatic screening in dried blood spots on filter paper. Clin Chim Acta 2004; 347:97–102.
5. Gort L, Coll MJ, Chabás A. Glycogen storage disease type II in Spanish patients: high frequency of c.1076-1G>C mutation. Mol Genet Metab 2007; 92:183–187.
6. Hawes ML, Kennedy W, O'Callaghan MW, et al. Differential muscular glycogen clearance after enzyme replacement therapy in a mouse model of Pompe disease. Mol Genet Metab 2007; 91:343–351.
7. Hirschhorn R, Reuser AJJ. Glycogen storage disease type II: acid alpha-glucosidase (acid maltase) deficiency. In: Childs B, et al., eds. The Metabolic and Molecular Bases of Inherited Disease, 8th ed. New York: McGraw-Hill; 2000: pp. 3389–3420.
8. Kishnani P, Corzo D, Nicolino M, et al. Recombinant human acid alpha-glucosidase: major clinical benefits in infantile-onset Pompe disease. Neurology 2007; 68:99–109.
9. Kishnani PS, Howell RR. Pompe disease in infants and children. J Pediatr 2004; 144:S35–S43.
10. Kishnani PS, Hwu WL, Mandel H, et al. A retrospective, multinational, multicenter study on the natural history of infantile-onset Pompe disease. J Pediatr 2006a; 148:671–676.
11. Kishnani PS, Nicolino M, Voit T, et al. Chinese hamster ovary cell-derived recombinant human acid-glucosidase in infantile-onset Pompe disease. J Pediatr 2006b; 149:89–97.
12. Koeberl DD, Kishnani PS, Chen YT. Glycogen storage disease types I and II: treatment updates. J Inherit Metab Dis 2007; 30:159–164.
13. Kroos MA, Pomponio RJ, Hagemans ML, et al. Broad spectrum of Pompe disease in patients with the same c.-32-13T->G haplotype. Neurology 2007; 68:110–115.
14. Müller-Felber W, Horvath R, Gempel K, et al. Late onset Pompe disease: clinical and neurophysiological spectrum of 38 patients including long-term follow-up in 18 patients. Neuromuscul Disord 2007; 17:698–706.
15. Slonim A, Boone L, Minikes J, et al. Benign course of glycogen storage disease type II in two brothers: nature or nurture? Muscle Nerve 2006; 33:571–574.
16. van Den HH, Reuser AJ, Vulto AG, et al. Recombinant human alpha-glucosidase from rabbit milk in Pompe patients. Lancet 2000; 356:397–398.
17. van den Hout HMP, Hop W, van Diggelen OP, et al. The natural course of infantile Pompe's disease: 20 original cases compared with 133 cases from the literature. Pediatrics 2003; 112:332–340.
18. Winkle LPF, Kamphoven JHJ, van den Hout HMP, et al. Morphological changes in muscle tissue of patients with infantile Pompe's disease receiving enzyme replacement therapy. Muscle Nerve 2003; 27:743–751.
19. Young SP, Corzo D, Kishnani P, et al. Diagnostic value of urinary and plasma glucose tetrasaccharides in infantile and late onset glycogen storage disease type II. Mol Genet Metab 2005; 84:241–242.

# Chapter 5
# Diseases of the Human Mitochondrial Oxidative Phosphorylation System

**Julio Montoya, Ester López-Gallardo, María Dolores Herrero-Martín, Íñigo Martínez-Romero, Aurora Gómez-Durán, David Pacheu, Magdalena Carreras, Carmen Díez-Sánchez, Manuel J. López-Pérez, and Eduardo Ruiz-Pesini**

**Abstract** Mitochondrial diseases, or diseases of the oxidative phosphorylation system, consist of a group of disorders originated by a deficient synthesis of ATP. This system is composed of proteins codified in the two genetic systems of the cell, the nuclear and the mitochondrial genomes, and, therefore, the mode of inheritance could be either mendelian or maternal. The diseases can also appear sporadically. Due to the central role that mitochondria play in cellular physiology, these diseases are a social and health problem of great importance. They are considered rare diseases; however, together they constitute a large variety of genetic disorders.

It is also believed that mitochondria are involved, directly or indirectly, in many other human diseases, mainly in age-related diseases. This review will focus mainly on describing the special characteristics of the mitochondrial genetic system and the diseases caused by mitochondrial DNA mutations. We will also note the difficulties in studying these pathologies, and the possible involvement of the genetic variability of the mitochondrial genome in the development of these diseases.

**Keywords:** Mitochondrial diseases · Oxidative phosphorylation system · Mitochondrial DNA · Mitochondrial genetics · Nuclear DNA

## 5.1 Introduction

The OXidative PHOSphorylation (OXPHOS) system, localized in the mitochondrial inner membrane, consists of the electron transport chain (ETC) and the ATP synthase. ETC is made up of the multienzymatic respiratory complexes I to IV. The reducing equivalents originating in ß-oxidation and in the Krebs cycle pass along the respiratory complexes, and the energy generated is used to pump protons from the mitochondrial matrix to the intermembrane space. The electrochemical

J. Montoya (✉)
Departamento de Bioquímica y Biología Molecular y Celular, Universidad de Zaragoza, Miguel Servet, 177 50013, Zaragoza, Spain
e-mail: jmontoya@unizar.es

C. Espinós et al. (eds.), *Inherited Neuromuscular Diseases*, Advances in Experimental Medicine and Biology 652, DOI 10.1007/978-90-481-2813-6_5,

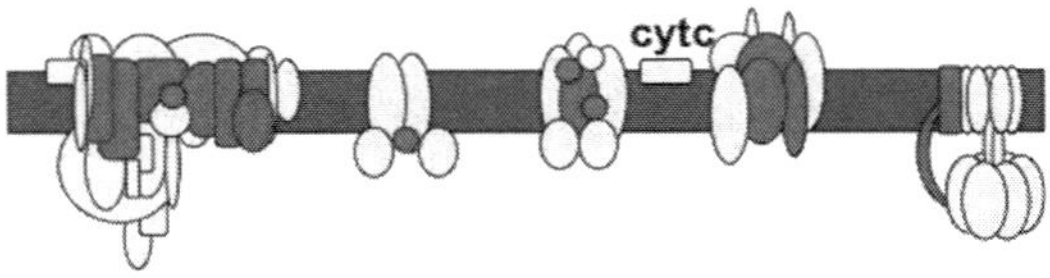


| | CI | CII | CIII | CIV | CV |
|---|---|---|---|---|---|
| mtDNA | 7 | 4 | 1 | 3 | 2 |
| nDNA | 39 | 0 | 10 | 10 | 14 |

**Fig. 5.1** Scheme of the oxidative phosphorylation system. *Green*: mtDNA-encoded subunits; *Yellow*: nDNA-encoded subunits; *Red dots*: Coenzyme Q. CI to V: OXPHOS Complexes I to V

proton gradient thus originated is utilized by complex V to produce ATP (Fig. 5.1). However, this system is also involved in the production of heat, in the process of protein import to mitochondria, in the regulation of apoptosis, in the maintenance of cellular calcium levels, and in the generation of reactive oxygen species (ROS). ROS, calcium, and ATP can trigger different biochemical pathways and alter the cellular metabolic status. All the information coming from outside (nutrients and oxygen) is integrated in OXPHOS, and it can modify the level of some cellular messengers and send signals to the nucleus.

Human cells contain two genetic systems: nuclear (nDNA) that encodes for the great majority of the cell proteins, and the mitochondrial (mtDNA) that presents characteristics very different from nDNA, probably due to its location in a subcellular organelle. OXPHOS and the mitochondrial ribosomes are unique since their biogenesis depends on the coordinated expression of both genomes. OXPHOS consists of around 100 structural proteins and only 13 of these are encoded into the mitochondrial genome; the others, together with those needed to assemble the complexes and to maintain and express the mitochondrial genome, are encoded in the nucleus.

Mitochondrial diseases, or diseases of the oxidative phosphorylation system, consist of a group of disorders originating in deficient synthesis of ATP. These diseases can be caused by mutations in mtDNA (maternally inherited) or in nDNA (Mendelian inherited). The first mutations in mtDNA were discovered in 1988 [22, 58, 61]. Since then, the number of described mutations in this genome has increased enormously, although many of them have only been described in one patient. Later on, mutations in mitochondrial protein genes encoded in nDNA were also discovered [47]. Also, due to the central role that mitochondria play in cellular physiology, these diseases have become a social and health problem of great importance. It is also considered that mitochondria are involved, directly or indirectly, in many other human diseases, mainly in aging, age-related diseases, diabetes, and cancer, etc. [57].

Mitochondrial diseases show large phenotypic variability with symptoms that affect different organs and tissues, and the participation of specialists from different areas – clinicians, anatomopathologists, biochemists and geneticists – is required for correct diagnoses. The special characteristics of mitochondrial genome, the maternal inheritance, the high number of copies present in a cell and the mitotic segregation confer to these diseases very particular properties. These disorders collectively affect approximately 1/5000 births [49]. In this review, the main characteristics of the diseases caused by mtDNA damage and some aspects of the nDNA mutations that affect OXPHOS will be summarized.

## 5.2 Mitochondrial Genetic System

Human mtDNA is a super-coiled double-strand closed circular molecule of approximately 16,569 base pairs [1, 2]. There are several copies per mitochondrion and many mitochondria per cell. This molecule encodes 37 genes: 2 rRNAs (12S and 16S rRNAs), 22 tRNAs and 13 polypeptide components of four of the five OXPHOS complexes; seven (ND1–ND6 and ND4L) of the 46 complex I (NADH:ubiquinone oxidoreductase) subunits; one (apocytochrome b) of the 11 complex III (ubiquinol:cytochrome c oxidoreductase) subunits; three (COI-COIII) of the 13 complex IV (cytochrome c oxidase) polypeptides, and two (ATP6 and ATP8) of the 16 complex V (ATP synthase) subunits (Fig. 5.2).

The basic features of the mitochondrial genetic system, the mode of replication and transcription, and the proteins that encode were described in the 1980s [1, 9, 12, 36, 37, 39, 40, 42, 43, 46]. The genetic organization of mtDNA is extremely compact: there are some overlapping genes, the 3'-end CCA of the tRNAs and some termination codons not encoded in the mtDNA; there are no intronic sequences and almost no non-coding nucleotides between genes. A non-coding region (control region) of approximately 1.1 kb, located between the $tRNA^{Phe}$ and $tRNA^{Pro}$, contains the origin of replication for the heavy strand ($O_H$), the transcription promoters (LSP and HSP, light and heavy strand promoters, respectively) and the regulatory elements for mtDNA expression (Fig. 5.2).

MtDNA replication starts at $O_H$, which is located downstream from the LSP in the D-loop region, and proceeds along the parental L-strand to produce a daughter H-strand circle. When H-strand replication reaches $O_L$, the parental H-strand is displaced, the initiation site for L-strand synthesis is exposed, and its replication starts and proceeds in the opposite direction, producing a daughter L-strand (Fig. 5.2) [12, 23]. This asymmetric replication of both strands allows the synthesis to be continuous. However, recently, a second mechanism of mtDNA replication has been described. According to this description, mtDNA replication occurs in a bidirectional and symmetric way, similar to replication of bacterial DNA [6, 60].

RNA is synthesized in human mitochondria by means of three transcription units that start at three different initiation points, one for the L-strand (L) and two for the H-strand ($H_1$ and $H_2$); these transcribe three polycistronic molecules [31, 36, 37] (Fig. 5.2). The $H_1$ transcription unit starts immediately upstream from the $tRNA^{Phe}$ gene and ends at the 3'-end of the 16S rRNA. This transcription unit is responsible for the synthesis of the two rRNAs, $tRNA^{Phe}$ and $tRNA^{Val}$. The second transcription unit starts at the initiation site $H_2$, 2 nt upstream from the 5'-end of the 12S rRNA gene [31, 37], and gives rise to a polycistronic molecule covering almost the whole

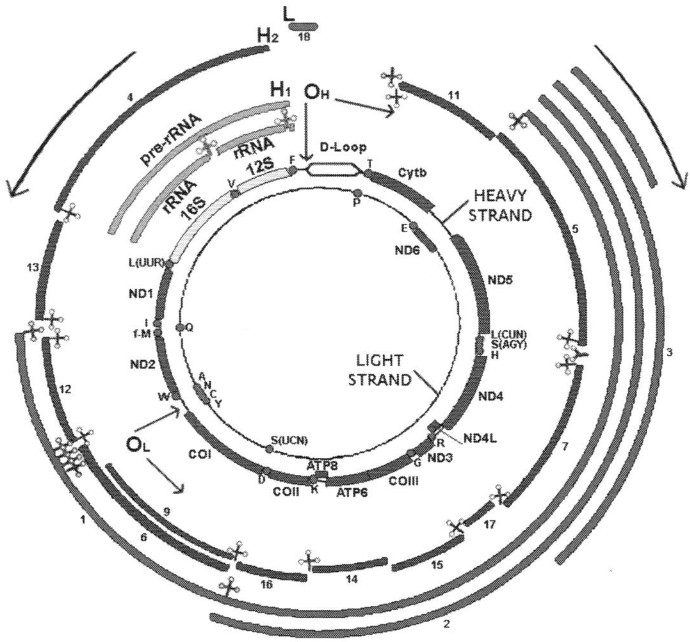


**Fig. 5.2** Genetic and transcription map of human mtDNA. The two inner circles represent both mtDNA strands with the encoded genes: rRNAs (*yellow*), tRNAs (*red dots*) and protein coding genes (*blue*). External circles represent the RNAs transcribed from the heavy strand: (*orange* or *blue* for the RNAs derived from the $H_1$ or $H_2$ transcription units) and light strand (*pink*). ND1 to ND6 are subunits 1–6 of NADH dehydrogenase (complex I); cyt b, cytochrome *b* subunit of complex III; CO I, CO II and CO III, are subunits of cytochrome *c* oxidase (complex IV) and ATP6 and ATP8, subunits of ATP synthase (complex V). $O_H$ and $O_L$ represent replication origins for the H- and L-strand, respectively, according to the classical model of replication. $H_1$, $H_2$ and L indicate initiation points for the three transcription units of the H- and L-strand, respectively. Arrows at the $O_H$ and $O_L$, and in the outside part of the figure, indicate the direction of replication and transcription of the two strands. tRNA genes are indicated by the one letter code of the corresponding amino acid (from Montoya et al. BBA-Bioenergetics 1757, 9–10, 2006)

H-strand. The mRNAs for 12 polypeptides and 14 tRNAs derive from the processing of this polycistron. The L-strand gives rise to a single polycistron starting at the 5'-end of the RNA 18, from which eight tRNAs, the ND6 mRNA and an RNA primer for initiation of replication are derived.

The polycistronic primary transcripts synthesized from the three initiation sites are processed, according to the "tRNA punctuation" model, to yield the mature rRNAs, mRNAs and tRNAs after the precise endonucleolytic cleavage on both sides of the tRNA molecules [37, 44]). Processing of the polycistronic molecules requires endonucleolytic cleavages at the 5'- and 3'-end of the tRNAs, the addition of the CCA to the 3'-end of the tRNAs, polyadenylation activity for mRNAs and rRNAs, and, finally, post-transcriptional modification of a subset of tRNA and rRNA nucleotides [39].

MtDNA transcription requires a mitochondrial RNA polymerase (mtRPOL or POLRMT) [54] and at least three transcription factors: TFAM (or mtTFA) and either TFB1M or TFB2M for initiation and mTERF for termination of the transcription unit starting at $H_1$ (Fig. 5.2) [39]. All of these are encoded by nDNA.

## 5.3 Mitochondrial Genetics

The location of mtDNA in a cytoplasmic organelle has conferred genetic features that differentiate it from nDNA The main genetic features of the mitochondrial genome are:

### *5.3.1 Maternal Inheritance*

mtDNA is maternally inherited. The mother will pass her mtDNA to all her children but only the daughters will transmit it to all members of the next generation. This is due to the high copy number of mtDNA within the oocyte and to the fact that mitochondria from the spermatozoon are eliminated in the first cellular divisions [51].

### *5.3.2 Polyplasmy*

The mtDNA copy number varies within the different tissues: from a very low number of molecules in platelets to more than 100,000 copies in the oocyte. The number of molecules in most tissues varies between 1000 and 10,000 per cell and each mitochondrion has between 2 and 10 copies organized in nucleoids. In normal tissues, all mtDNA molecules are identical (homoplasmy). However, the appearance of a mutation can give rise to the existence of two different types of molecules, normal and mutated (heteroplasmy).

### *5.3.3 Mitotic Segregation*

In the case of heteroplasmy, mtDNA distributes randomly at cell division, giving rise to three possible genotypes: homoplasmic for normal mtDNA, homoplasmic for mutated mtDNA, and heteroplasmic with diverse percentages of the mutated molecule. As a consequence of this, the phenotype of a cell will depend on the percentage of mutated mtDNA that it contains. One of the main features of the mitochondrial diseases is that they are, in general, multisystemic.

### *5.3.4 Threshold Effect*

In the case of heteroplasmy, there is a threshold level of the pathogenic mutation that is critical for the clinical expression of the disease. The percentage of mtDNA required to cause the dysfunction of an organ or tissue varies depending on their

ATP needs. The minimum critical number of mutated mtDNA is usually very high but there are cases in which a lower percentage will give rise to a clinical pathologic state. In many cases, the mutation is found in low levels in blood cells but very likely this percentage will be higher in the most affected tissues. For this reason, it is always advisable to use the appropriate tissue to carry out the test. If this is not possible, nowadays it is advisable to use cells collected from urine that have a higher percentage of the mutation than blood cells.

In the case of homoplasmic pathogenic mutations, they are transmitted in homoplasmy to all maternal children but in many cases not all the offspring will develop the disease. This is very frequently found in the Leber hereditary optic neuropathy (LHON). It is quite possible that nuclear or environmental factors may be important in the expression of the disorder. All this makes it very difficult to give genetic counselling.

### *5.3.5 High Mutational Rate*

mtDNA is very vulnerable and its mutational rate is 10–20 times higher than that of nDNA genes with related function. There are several explanations for this fact. One of them is the accumulation of reactive oxygen species (ROS), produced as a consequence of respiratory transport chain activity, in the same compartment. ROS can damage the highly compact gene organization of mtDNA and affect different genes. Also, although there are mtDNA repair systems, these may not be efficient, resulting in high mutational rate.

It is generally thought that a normal individual is born with an initial homoplasmic mtDNA and an energy capacity that is higher than the bioenergetic threshold. Therefore, all the organs function normally. However, with age mtDNA mutations accumulate in different tissues and reduce the bioenergetic capacity to below the threshold levels and symptoms of a disease may appear. If the individual in question harbors a pathogenic mutation or mutations that give rise to a lower initial bioenergetic capacity, the organs and tissues may fail at an earlier age. In 1990, some evidence was found to support this theory, when it was shown that some mutations accumulated with age [34].

## 5.4 Diseases Caused by Mutations in mtDNA

Since all proteins encoded in the mitochondrial genome are components of OXPHOS, all the genes encoded in mtDNA (mRNA, tRNA and rRNA genes) participate in their synthesis and, therefore, the diseases caused by mutations in this genome present the common characteristic of having a deficient synthesis of ATP. As mitochondria are vital components of all types of nucleated cells, it is not surprising that these diseases are in general multisystemic, and that the clinical features are very variable (Table 5.1). Sometimes it is possible to define specific syndromes,

**Table 5.1** Some of the common clinical characteristics for the diagnoses of mtDNA-associated disorders

| Organ/tissue | Clinical features |
|---|---|
| **Nervous system** | Encephalopathy |
| | Stroke |
| | Epilepsy |
| | Cerebellar ataxia |
| | Seizures |
| | Myoclonus |
| | Mental and psychomotor retardation |
| | Migraine |
| | Cortical blindness |
| | Peripheral neuropathy |
| | Dementia |
| | Speech disturbances |
| | Sensorineural deafness |
| | Dystonia |
| **Muscle** | Myopathy |
| | Exercise intolerance |
| | Weakness |
| | Ophthalmoplegia |
| | Ptosis |
| | Myoglobinuria |
| **Blood** | Syderoblastic anaemia |
| **Eye** | Optic atrophy |
| | Pigmentary retinopathy |
| | Cataracts |
| **Ear** | Deafness |
| **Heart** | Cardiomyopathy |
| | Conduction blocks |
| **Endocrine system** | Diabetes mellitus |
| | Diabetes insipidus |
| | Hypoparathyroidism |
| | Short stature |
| **Intestine** | Intestinal pseudo-obstruction |
| | Vomiting |
| **Pancreas** | Exocrine pancreatic dysfunction |
| **Liver** | Liver failure |
| **Kidney** | Fanconi syndrome |
| | Kidney failure |
| **Muscle morphology** | Ragged-red fibers |
| | Paracrystaline inclusions in mitochondria |
| **Histochemistry** | COX negative fibers |
| **Biochemistry** | Decreased activity of the respiratory complexes in muscle biopsy, fibroblasts, etc. |
| | OXPHOS complex assembly defects |
| | Decreased ATP levels |
| **Laboratory** | Lactic acidosis in blood and/or cerebrospinal fluid |
| | Hypoglycemia |
| **Genetics** | Maternal inheritance |
| | Sporadic |
| **Cellular models (cybrids)** | Metabolic defects transmitted by the mtDNA mutation |

**Table 5.2** Diseases caused by mtDNA damage

| |
|---|
| *Diseases associated with mtDNA point mutations in protein-coding genes* |
| Leber hereditary optic neuropathy (LHON) |
| NARP syndrome (Neuropathy, ataxia and retinitis pigmentosa) |
| Maternal inherited Leigh syndrome (MILS) |
| Exercise intolerance |
| *Diseases associated with mtDNA point mutations in tRNA genes* |
| MELAS syndrome (Mitochondrial Encephalopathy, Lactic Acidosis and stroke-like episodes) |
| MERRF syndrome (Myoclonus Epilepsy and Ragged-Red Fibers) |
| Diabetes and deafness |
| Neurosensorial deafness |
| *Diseases associated with mtDNA point mutations in rRNA genes* |
| Aminoglycoside-induced hearing loss |
| *Diseases associated with mtDNA deletions* |
| Chronic progressive external ophthalmoplegia (CPEO) |
| Kearns-Sayre syndrome (KS) |
| Pearson syndrome |
| *Diseases associated with mtDNA depletion* |

but in general we find overlapping symptoms and a large variety of phenotypes. At other times, and mainly in children, the symptoms are not so specific and it is very difficult to define the disorder. In many cases diagnosis depends on the results of several types of analysis (clinical, laboratory, histochemical, biochemical and genetic). However, some of the mtDNA diseases affect just a single tissue, such as the optical nerve in LHON (Leber Hereditary Optic Neuropathy) and the coclear cells in mitochondrial deafness.

The number of mtDNA mutations found up to now is very large (more that 100 point mutations, although some of them have been described only once; and there are many deletions). For this reason, in this chapter we will only mention the main disorders known up to now (Table 5.2), with the main characteristics, the most common mutations and the hot-spot genes that cause the diseases [13, 15] (Table 5.3).

Besides the mutations mentioned above, there might be other ones in mtDNA sequences involved in DNA and RNA metabolism that might also cause diseases. In this sense, a high proportion of the somatic T414G transversion is found in the fibroblast mtDNA molecules of the aged but it is absent in younger individuals [34]. This mutation is located in the middle of the promoter for the L-strand transcription, at a position adjacent to a segment with high affinity for TFAM transcription factor. But these mtDNA elements related to the initiation of transcription are not the only ones to be associated with particular phenotypes. Thus, $tRNA^{Leu(UUR)}$ is a mutational hot spot and several mutations have been described in the region where the termination factor mTERF binds [27]. This suggests that other functions, besides those related to normal tRNA, might be affected by these mutations. Thus, the A3243G transition located in mTERF binding site for the $H_1$ transcription

**Table 5.3** Mitochondrial phenotypes associated with mtDNA damage

| Disease (syndrome). Clinical and genetic features | Mutation | Gene |
|---|---|---|
| *LHON (Leber hereditary optic neuropathy).* Acute or subacute loss of vision due to severe bilateral optic atrophy. Onset between 20 and 30 years. Telangiectatic microangiopathy, pseudoedema of the optic disc. Sometimes accompanied by cerebellar ataxia, periferal neuropathy, or cardiac conduction abnormalities. Maternal inheritance, hetero- or homoplasmic | G3460A<br>G11778A<br>T14484C | ND1<br>ND4<br>ND6 |
| *NARP (Neuropathy, Ataxia Retinitis Pigmentosa).* Usually accompanied by developmental delay, dementia, muscle weakness, optic atrophy. Maternal inheritance, heteroplasmic | T8993G<br>T8993C | ATP6<br>ATP6 |
| *Leigh (MILS). Maternally inherited Leigh syndrome.* Severe encaphalomyopathy with onset during the first months of life, with signs of progressive brain stem dysfunction with abnormal respiration (hyperapnea), nistagmus, developmental delay, psychomotor regression, hypotonia, seizures, ophthalmoparesis, lactic acidosis. Maternal inheritance, heteroplasmic | T8993G<br>T8993C<br>T9176C<br>T9176G | ATP6<br>ATP6<br>ATP6<br>ATP6 |
| *Striatum bilateral necrosis* | T14487C | ND6 |
| *Dystonia* | G14459A | ND6 |
| *Exercise intolerance.* Fatigue, muscle weakness, myalgia with or without myoglobinuria, ragged-red fibers. Sporadic, somatic, heteroplasmic | hot spot | Cytochrome b |
| *MELAS. Mitochondrial Encephalopathy, Lactic Acidosis and stroke-like episodes.* Accompanied by focal brain abnormalities, focal or generalized seizures, recurrent headache, vomiting, dementia, limb weakness, short stature, deafness, ragged-red fibers. Maternal inheritance, heteroplasmic | A3243G<br>Hot spot | $tRNA^{Leu(UUR)}$<br>$tRNA^{Leu(UUR)}$ |
| *MERRF. Myoclonus Epilepsy and Ragged-Red Fibers.* Myoclonus, seizures, mitochondrial myopathy, cerebellar ataxia. Additional signs may include optic atrophy, deafness, dementia, peripheral neuropathy, pigmentary retinopathy, multiple lipomatosis, dementia. Maternal inheritance, heteroplasmic | A8344G<br>G8363A<br>Hot spot | $tRNA^{Lys}$<br>$tRNA^{Lys}$<br>$tRNA^{Lys}$ |
| *Diabetes and deafness.* Maternal inheritance, heteroplasmic | A3243G | $tRNA^{Leu(UUR)}$ |
| *Neurosensorial deafness.* Maternal inheritance, hetero- or homoplasmic | hot spot | $tRNA^{Ser(UCN)}$ |
| *Symmetric multiple lipomatosis* | hot spot | $tRNA^{Lys}$ |
| *Exercise intolerance* | G12334A | $tRNA^{Leu(CUN)}$ |
| *Deafness induced by aminoglycosides* | A1555G<br>C1494T | rRNA 12S<br>rRNA 12S |

**Table 5.3** (continued)

| Disease (syndrome). Clinical and genetic features | Mutation | Gene |
|---|---|---|
| *Pearson syndrome.* Refractory sideroblastic anemia, vacuolization of marrow precursors and exocrine pancreatic dysfunction, short stature and lactic acidosis. Affects children, who die in early infancy. Survivors usually generate Kearns-Sayre syndrome. Usually sporadic and heteroplasmic | Large-scale single deletions | Deletion of several genes |
| *CPEO.- Chronic Progressive External Ophthalmoplegia.* Benign clinical condition presenting ophthalmoplegia, ptosis, myopathy. Ragged-red and COX-negative fibers. Usually sporadic and heteroplasmic | Large-scale single or multiple deletions | Deletion of several genes |
| *Kearns-Sayre.* Invariant triad of onset before 20 years, CPEO and pigmentary retinopathy, plus at least one of the following symptoms: cerebellar syndrome, heart block, CSF protein level above 100 mg/ml. Usually also accompanied by ataxia, deafness, dementia, endocrine failures. Ragged-red and COX-negative fibers. Usually sporadic and heteroplasmic | Large-scale single deletions | Deletion of several genes |
| *MNGIE.- Myoneurogastrointestinal encephalopathy.* Progressive external ophthalmoplegia, neuropathy, leukoencephalopathy and intestinal dysmotility with cachexia. Affects young adults. It originates in mutations in the thymidin phosphorylase (TP) gene that produce autosomic recessive disease with multiple deletions in mtDNA | Multiple deletions | Deletion of several genes |
| *DIDMOAD* (or Wolfram syndrome if it is of early onset)**:** Diabetes Insipida, Diabetes Mellitus, optic atrophy, neurosensorial deafness | Multiple deletions | Deletion of several genes |
| *Toni-Debré-Fanconi.* Renal proximal tubulopathy in the presence of normal, or nearly normal, glomerular filtration. Defect in the reabsorption of glucose, amino acids, phosphate and potassium | Large-scale single deletions | Deletion of several genes |
| *Aminoaciuria.* Rickets, polyuria, short stature, microcephaly, mental retardation. Skin hyperpigmentation | | |
| *MtDNA depletion.* Large decrease in mtDNA levels. Variable clinical spectra: myopathic, hepatocerebral forms, etc. Disorder of infancy or childhood. Autosomic recessive | mtDNA depletion | Nuclear genes involved in nucleotide metabolism |
| *Alpers.* Progressive sclerosing poliodystrophy, astrocytosis and espongiosis of the cerebral grey matter. Accompanied by developmental delay, progressive mental retardation, hypotonia, spasticity, seizures. Signs of chronic liver dysfunction spongiosis, tubulopathy, myopathy | mtDNA depletion | DNA polymerase gamma |

A list of mtDNA mutations associated with different phenotypes can be found at MITOMAP: A Human Mitochondrial Genome Database, 2005 http://www.mitomap.org. Most of the mutations not cited in the table have been found in single cases.

unit substantially decreases binding of this factor [10, 41]), resulting in severe impairment of transcription termination at the end of the 16S rRNA gene [21].

tRNAs are considered processing signals and many tRNA mutations have been shown to affect RNA metabolism. For example, the MELAS A3243G mutation provokes a small but consistent increase in the steady-state levels of new RNA (RNA 19), which contains the sequences of 16S rRNA + $tRNA^{Leu(UUR)}$ + ND1, suggesting a defect in the $tRNA^{Leu(UUR)}$ processing [25]. Other pathologic $tRNA^{Leu(UUR)}$ mutations, such as 3256T, 3271C and 3302G, also increase the levels of RNA 19 [4, 19, 26].

Mutations in other mitochondrial tRNA genes also affect the RNA processing efficiency. Thus, mutations at positions 7510G, 7511G and 7512G in $tRNA^{Ser(UCN)}$ decrease this efficiency [59]. However, mutations at positions 7442 Ci, 7443C, 7444T and 7445C and 7445G, adjacent to the 3'-end of the $tRNA^{Ser(UCN)}$ gene, have not only quantitative but also qualitative effects on the processing of this tRNA. Thus, processing occurs at a position located one nucleotide downstream from the normal processing site [55, 59]. This qualitative defect was not found in another study of 7445C [18].

RNA stability is also affected by a microdeletion ($\mu\Delta 9205$) removing two residues at the ATP6 termination codon. This was found in a patient with mitochondrial disease. The polyadenylation form of the RNA 14 revealed substantial abnormalities. Most of this mutated RNA terminated with short poly(A) extensions and the steady-state levels of the RNA 14 were markedly decreased. The enhanced turnover of RNA 14 was due to a translation-dependent deadenylation decay mechanism [53].

MtDNA diseases are also unique inasmuch as different mutations in the same or in different genes might give rise to the same phenotype, just as the same mutation can give rise to very different phenotyes. The spectrum of phenotypes related to mtDNA mutations is increasing constantly, and it is also possible that very mild phenotypes may be missed if they do not resemble a true mitochondrial disease [20]. It is also quite likely that we will also see how different damage in mtDNA will play a role in other very important neurodegenerative diseases such as Parkinson, Alzheimer, and other age-related diseases, as well as in cancer, diabetes, etc. Sequencing of the whole mtDNA will be necessary in order to find new mutations associated with well-described and new phenotypes.

## 5.5 New mtDNA Point Mutations: Pathogenicity Criteria

Since mtDNA presents a high mutational rate it is quite possible that, besides the high number of point mutations that have already been described, we may still find new mutations related to specific syndromes or to new phenotypes. Thus, very recently our group has found a new mutation in the subunit I of cytochrome c oxidase associated with a very mild phenotype characterized by moderate mental retardation, weakness and fatigability [20]. As sequencing of the whole mitochondrial genome is no longer a problem, the probability of finding new mutations has increased. However, we have to keep in mind that we should sequence the mtDNA from the most appropriate tissue since many disorders may be caused by somatic mutations. Because of the high phenotypic variability, we also have to think that we may be missing some mild phenotypes not viewed as mitochondrial phenotypes. We must bear in mind that the field of mitochondrial pathology is broader than previously thought.

Because of the abundance of different mitochondrial DNA variants that we can find, it is not easy to determine whether a new mutation is pathogenic. To facilitate this task, different criteria have been proposed. However, not every mutation fufills all the criteria; therefore, we should be extremely careful in their application and maintain an open mind in defining the pathogenicity of a new mutation, in order to make progress in this field [38]. The different criteria are:

### *5.5.1 The Mutation must Be Present in Patients and Absent in Controls*

A pathologic mutation should not be found in normal individuals. However, there are clear pathologic mtDNA mutations that present incomplete penetrance and they can be found in healthy members of the same pedigree (LHON and non-syndromic deafness mutations); they probably require other factors for expression. How many controls should we check? Nowadays, there are more than 3500 published mtDNA coding region sequences and it is very easy to check all of them. Thus, this is a good number of sequences in order to start to rule out the absence of a new mutation in control individuals.

It may also be true that mutations described in population studies could happen to be pathological in a patient because of intra- or intermolecular compensations in the control individual.

### *5.5.2 The Mutation Must Be Found in Different Mitochondrial Genetic Backgrounds*

This criterion excludes the possibility of a new mutation as an ancient polymorphism, meaning that there is an independent association with the phenotype. New mutations cannot fulfill this criterion.

### *5.5.3 The Percentage of the Mutation Must Correlate with the Phenotype*

As previously mentioned, the number of mtDNA molecules is usually very high, and therefore gives rise to the possibility of heteroplasmy and homoplasmy within the cell, and to the existence of correlation between genotype and phenotype. Thus, the phenotype of a heteroplasmic mutation will depend on the nature and percentage of mutated molecules that are present in a specific tissue. A pathological mutation is usually, although not always, associated with heteroplasmy. However, this does not mean that heteroplasmy is synonymous with pathogenicity, since any mtDNA variant will have passed through a heteroplasmic state before being fixed in the general population. This criterion is not applicable to homoplasmic pathological mutations.

### *5.5.4 The Mutation Should Be the Best mtDNA Candidate Variant to Be Pathologic*

The discovery of a good candidate for pathologic mutation does not preclude the need to resequence the whole mtDNA to rule out the possible existence of a second and more important mutation, thereby demonstrating the importance of whole mitochondrial genome sequencing of individuals suffering an mtDNA disease. Furthermore, there is also the possibility of the co-existence of more than one pathological mutation [7].

### *5.5.5 The Mutation Must Affect Highly Evolutionarily Conserved Nucleotides*

A pathologic mutation usually affects a highly conserved nucleotide. The degree of conservation frequently depends on the functional importance of the nucleotide. However, we can also find pathologic mutations that are not very well evolutionarily conserved, probably because of the existence of compensatory mutations.

### *5.5.6 The Mutation Must Affect Functionally Important Domains*

This is an important issue, although the structure and function of some OXPHOS subunits and the rRNA and tRNA functional domains are still unknown. It is also possible that a mutation may affect important positions involved in the replication, transcription and processing or modification of RNA.

### *5.5.7 Transfer of the mtDNA with a Mutation to Other Cell Lines Must Be Accompanied By Transfer of the Cellular and Molecular Defect: Transmitochondrial Cybrids as a Model to Confirm the Pathogenicity of a New Mutation.*

The previous criteria give some evidence about the pathogenicity of a specific point mutation. However, it is necessary to have functional evidence if we want to confirm the pathogenicity of a new mutation. In this sense, the transmitochondrial cell line model is nowadays considered to provide the strongest evidence to define a candidate mutation as pathologic. The transfer of a particular phenotype by repopulating human cell lines emptied of their mtDNA ($\sigma^0$–cells) with another nuclear background is very good evidence of the involvement of the particular mutated mtDNA [24]. This technique has been used to determine the pathogenicity of human mtDNA point mutations since 1991 [11]. However, the results are not always straightforward since other mitochondrial, nuclear or environmental factors can act as modifying factors, and the ability to detect relevant effects depends on the parameters searched and experimental conditions used.

Although there are some animal models for mtDNA pathologic mutations, none of them reproduce a described human mtDNA pathologic mutation.

## 5.6 mtDNA Haplogrupos and Diseases

As mentioned above, the location of the mtDNA in the mitochondrial matrix and close to the inner membrane makes it vulnerable to ROS and, in consequence, to

**Table 5.4** Mitochondrial haplogroup-associated phenotypes

| mtDNA haplogroups | Phenotype |
|---|---|
| J | Longevity (R) |
| | LHON (S) |
| | Parkinson Disease (R) |
| | Knee Osteoarthritis (R) |
| | Age-related Maculopathy (S) |
| T | DIDMOAD (S) |
| | Astenozoospermia (S) |
| | Alzheimer Disease (R) |
| | NRTIs-Associated Peripheral Neuropathy (S) |
| | Endurance Training |
| H | Lewy Body Dementia (S) |
| H1 | Ischemic Stroke (R) |
| H5 | Alzheimer Disease (S) |
| | Hearing loss and migraine (S) |
| U | Alzheimer Disease (S) |
| Uk | Longevity (R) |
| | Parkinson Disease (R) |
| | Multiple Sclerosis (S) |
| | Age Related Hearing Loss (S) |
| | LHON (S) |
| D | Longevity (R) |
| | Heart attack (R) |
| R | Sepsis (R) |
| N9a | Type 2 Diabetes (R) |

S and R denote susceptibility and resistance, respectively. NRTIs: Nucleotide Reverse Transcriptase inhibitors.

generating mutations at a higher rate than the nDNA. Each new mutation in mtDNA gives rise to a new mitochondrial variant or haplotype. Human population can be classify into haplogroups that are defined as a group of phylogenetically related haplotypes. These mitochondrial variants might modulate oxidative phosphorylation and predispose to or protect from common diseases (Alzheimer, Parkinson, diabetes, age-related diseases, etc.) or to the penetrance of specific phenotypes (Table 5.4).

## 5.7 Oxphos Disorders Without mtDNA Mutations

Besides the 13 polypeptides encoded in mtDNA, most of the protein components of OXPHOS, as well as the enzymes and factors involved in the maintenance, transcription and translation of the mitochondrial genome, as well as the proteins of the import machinery, are encoded by nDNA. Therefore, one can easily think that most of the pathological mutations that cause mitochondrial diseases should affect nuclear genes. In general, most of these mutations produce severe mitochondriopathies with an early onset of the disease. In contrast to mtDNA, nDNA mutations are Mendelian inherited.

The first reported mutation on a nuclear encoded OXPHOS gene was in a patient with Leigh syndrome and complex II deficiency [5]. However, the most common cause of OXPHOS disease is Complex I deficiency. In fact mutations in this gene are responsible for approximately 30% of very variable phenotypes, although the great majority of them also cause Leigh syndrome. However, the percentage of mutations in this complex is not as high as expected, suggesting that they are incompatible with life. Deficiencies of complexes III, IV and V with mutations in nDNA coding genes are, to date, very rare. Thus, only two mutations in the subunit VII gene of complex III, ubiquinol-cytochrome c reductase (UQCRQ) [3], and one in the COX6B1 gene of complex IV [32], causing severe infantile phenotypes, have been reported up to now. As mentioned before the lack of mutations in these genes could be due to their lethal nature [30, 47].

OXPHOS complex assembly requires the participation of nuclear encoded proteins, and mutations in these genes have been associated with important OXPHOS diseases. Assembly proteins have been described for Complexes I, III, IV and V, although the majority of diseases related to these genes have been found in complex III (BCS1L) and complex IV (SURF1, SCO1, SCO2, COX10) [14, 16, 47].

Up to now, only a small number of patients with OXPHOS diseases have presented mutations in nuclear genes coding for structural or assembly genes. However, recent advances open new possibilities. Thus, mutations in nuclear-encoded genes involved in maintenance, RNA metabolism and translation of the mitochondrial genome have been described as causing important disorders. Multiple deletions and depletion of mtDNA have been associated with mutations in the genes for mitochondrial DNA polymerase gamma (POLG and POLG2) and Twinkle helicase, and in thymidine phosphorylase (TP), adenosin nucleoside translocator (ANT1), desoxyguanosin kinase (DGUOK), thymidin kinase (TK2), succinyl-CoA synthetase (SUCLA2), cytosolic p53-inducible ribonucleotide reductase (RRM2B), and

MPV17, which controls the nucleotide pool inside the mitochondria. Also, mutations in ribosomal proteins or elongation factors have been found in very severe multiple OXPHOS defects that impair translation of mtDNA-encoded proteins [35, 50, 56]).

As mentioned above, in relation to mtDNA, mutations in nuclear genes involved in RNA metabolism might also cause diseases. Thus, trans-acting elements can also be related to mitochondrial diseases. TFB1M and TFB2M initiation factors are closely related to rRNA methyltransferases [33]. These transcription factors may play other roles in mitochondrial gene expression. In fact, TFB1M expression in *E. coli* lacking the KsgA rRNA methyltransferase restores dimethylation of two adjacent residues in a stem-loop of bacterial 16S rRNA. Moreover, resistance to the antibiotic kasugamycin is reversed by expression of TFB1M. The target adenine residues of this stem-loop are evolutionarily conserved in the mitochondrial 12S rRNA [48]. This mitochondrial stem-loop is close to the 3'-end of the 12S rRNA and 28 nucleotides downstream from a pathogenic mtDNA mutation (A1555G) that predisposes individuals to deafness. Interestingly, there is evidence that nuclear genetic background influences the phenotypic expression of this mutation. A polymorphism on chromosome 6 near the TFB1M gene, but located outside the coding region, was implicated as a nuclear modifying locus of the 1555G mutation in several pedigrees of mitochondrial deafness [8]. Several clues imply that altered TFB1M expression affects the disease at rRNA methylation. Thus, the methylation status of the analogous bacterial 16S stem-loop determines sensitivity to amynoglycosides, the same antibiotics inducing deafness in patients with the mitochondrial mutation. The bacterial stem-loop structure is altered by the dimethylation of the conserved adenines, which suggests that the mitochondrial homologue structure could also be affected [45].

Mitochondrial mutations in tRNA genes have been associated with different pathologic phenotypes. Recently, a candidate gene for prostate cancer susceptibility, ELAC2, was identified. Several variants of this gene are significantly associated with the occurrence of this disease. Surprisingly, the prostate cancer susceptibility gene encodes tRNase Z, the enzyme required for 3'-end processing of the mitochondrial tRNAs [52].

## 5.8 Oxphos Defects Due to Environmental Factors

Frequently, mitochondrial deafness appears after amynoglycoside treatment. Two mtDNA mutations have been described as associated with mitochondrial deafness. The first one is an A1555G transition in the 12S rRNA and the second one at C1494T. Both of these two mutations rebuild a new Watson-Crick base pair that facilitates amynoglycoside binding to this region of 12S rRNA, important for decoding. Then, mitochondrial protein synthesis is impaired. Thus, environmental factors can interact with particular genotypes and modify the expression of pathologic mutations.

Other environmental factors may contribute, along with mtDNA variants, to the development of mitochondrial diseases. Thus, MPTP, a by-product of the illegal manufacture of abused drugs, acts as an inhibitor of respiratory complex I, producing a phenotype similar to that of Parkinson disease [28]. Annonacine, another complex I inhibitor, found in tropical plants frequently used as food and with medicinal purposes as well, causes Guadalupe Parkinson [29]. Statines, used to decrease cholesterol levels, have been associated with mitochondrial myopathies. Cholesterol and ubiquinone share the first steps of the biosynthetic pathway [17]. Thus, the inhibition of this biochemical pathway could decrease ubiquinone levels and electron transport between respiratory complexes I and II on the one hand and complex III on the other.

## 5.9 mtDNA in Multifactorial Disorders

In recent years, many results have associated the mtDNA variation to multifactorial disorders, such aging and age-linked diseases. These associations support the common variant-common disease hypothesis for multifactorial disorders. According to this hypothesis, the disorder would be due to a rare combination of common genetic variants. Mitochondrial haplogroups are defined by ancient mutations and are therefore relatively well represented in populations. Nonetheless, more recent deleterious variants that could contribute to susceptibility to some pathologies would be rare in populations, lending support to the rare allele hypothesis for multifactorial disorders.

## 5.10 Conclusions

Oxidative phosphorylation diseases are a group of disorders due to deficient production of ATP. The most surprising clinical feature is that they are multisystemic and affect organs unrelated. For many years, mtDNA has been considered the only agent responsible for these diseases, because mtDNA genes only encode for OXPHOS subunits. However, all the proteins involved in the maintenance and expression of this genetic system, most of the subunits of the OXPHOS system, and many other proteins involved in the assembly are encoded in nuclear genes. Thus, mutations in these genes can also produce mitochondrial disorders. mtDNA mutations are maternally inherited or sporadic. They can also be produced by nuclear mutations in genes that regulate mtDNA biosynthesis and, therefore, Mendelian inheritance.

**Acknowledgements** This project was supported by grants from the Instituto de Salud Carlos III-FIS (PI07-0045 and PI05-0647) and the Diputación General de Aragón (Grupos Consolidados B33 and PM063-2007). CIBER de Enfermedades Raras (CIBERER) is an initiative of the ISCIII.

## References


1. Anderson, S., Bankier, A. T., Barrell, B. G., de-Bruijn, M. H. L., Coulson, A. R., Drouin, J., Eperon, I. C., Nierlich, D. P., Roe, B. A., Sanger, F., Schreier, H. P., Smith, A. J. H., Stader, R. and Young, I. G. *Sequence and organization of the human mitochondrial genome*. Nature (1981) 290: 427–465.
2. Andrews, R. M., Kubacka, I., Chinnery, P. F., Lightowlers, R. N., Turnbull, D. M. and Howell, N. Reanalysis and revision of the Cambridge reference sequence for human mitochondrial DNA. Nat Genet (1999) 23: 147.
3. Barel, O., Shorer, Z., Flusser, H., Ofir, R., Narkis, G., Finer, G., Shalev, H., Nasasra, A., Saada, A. and Birk, O. S. Mitochondrial complex III deficiency associated with a homozygous mutation in UQCRQ. Am J Hum Genet (2008) 82: 1211–1216.
4. Bindoff, L. A., Howell, N., Poulton, J., Mccullough, D. A., Morten, K. J., Lightowlers, R. N., Turnbull, D. M. and Weber, K. Abnormal RNA processing associated with a novel transfer RNA mutation in mitochondrial DNA - a potential disease mechanism. J Biol Chem (1993) 268: 19559–19564.
5. Bourgerom, T., Rustin, P., Chretien, D., Birch-Machin, M., Bourgeois, M., Viegas-Pequignot, E., Munnich, A. and Rotig, A. Mutation of a nuclear succinate dehydrogenase gene results in mitochondrial respiratory chain deficiency. Nat Genet (1995) 11: 144–149.
6. Bowmaker, M., Yang, M. Y., Yasukawa, T., Reyes, A., Jacobs, H. T., Huberman, J. A. and Holt, I. J. Mammalian mitochondrial DNA replicates bidirectionally from an initiation zone. J Biol Chem (2003) 278: 50961–50969.
7. Brown, M. D., Allen, J. C., Van Stavern, G. P., Newman, N. J. and Wallace, D. C. Clinical, genetic, and biochemical characterization of a Leber hereditary optic neuropathy family containing both the 11778 and 14484 primary mutations. Am J Med Genet (2001) 104: 331–338.
8. Bykhovskaya, Y., Mengesha, E., Wang, D., Yang, H., Estivill, X., Shohat, M. and Fischel-Ghodsian, N. Human mitochondrial transcription factor B1 as a modifier gene for hearing loss associated with the mitochondrial A1555G mutation. Mol Genet Metab (2004) 82: 27–32.
9. Chomyn, A., Mariottini, P., Cleeter, M. W. J., Ragan, C. I., Doolittle, R. F., Matsuno-Yagi, A., Hatefi, Y. and Attardi, G. Functional assignment of the products of the unidentified reading frames o human mitochondrial DNA. In: Achievements and Perspectives of Mitochondrial Research (Quagliarello, E., Slater, E. C., Plamieri, F., Saccone, C. and Kroon, A. M. Amsterdam, eds). Volume II Elsevier Sciences. 1985, 259–275.
10. Chomyn, A., Martinuzzi, A., Yoneda, M., Daga, A., Hurko, O., Johns, D., Lai, S. T., Nonaka, I., Angelini, C. and Attardi, G. MELAS mutation in mtDNA binding site for transcription termination factor causes defects in protein synthesis and in respiration but no change in levels of upstream and downstream mature transcritps. Proc Natl Acad Sci U S A (1992) 89: 4221–4225.
11. Chomyn, A., Meola, G., Bresolin, N., Lai, S. T., Scarlato, G. and Attardi, G. In vitro genetic transfer of protein synthesis and respiration defects to mitochondrial DNA-less cells with myopathty-patient mitochodria. Mol Cell Biol (1991) 11: 2236–2244.
12. Clayton, D. A. Replication of animal mitochondrial DNA. Cell (1982) 28: 693–705.
13. DiMauro, S. and Bonilla, E. Mitochondrial encephalomyopathies. In: The molecular and Genetic Basis of neurological diseases (Rosenberg, R. N., Prusiner, S. B., DiMauro, S. and Barchi, R. L., eds). Boston, Butterworth-Heinemann: 1998, 201–235.
14. Dimauro, S. and Hirano, M. Mitochondrial encephalomyopathies: an update. Neuromuscul Disord (2005) 15: 276–286.
15. DiMauro, S., Hirano, M. and Schon, E. A., Eds.. Mitochondrial Medicine. Abingdon, Informa Health Care: 2006.
16. Dimauro, S. and Schon, E. A. Mitochondrial disorders in the nervous system. Annu Rev Neurosci (2008) 31: 91–123.

17. Evans, M. and Rees, A. Effects of HMG-CoA reductase inhibitors on skeletal muscle: are all statins the same? Drug Safety (2002) 25: 649–663.
18. Guan, M. X., Enriquez, J. A., Fischel-Ghodsian, N., Puranam, R. S., Lin, C. P., Maw, M. A. and Attardi, G. The deafness-associated mitochondrial DNA mutation at position 7445, which affects tRNASer(UCN) precursor processing, has long-range effects on NADH dehydrogenase subunit ND6 gene expression. Mol Cell Biol (1998) 18: 5868–5879.
19. Hao, H. L. and Moraes, C. T. Functional and molecular mitochondrial abnormalities associated with a C->T transition at position 3256 of the human mitochondrial genome – the effects of a pathogenic mitochondrial tRNA point mutation in organelle translation and RNA processing. J Biol Chem (1996) 271: 2347–2352.
20. Herrero-Martin, M. D., Pineda, M., Briones, P., Lopez-Gallardo, E., Carreras, M., Benac, M., Angel Idoate, M., Vilaseca, M. A., Artuch, R., Lopez-Perez, M. J., Ruiz-Pesini, E. and Montoya, J. A new pathologic mitochondrial DNA mutation in the cytochrome oxidase subunit I (MT-CO1). Hum Mutat (2008) 29: E103–E111.
21. Hess, J. F., Parisi, M. A., Bennett, J. L. and Clayton, D. A. Impairment of mitochondrial transcription termination by a point mutation associated with the MELAS subgroup of mitochondrial encephalomyopathies. Nature (1991) 351: 236–239.
22. Holt, I. J., Harding, A. E. and Morgan-Hughes, J. A. Deletions of muscle mitochondrial DNA in patients with mitochondrial myopathies. Nature (1988) 331: 717–719.
23. Kasamatsu, H., Robberson, D. L. and Vinograd, J. A novel closed-circular mitochondrial DNA with properties of a replicating intermediate. Proc Natl Acad Sci U S A (1971) 68: 2252–2257.
24. King, M. P. and Attardi, G. Human cells lacking mtDNA: repopulation with exogenous mitochondria by complementation. Science (1989) 246: 500–503.
25. King, M. P., Koga, Y., Davidson, M. and Schon, E. A. Defects in mitochondrial protein synthesis and respiratory chain activity segregate with the transfer RNA(Leu)(UUR) mutation associated with mitochondrial myopathy, encephalopathy, lactic acidosis, and strokelike episodes. Mol Cell Biol (1992) 12: 480–490.
26. Koga, Y., Davidson, M., Schon, E. A. and King, M. P. Fine mapping of mitochondrial RNAs derived from the mtDNA region containing a point mutation associated with MELAS. Nucleic Acids Res (1993) 21: 657–662.
27. Kruse, B., Narasimhan, N. and Attardi, G. Termination of transcription in human mitochondria: identification and purification of a DNA binding protein factor that promotes termination. Cell (1989) 58: 391–397.
28. Langston, J. W., Ballard, P., Tetrud, J. W. and Irwin, I. Chronic Parkinsonism in humans due to a product of meperidine-analog synthesis. Science (1983) 219: 979–980.
29. Lannuzel, A., Michel, P. P., Hoglinger, G. U., Champy, P., Jousset, A., Medja, F., Lombes, A., Darios, F., Gleye, C., Laurens, A., Hocquemiller, R., Hirsch, E. C. and Ruberg, M. The mitochondrial complex I inhibitor annonacin is toxic to mesencephalic dopaminergic neurons by impairment of energy metabolism. Neuroscience (2003) 121: 287–296.
30. Loeffen, J. L. C. M., Smeitink, J. A. M., Trijbels, J. M. F., Janssen, A. J. M., Triepels, R. H., Sengers, R. C. A. and van den Heuvel, L. P. Isolated complex I deficiency in children: clinical, biochemical and genetic aspects. Hum Mutat (2000) 15: 123–134.
31. Martin, M., Cho, J., Cesare, A. J., Griffith, J. D. and Attardi, G. Termination factor-mediated DNA loop between termination and initiation sites drives mitochondrial rRNA synthesis. Cell (2005) 123: 1227–1240.
32. Massa, V., Fernandez-Vizarra, E., Alshahwan, S., Bakhsh, E., Goffrini, P., Ferrero, I., Mereghetti, P., D'Adamo, P., Gasparini, P. and Zeviani, M. Severe infantile encephalomyopathy caused by a mutation in COX6B1, a nucleus-encoded subunit of cytochrome C oxidase. Am J Hum Genet (2008) 82: 1281–1289.
33. McCulloch, V., Seidel-Rogol, B. L. and Shadel, G. S. A human mitochondrial transcription factor is related to RNA adenine methyltransferases and binds s-adenosylmethionine. Mol Cell Biol (2002) 22: 1116–1125.

34. Michikawa, Y., Mazzucchelli, F., Bresolin, N., Scarlato, G. and Attardi, G. Aging-dependent large accumulation of point mutations in the human mtDNA control region for replication. Science (1999) 286: 774–779.
35. Miller, C., Saada, A., Shaul, N., Shabtai, N., Ben-Shalom, E., Shaag, A., Hershkovitz, E. and Elpeleg, O. Defective mitochondrial translation caused by a ribosomal protein (MRPS16) mutation. Ann Neurol (2004) 56: 734–738.
36. Montoya, J., Christianson, T., Levens, D., Rabinowitz, M. and Attardi, G. Identification of initiation sites for heavy strand and light strand transcription in human mitochondrial DNA. Proc Natl Acad Sci U S A (1982) 79: 7195–7199.
37. Montoya, J., Gaines, G. L. and Attardi, G. The pattern of transcription of the human mitochondrial rRNA genes reveals two overlapping transcription units. Cell (1983) 34: 151–159.
38. Montoya, J., Lopez-Gallardo, E., Diez-Sanchez, C., Lopez-Perez, M. J. and Ruiz-Pesini, E. 20 years of human mtDNA pathologic point mutations: Carefully reading the pathogenicity criteria. Bba Bioenergetics (2009) 1787: 476–483.
39. Montoya, J., Lopez-Perez, M. J. and Ruiz-Pesini, E. Mitochondrial DNA transcription and diseases: past, present and future. Biochim Biophys Acta (2006) 1757: 1179–1189.
40. Montoya, J., Ojala, D. and Attardi, G. Distinctive features of the 5`-terminal sequences of the human mitochondrial mRNAs. Nature (1981) 290: 465–470.
41. Nam, S. C. and Kang, C. DNA light-strand preferential recognition of human mitochondria transcription termination factor mTERF. J Biochem Mol Biol (2005) 38: 690–694.
42. Ojala, D., Merkel, C., Gelfand, R. and Attardi, G. The tRNA genes punctuate the reading of genetic information in human mitochondrial DNA. Cell (1980) 22: 393–403.
43. Ojala, D., Montoya, J. and Attardi, G. The putative mRNA per subunit II of human cytochrome c starts directly at the translation initiation codon. Nature (1980) 287: 79–82.
44. Ojala, D., Montoya, J. and Attardi, G. tRNA punctuation model of RNA processing in human mitochondria. Nature (1981) 290: 470–474.
45. Rife, J. P. and Moore, P. B. The structure of a methylated tetraloop in 16S ribosomal RNA. Structure (1998) 6: 747–756.
46. Robberson, D. L., Kasamatsu, H. and Vinograd, J. Replication of mitochondrial DNA. Circular replicative intermediates in mouse L cells. Proc Natl Acad Sci U S A (1972) 69: 713–741.
47. Ruiz-Pesini, E., Lopez-Gallardo, E., Dahmani, Y., Herrero, M. D., Solano, A., Diez-Sanchez, C., Lopez-Perez, M. and Montoya, J. Diseases of the human mitochondrial oxidative phosphorylation system. Rev Neurol (2006) 43: 416–424.
48. SeidelRogol, B. L., McCulloch, V. and Shadel, G. S. Human mitochondrial transcription factor B1 methylates ribosomal RNA at a conserved stem-loop. Nat Genet (2003) 33: 23–24.
49. Skladal, D., Halliday, J. and R., T. D. Minimum birth prevalence of mitochondrial respiratory chain disorders in children. Brain: J Neurol (2003) 126: 1905–1912.
50. Smeitink, J. A., Elpeleg, O., Antonicka, H., Diepstra, H., Saada, A., Smits, P., Sasarman, F., Vriend, G., Jacob-Hirsch, J., Shaag, A., Rechavi, G., Welling, B., Horst, J., Rodenburg, R. J., van den Heuvel, B. and Shoubridge, E. A. Distinct clinical phenotypes associated with a mutation in the mitochondrial translation elongation factor EFTs. Am J Hum Genet (2006) 79: 869–877.
51. Sutovsky, P., Neuber, E. and Schatten, G. Ubiquitin-dependent sperm quality control mechanism recognizes spermatozoa with DNA defects as revealed by dual ubiquitin-TUNEL assay. Mol Reprod Dev (2002) 61: 406–413.
52. Takaku, H., Minagawa, A., Takagi, M. and Nashimoto, M. A candidate prostate cancer susceptibility gene encodes tRNA 3′ processing endoribonuclease. Nucleic Acids Res (2003) 31: 2272–2278.
53. Temperley, R. J., Seneca, S. H., Tonska, K., Bartnik, E., Bindoff, L. A., Lightowlers, R. N. and Chrzanowska-Lightowlers, Z. M. Investigation of a pathogenic mtDNA microdeletion

reveals a translation-dependent deadenylation decay pathway in human mitochondria. Hum Mol Genet (2003) 12: 2341–2348.
54. Tiranti, V., Savoia, A., Forti, F., DApolito, M. F., Centra, M., Racchi, M. and Zeviani, M. Identification of the gene encoding the human mitochondrial RNA polymerase (h-mtRPOL) by cyberscreening of the expressed sequence tags database. Hum Mol Genet (1997) 6: 615–625.
55. Toompuu, M., Levinger, L. L., Nadal, A., Gomez, J. and Jacobs, H. T. The 7472insC mtDNA mutation impairs 5′ and 3′ processing of tRNA(Ser(UCN)). Biochem Biophys Res Commun (2004) 322: 803–813.
56. Valente, L., Tiranti, V., Marsano, R. M., Malfatti, E., Fernandez-Vizarra, E., Donnini, C., Mereghetti, P., De Gioia, L., Burlina, A., Castellan, C., Comi, G. P., Savasta, S., Ferrero, I. and Zeviani, M. Infantile encephalopathy and defective mitochondrial DNA translation in patients with mutations of mitochondrial elongation factors EFG1 and EFTu. Am J Hum Genet (2007) 80: 44–58.
57. Wallace, D. C., Ruiz-Pesini, E. and Mishmar, D. mtDNA variation, climatic adaptation, degenerative diseases, and longevity. Cold Spring Harb Symp Quant Biol (2003) 68: 479–486.
58. Wallace, D. C., Singh, G., Lott, M. T., Hodge, J. A., Schurr, T. G., Lezza, A. M. S., II, L. J. E. and Nikoskelainen, E. K. Mitochondrial DNA mutation associated with Leber′s hereditary optic neuropathy. Science (1988) 242: 1427–1430.
59. Yan, H., Zareen, N. and Levinger, L. Naturally occurring mutations in human mitochondrial pre-tRNASer(UCN) can affect the tRNase Z cleavage site, processing kinetics and substrate secondary structure. J Biol Chem (2006) 281: 3926–3935.
60. Yasukawa, T., Yang, M. Y., Jacobs, H. T. and Holt, I. J. A bidirectional origin of replication maps to the major noncoding region of human mitochondrial DNA. Mol Cell (2005) 18: 651–662.
61. Zeviani, M., Moraes, C. T., DiMauro, S., Nakase, H., Bonilla, E., Schon, E. and Rowland, L. P. Deletions of mitochondrial DNA in Kearns-Sayre syndrome. Neurology (1988) 38: 1339–1346.

# Chapter 6
# Mitochondrial Diseases: A Cross-Talk Between Mitochondrial and Nuclear Genomes

**Antonella Spinazzola and Massimo Zeviani**

**Abstract** More than one billion years ago, mitochondria were free-living prokaryotic organisms with their own DNA. However, during the evolution, ancestral genes have been transferred from the mitochondrial to the nuclear genome so that mtDNA became dependent on numerous nucleus-encoded factors for its integrity, replication and expression. Mutations in any of these factors may alter the cross-talk between the two genomes and cause Mendelian diseases that affect mtDNA integrity or expression.

**Keywords:** Mitochondrial disease · Mitochondrial DNA · mtDNA depletion · mtDNA multiple deletions · Nucleus · Mitochondrion cross-talk

## 6.1 Introduction

Mitochondrial disorders are clinical phenotypes associated with abnormalities of the terminal component of aerobic energy metabolism, i.e. oxidative phosphorylation (OXPHOS). The main metabolic consequences of OXPHOS impairment include accumulation of metabolic intermediates, increased generation of reactive oxygen species (ROS) and decreased ATP production. OXPHOS is carried out in the inner mitochondrial membrane by a series of multiheteromeric enzyme complexes plus electron-shuttle molecules, collectively constituing the respiratory chain. The formation of the respiratory chain is under the control of two separate genetic systems, the nuclear genome and the mitochondrial genome (mitochondrial DNA, mtDNA). In humans, thirteen of the approximately 85 proteins composing the five major OXPHOS complexes are encoded by the 16.6 kb mtDNA. Seven mtDNA-encoded polypeptides are subunits of complex I (NADH:ubiquinone oxidoreductase, cI), three are subunits of complex IV (cytochrome *c* oxidase, cIV),

M. Zeviani (✉)
Unit of Molecular Neurogenetics, Foundation IRCCS Neurological Institute "C. Besta", via Libero Temolo 4, 20126 Milano, Italy
e-mail: zeviani@istituto-besta.it

C. Espinós et al. (eds.), *Inherited Neuromuscular Diseases*, Advances in Experimental Medicine and Biology 652, DOI 10.1007/978-90-481-2813-6_6,

two are subunits of the $F_O$ (integral membrane) domain of $F_1F_O$-ATP synthase, cV, and one, cytochrome *b*, is part of complex III (ubiquinol-cytochrome *c* oxidoreductase, cIII). In addition, mtDNA encodes 2 rRNAs and 22 tRNAs that contribute to the autonomous translational apparatus of mitochondria, dedicated to the synthesis of the 13 mtDNA-encoded polypeptides. All other structural subunits, including all four components constituting complex II (succinate-ubiquinone oxidoreductase, cII), the assembly factors of the OXPHOS complexes, the enzymatic apparatus providing the biosynthesis of cofactors and non-protein components of the respiratory chain (e.g. CoQ biosynthesis), as well as all of the proteins involved in the replication, transcription and translation of mtDNA, are nuclear-coded and targeted either co- or post-translationally to the mitochondria.

According to recent data obtained from integrative analysis, the mitochondrial proteome is nowadays estimated to consist of approximately 1500 gene products [6]. The number of candidate genes potentially able to determine mitochondrial disease is believed to outnumber this figure.

## 6.2 Genetics and Pathophysiology

A clinical-genetic classification of mitochondrial disorders due to mutations in nuclear genes includes: (i) disorders due to mutations in respiratory chain subunits; (ii) disorders due to mutations in ancillary proteins; (iii) disorders due to faulty intergenomic communication affecting the maintenance and expression of mtDNA; (iv) disorders due to defects in biosynthetic enzymes for lipids or cofactors; (v) coenzyme Q-deficiency and disorders of the lipid milieu of the inner mitochondrial membrane; (vi) disorders due to defective mitochondrial trafficking and transport machinery; (vii) disorders due to gene defects encoding proteins involved in mitochondrial biogenesis.

In this chapter we focus on the Mendelian mitochondriopathies affecting mtDNA stability and integrity.

### *6.2.1 Stability and the Integrity of mtDNA: Pathophysiological Aspects*

Mitochondrial mtDNA maintenance is a complex process. In the yeast *Saccharomyces cerevisiae* over a hundred genes have been identified, deletion of which causes loss or instability of mtDNA [9]. However, both factors and mechanisms regulating mtDNA stability are only beginning to be decoded in mammals. The progress made in yeast, as well as in bacteria and bacteriophages, can be very effective to understanding human pathologies, since orthologs of mammalian proteins related to mtDNA maintenance are conserved in different phyla of protist organims.

Two models of mtDNA replication have been proposed. According to the orthodox "strand displacement" or "asynchronous" model, there are two spatially and temporally separated origins of replication, one for each mtDNA strand. The process of mtDNA replication begins at the heavy strand origin ($O_H$), located in the mtDNA non-coding region (displacement loop, D-loop), and proceeds in an anticlockwise direction until 2/3 of the mtDNA circle, where the origin of light-strand replication ($O_L$) is exposed. This allows light-strand synthesis to proceed in the clockwise direction until the entire molecule is copied [8]. The resulting concatamer is then resolved by the action of a topoisomerase. A few years ago, Holt and colleagues [27] proposed a "strand-synchronous" model of mtDNA replication, based on experimental evidence from 2-dimension gel electrophoresis of mtDNA replication intermediates. According to this model, replication is initiated from multiple origins, distributed across a 4 Kb fragment 3′ from the non-coding D-loop and proceeds in both directions in replication "bubbles". A relevant role is played by mt-RNA complementary strands that intervene in, and are possibly acting as primers of, mtDNA replication.

Irrespective of the model, mtDNA replication requires a number of nuclear-encoded proteins, few of which have been identified in humans, including a DNA polymerase activity provided by pol-g, the only DNA polymerase found in mitochondria, a mitochondrial single-strand binding-protein (mtSSB), a DNA-RNA primase/helicase, denominated Twinkle, a DNA ligase and several topoisomerases. As we alluded to earlier, replication-initiation is provided by RNA primers that couple replication with transcription, a complex process which is in turn carried out by a set of factors that include the mtDNA-specific RNA polymerase, and several transcription factors and regulators: TFAM TFB1M, TFB2M, mTERF, etc. The supply of deoxy-nucleotides and ribonucleotides, the building blocks of mtDNA replication and transcription, is essential for both processes and is carried out by complex metabolic pathways partly distinct for purines and pyrimidines. Finally, a number of proteins, including some of those mentioned above, are involved in the formation of the mitochondrial nucleoid, a dynamic structure containing one or more copies of mtDNA, which represents the fundamental segregation unit of mtDNA inheritance.

In principle, any defective protein involved in mtDNA replication, maintenance, and integrity could precipitate loss or instability of mtDNA, causing either qualitative (multiple deletions) or quantitative (depletions) mtDNA molecular lesions. So far, only a few such proteins have been identified as responsible of human mitochondrial disease.

#### 6.2.1.1 Clinical Manifestations

Autosomal disorders classified as defects of nuclear-mitochondrial intergenomic communication can be associated with the accumulation of mtDNA large-scale rearrangements (mtDNA breakage syndromes) or characterized by severe reduction of the mtDNA copy number. The most frequent clinical presentations are:

(1) an adult-onset encephalomyopathy, defined clinically by PEO, genetically by the autosomal dominant (ad) or, more rarely, recessive (ar) transmission of the trait, and molecularly by the presence of multiple deletions of mtDNA.

(2) a spectrum of recessive neurological syndromes ranging from typical infantile hepatopathic poliodystrophy (Alpers-Huttenlocher syndrome) to juvenile onset sensory-ataxia neuropathy, dysarthria and ophthalmoplegia to a combination of spinocerebellar ataxia and epilepsy with or without external ophthalmoplegia..
(3) an autosomal recessive multisystem disorder known as mitochondrial neuro-gastrointestinal encephalomyopathy, characterized by combined accumulation of multiple deletions and partial depletion of mtDNA
(4) early-onset, organ-specific autosomal recessive syndromes associated with profound mtDNA depletion.

*Autosomal dominant progressive external ophthalmoplegia* (adPEO) is characterized by the accumulation of multiple mtDNA deletions in postmitotic tissues. The typical clinical feature of adPEO is progressive muscle weakness, most severely affecting extraocular muscles, determining the progression of bilateral eyelid ptosis and severe limitation of eye movements. The disease has adult-onset between 20 and 40 years of age. Additional features vary among families; they may include ataxia, sensorineural hearing loss, cataracts, hypogonadism, parkinsonism, and psychiatric abnormalities consisting of severe depression and avoidant personality. Dysphagia, dysphonia, weakness of facial muscles, and peripheral neuropathy may be prominent symptoms in selected families [60, 80]. At rest, elevated levels of plasma lactate are detected only in severely affected patients. Symptoms seem to progress along with the age of the patients. Skeletal muscle shows RRFs and a mild reduction in the activities of respiratory chain enzymes [24, 60].

CPEO has been reported in association with deafness, ataxia, axonal sensory-motor polyneuropathy, and mitochondrial myopathy with cytochrome c oxidase negative and Ragged Red Fibres to delineate a syndromic form of dominant optic atrophy (DOA).

*Autosomal recessive progressive ophthalmoplegia (arPEO)*. In 1989 the presence of multiple mtDNA deletions was reported in muscle specimens from 2 siblings with PEO, optic atrophy, muscle weakness, and peripheral neuropathy [79]. However, in contrast to autosomal dominant PEO families, the 2 siblings were the only affected members of the pedigree and were born from consanguineous, apparently healthy parents, suggesting an autosomal recessive transmission [44]. Since then, multiple deletions of mtDNA have been reported in numerous sporadic PEO cases or in other families in which PEO was clearly transmitted as a recessive trait (up to 11% in our series) [1, 37].

*Sensory-ataxia neuropathy, dysarthria and ophthalmoplegia; spinocerebellar ataxia-epilepsy syndrome; Alpers' syndrome*. These syndromes are all characterized by an association with recessive mutations of *POLG1*, the master gene of mtDNA replication, and form a continuum ranging from Alpers' syndrome (hepatopathic poliodystrophy, see below), to childhood- or juvenile-onset autosomal recessive, progressive sensory-ataxic syndromes, with or without epilepsy. Sensory ataxic neuropathy with dysarthria and ophthalmoplegia (SANDO) is used to classify some of these families, but in other families, cerebellar signs, myoclonus, and seizures are additional prominent findings.

*Mitochondrial neurogastrointestinal encephalomyopathy* (MNGIE) is an autosomal recessive disorder characterized by the combination of 6 features: (1) progressive external ophthalmoplegia, (2) severe gastrointestinal dysmotility, (3) cachexia, (4) peripheral neuropathy predominantly demyelinating, (5) diffuse leukoencephalopathy, (6) evidence of mitochondrial dysfunction (histological, biochemical and or genetic abnormalities of mitochondria) [25, 51]).

*MtDNA depletion syndromes.* In contrast to other types of mtDNA defects, mtDNA depletion syndrome (MDS) is a quantitative abnormality: there is paucity of mtDNA, but the remaining mtDNA does not harbor any mutations or rearrangements. MDS is transmitted as an autosomal recessive trait, and is phenotypically heterogeneous [65].

Some children present with myopathy, others with liver failure in infancy, and some with multisystem involvement. Consistent with the different phenotypes, mtDNA depletion may affect either a specific tissue (most commonly muscle or liver) or multiple organs, including heart, brain, and kidney. However, three well-established forms of MDS are known: myopathic (OMIM #609560), encephalomyopathic (OMIM #612073) and hepatocerebral (OMIM #251880).

In the myopathic form, affected children are typically born after uncomplicated pregnancy, although decreased fetal movements are noted in some cases. The patient usually present in the first year of life with feeding difficulty, failure to thrive, hypotonia, weakness, and occasionally PEO. Death is usually due to pulmonary insufficiency and infections, but some patients survive into their teens [45, 52, 70] . The clinical spectrum includes (a) a spinal muscular atrophy type 3, (b) rigid spine syndrome, (c) a severe muscle weakness with marked dystrophic alterations, encephalopathy and seizures [19], (d) a milder myopathic phenotype with no motor regression and longer survival [52]. A severe variant form that combines a floppy-infant syndrome with renal proximal tubulopathy and nephrocalcinosis has been also described [4, 5]. Muscle biopsy may show proliferation of mitochondria (ragged-red fibers) and patchy or diffuse COX deficiency. Biochemical defects of all mtDNA-related respiratory chain complexes are always present in muscle mitochondria. Patients with mtDNA depletion in muscle tend to have elevated serum creatine kinase (CK) levels, ranging from 2 to 30 times the upper limit of normal. This is an important clue for the diagnosis, because increased serum CK is relatively uncommon in patients with other mitochondrial myopathies.

Encephalomyopathic forms. Two forms have been reported. The first is characterized by high lactate in blood, severe psychomotor retardation with muscle hypotonia, impaired hearing and generalized seizures, followed by knee and hip contractures, finger dystonia and mild ptosis. Brain MRI is suggestive of Leigh syndrome. Analysis of muscle biopsy revealed moderate mtDNA depletion (about 32%) [7, 14, 54]. The second form is associated with extremely severe combination of muscle and liver dysfunction, dysmorphic features, connatal lactic acidosis and death in the first days of life [54]. Both syndromes are hallmarked by methylmalonic aciduria, due to the accumulation of succinyl-CoA that lead to an increased concentration of methylmalonyl-CoA and therefore the methylmalonic acid.

Hepatocerebral form. In this form onset of symptoms is between birth and 6 months and death usually occurs within 1 year of age. The most common symptoms and signs include persistent vomiting, failure to thrive, hypotonia, and hypoglycemia associated with progressive neurological symptoms. Histological changes on liver biopsy include fatty degeneration, bile duct proliferation, fibrosis, and collapse of lobular architecture. Reduced COX histochemistry and combined deficiency of mtDNA encoded respiratory chain complexes were found in the liver of a few patients. A variant form of hepatocerebral MDS affects the Navajo people, with a prevalence of 1:1600 live births, hence the term Navajo Neuro-Hepatopathy (NNH). The major clinical features are hepatopathy, peripheral neuropathy, corneal anesthesia and scarring, acral mutilation, cerebral leukoencephalopathy, failure to thrive, and recurrent metabolic acidosis with intercurrent infections [74]. A peculiar form of hepatocerebral MDS is Alpers-Huttenlocher syndrome, an early onset, fatal disease, characterized by hepatic failure, intractable seizures, evolving into *epilepsia partialis continua*, and global neurological deterioration. The liver dysfunction is usually progressive as well, evolving from miscrovesicular staetosis with bile duct proliferation into cirrhosis and chronic liver failure. Using of valproic acid as a therapy for epilepsy can precipitate acute liver failure.

### 6.2.1.2 Etiology

Autosomal dominant PEO is a genetically heterogeneous clinical entity. Most of the adPEO families carry heterozygous mutations in one of three genes: *ANT1*, encoding the muscle-heart-specific mitochondrial adenine nucleotide translocator [33], *Twinkle*, encoding a mtDNA helicase [62], and *POLG1*, encoding the catalytic subunit of the mtDNA-specific polymerase gamma [71] (Table 6.1).

#### *ANT1* Mutations

The gene responsible for the adPEO form linked to the 4p locus encodes the muscle-heart specific isoform of the mitochondrial adenine nucleotide translocator [33]. Dominant missense mutations have been found in families with adPEO and in sporadic patients with mild, slowly progressive myopathy and little or no extramuscular symptoms. In 2005, Palmieri et al. [55] reported the first recessive mutation in the *ANT1* gene in a patient who presented with hypertrophic cardiomyopathy, mild myopathy with exercise intolerance, RRF and lactic acidosis, but no ophthalmoplegia. Southern blot analysis disclosed multiple deletions of muscle mitochondrial DNA and virtually no ATP uptake was measured in proteoliposomes reconstituted with protein extracts from muscle of this patient. Recently Galassi et al. [18] reported a patient with a clinical presentation characterized initially by PEO with mtDNA multiple deletions lately evolving into a severe neurological syndrome, which included sensory and cerebellar ataxia, peripheral neuropathy, parkinsonism, and depression. This complex phenotype is the result of mutations in two distinct proteins, ANT1 and Polgamma A, which cause additive, deleterious

**Table 6.1** Genes associated with disorders of integenomic communication

| Gene | Phenotype | Molecular features | Gene product | Inheritance |
|---|---|---|---|---|
| *mtDNA replication* | | | | |
| *POLG1* | PEO, PEO plus, Alpers' syndrome, SCAE, SANDO | multiple mtDNA deletions, mtDNA depletion | Catalytic subunit mt polymerase gamma | AR, AD |
| *POLG2* | PEO | multiple mtDNA deletions | Accessory subunit mt polymerase gamma | AD |
| *PEO1 (TWINKLE)* | PEO, IOSCA, Alpers-like phenotype | mtDNA depletion, multiple mtDNA deletions | mtDNA helicase | AD, AR |
| *Nucleotide metabolism and other functions* | | | | |
| *TYMP* | MNGIE | multiple mtDNA deletions, mtDNA depletion | Thymidine phosphorylase | AR |
| *DGUOK* | MDS (hepatocerebral) | mtDNA depletion | Deoxyguanine kinase | AR |
| *TK2* | MDS (myopathic) | mtDNA depletion | Thymidine kinase | AR |
| *SUCLA2* | MDS (encephalomyopathic) | mtDNA depletion | β-subunit of ADP-forming succinyl Coenzime A synthetase ligase | AR |
| *SUCLG1* | MDS (encephalomyopathic) | mtDNA depletion | β-subunit of succinyl-Coenzime A ligase | AR |
| *RRM2B* | MDS (myopathic +renal) | mtDNA depletion | p53-inducible ribonucleotide reductase | AR |
| *SLC25A4 (ANT1)* | PEO, cardiomyopathy | multiple mtDNA deletions | Adenine nucleotide translocator | AD, AR |
| *MPV17* | MDS (hepatocerebral) | mtDNA depletion | Unknown function | AR |
| *OPA1* | Optic atrophy, ADOA plus | multiple mtDNA deletions | Dynamin-related GTPase | AD |

effects on mtDNA maintenance and integrity. *ANT1* mutations are responsible for approximately 7% of the adPEO cases in our series.

## Twinkle Mutations

The gene responsible for the adPEO form linked to the 10q locus encodes the mitochondrial DNA and RNA helicase involved in replication of the mitochondrial genome [62]. Mutations in *TWINKLE*, also known as *PEO1,* are mainly associated with "pure" PEO. However, some patients can develop L-dopa responsive parkinsonism. In muscle, multiple mtDNA deletions are not always detectable [12]. Mouse transgenic models that overexpressed several of the PEO mutations in *TWINKLE* recapitulated many of the characteristic of human PEO, including multiple mtDNA deletions, progressive respiratory chain dysfunction, and cytochrome c oxidase deficiency. A specific, recessive twinkle mutation, Y508C change, causes Infantile Onset SpinoCerebellar Ataxia (IOSCA) [48] characterized by a combination of ataxia, athetosis, areflexia, muscle hypotonia and severe epilepsy. Ophthalmoplegia, hearing loss, and optic atrophy may appear later in the disease course. Recently, Hakonen et al. studied the consequences of IOSCA for the central nervous system, as well as the in vitro performance of the IOSCA mutant protein. As it has been already shown IOSCA brain does not harbour mtDNA deletions or increased amount of mtDNA point mutations. However, IOSCA show mtDNA depletion in the brain and the liver and especially large neurons show respiratory chain complex I and IV deficiency [22], in spite that helicase activity, hexamerization and nucleoid structure are unaffected. The latter result suggests that the Twinkle IOSCA mutation affects mtDNA maintenance in a highly context and cell-type specific manner.

## POLG1 Mutations

In humans mtDNA is copied by a 195 kDa heterotrimer consisting of a catalytic subunit, encoded by the *POLG1* gene, located on chromosome 15q25, and two identical accessory subunits, encoded by *POLG2*, located on chromosome 17q. Several studies performed in the last 5 years have demonstrated that mutations in *POLG1* are a major cause of human mitochondrial disease. So far more than 100 mutations in *POLG1* have been reported, and can be seen at http://dir-apps.niehs.nih.gov/polg/index.cfm . This gene is the most frequent cause of adPEO (50% of the cases in our series) [37, 71]. In adPEO due to *POLG1* mutations, prominent features are severe dysphagia and dysphonia and, occasionally, a movement disorder including parkinsonism, cerebellar dysfunction, or chorea [41]. Mental retardation, hypogonadism, including precocious menopause, and gastrointestinal dysmotility may be additional findings [17, 41].

Recessive mutations of *POLG1* are also responsible for a variety of other syndromes, including most of the (i) autosomal recessive PEO cases [37], (ii) the apparently sporadic PEO cases associated with the accumulation of multiple mtDNA deletions [1], (iii) different recessive disorders such as sensory ataxic neuropathy, dysarthria, ophthalmoplegia, SANDO, [73] or its variants (e.g. spinocerebellar ataxia and epilepsy, SCAE), (iv) infantile Alpers-Huttenlocher syndrome, characterized by liver insufficiency and gray matter disorder (poliodystrophy) [10, 15, 47]. Two mutant alleles carrying mutations in the spacer region of the *POLG1* protein (A467T and S748W) are recurrent in all these conditions [28, 72, 75], which may help the diagnostic workout in suspected cases. Recently two novel mutations in POLG1 were reported in a compound heterozygous patient with autosomal recessive PEO, followed by pseudo-orthostatic tremor evolving into levodopa-responsive parkinsonism. This observation support the hypothesis that mtDNA dysfunction is engaged in the pathogenesis of idiopathic Parkinson′s disease [23].

The molecular basis of this clinical heterogeneity can be explained in part by the structural and functional complexity of the enzyme. Pol-gA, the 145 kDa catalytic subunit encoded by POLG1, comprises an N-terminal exonuclease domain, with predominantly proofreading functions, and a polymerase domain, which performs the template-directed synthesis of the nascent mtDNA strands. Interestingly, recombinant mice carrying proofreading-deficient POLG enzymes show accumulation of mtDNA mutations, ultimately leading to precocious aging [69].

A single heterozygous dominant mutation in the gene encoding the accessory subunit *POLG2* was reported in a 60–year-old woman with adult-onset PEO, cardiac conduction defect, and increased CK [40].

The search for nuclear genes associated with PEO is not over, since several families or sporadic PEO cases with multiple mtDNA deletions have failed to show mutations in the above genes.

### OPA1 Mutations

OPA1 is a dynamin-like GTPase located in the inner mitochondrial membrane involved in mitochondrial fusion, cristae organization [13] and control of apoptosis [39]. *OPA1* is linked to non-syndromic autosomal dominant optic atrophy (ADOA) [2, 11], a condition characterized by slowly progressive visual loss starting in childhood, first described by the Danish ophthalmologist Paul Kjer in 1959 [35]. However, OPA1 missense mutations have been identified in different families with a more complex phenotype, consisting of a combination of ADOA with PEO, peripheral neuropathy, ataxia and deafness [3, 29]. Most of these patients had ragged-red and cytochrome-c-oxidase negative muscle fibres, with paracrystalline inclusions filling abnormally shaped mitochondria. Remarkably, all patients harbored multiple deletions of mitochondrial DNA (mtDNA) in their skeletal muscle, thus revealing an unrecognized role of the OPA1 protein in mtDNA stability.

### Thymidine Phosphorylase Mutations

The gene responsible for MNGIE, identified in 1999 [50], encodes the enzyme thymidine phosphorylase (*TYMP*), which is involved in the catabolism of pyrimidines, by promoting the phosphorolysis of thymidine into thymine and deoxyribose-phosphate. Defects of thymidine phosphorylase result in systemic accumulation of thymidine and deoxyuridine [43, 63], which leads to deoxynucleotide pool imbalance [16] that cause mtDNA depletion, multiple deletions and mtDNA point mutation. More than 80% of mtDNA mutations found in tissues from MNGIE patients are T-to-C transitions preceded by a short run of As. This signature mutation suggests a "next-nucleotide effect" caused by the more common misinsertion T:dGMP , which is quickly extended by the elevated dTTP concentration resulting from *TYMP* deficiency in the mitochondria of MNGIE cells [48]. HeLa cells grown in media supplemented with 50 μM thymidine demonstrated mtDNA deletions and elevated mitochondrial pools of dTTP and dGTP, a result that recapitulated many of the genetic effects seen in MNGIE [61].

Interestingly, there are potential treatments for MNGIE. Hemodialysis has been shown to transiently reduce thymidine levels in blood [63, 78]). Allogeneic stem cell transplantation has had some success in restoring *TYMP* activity and lowering plasma thymidine levels [26]. In addition, repeated platelet infusions can reduce thymidine levels in blood in MNGIE patients [38].

mtDNA Depletion Syndromes

Since 2001, MDS has been linked to mutations in 8 nuclear genes. Mutations in TK2 (OMIM ∗188250) and RRM2B (OMIM ∗604712) are associated with early-onset myopathy with or without renal proximal tubulopathy [4, 5, 57]. Mutations in SUCLA1 (OMIM ∗611224) and SUCLA2 (OMIM ∗603921), encoding isoforms of succinyl–coenzyme A lyase (a Krebscycle enzyme), have been associated with encephalomyopathy [14, 54] while mutations in *Twinkle* (OMIM ∗60675) [21, 59], *POLG* (OMIM ∗174763) [15, 47] *DGUOK* (OMIM ∗601465) [42, 58, 68]), and *MPV17* [64, 67, 77]. Remarkably, the same pathogenic mutation in *MPV17* that was previously identified in an Italian family [64] was later found to be responsible for NNH [31] raising the possibility of a common founder effect. Haplotype analysis of the MPV17 locus in the Italian MDS and in several NNH families demonstrated that the mutation occurred independently in the two populations [66].

#### 6.2.1.3 Pathogenesis and Pathophysiology

A pathogenetic role of multiple mtDNA deletions in autosomal dominant or recessive PEO is supported by the tight segregation of the molecular lesions with the onset and severity of the disease. For instance, Lamantea and colleagues have reported that, in a series of *POLG1*-mutant PEO families, some of which were consanguineous, homozygous mutant individuals appeared more severely affected and showed the presence of much higher amounts of multiple mtDNA deletions in muscle, than their heterozygous relatives [37]. Moslemi and colleagues [46] have demonstrated close correlation between the accumulation of deletions and the segmental ragged-red COX-negative transformation of muscle fibers. These authors showed that within a single COX-deficient muscle fiber segment, only one single deletion could be detected. These results support the idea that a single deleted mtDNA is clonally expanded in each COX-deficient muscle fiber segment.

Deletions are absent in cultured fibroblasts, peripheral blood cells, and cultured myoblasts, but can be detected in stable tissues including (besides skeletal muscle), the brain, heart, and, in lesser amount, kidney and liver. Rearranged mtDNA molecules, similar if not identical to those found in the multiple deletions syndromes, are present at low levels (from less than 0.1 copies per cell to greater than 100 copies per cell) in the stable tissues of normal adult individuals [30]. In normal conditions, rearranged and wild-type molecules seem to coexist in an equilibrium condition in which rearrangements are continuously lost and reformed but never accumulate to physiologically damaging levels (except under highly abnormal conditions). The latter mechanisms could include (1) an enhancement of the processes by which rearrangements are continuously generated, (2) inhibition of the processes by which they are usually lost, and (3) a (possibly transient) alteration in their selective value, whether at the level of phenotype or replicative advantage. In support of the first mechanism stands the observation that the genes responsible for mitochondrial neurogastrointestinal encephalomyopathy and one form of autosomal dominant PEO, namely thymidine phosphorylase and the muscle-specific mitochondrial adenine translocator, are both involved in nucleotide metabolism. Disturbances of the nucleotide pool available for mtDNA replication, as well as abnormalities in either the mitochondrial helicase or DNA polymerase, are likely to affect the rate or process of DNA replication, which could ultimately lead to the exaggerated production of rearranged mtDNA molecules [20]. In addition to large-scale rearrangements, increased frequency of mtDNA point mutations have been reported in MNGIE [48]. Several somatic mutations, mostly T>C transitions preceded by $5'\text{-}A_{(n)}$ sequences, were identified in the mtDNA molecule of tissues and cultured cells from MNGIE patients. Some mutations were clearly pathogenic, as they predict the loss of or an abnormality in the function of mtDNA-encoded proteins or tRNA. The accumulation of these mutations is likely to be due to next-nucleotide effects and dislocation mutagenesis, as a consequence of increased levels of mitochondrial deoxythymidine and deoxyuridine pools. Finally, the mtDNA damage caused by *POLG1* mutations in Alpers-Huttenlocher syndrome, sensory-ataxia neuropathy, dysarthria and ophthalmoplegia, and ataxia-epilepsy syndrome is unclear. Depletion of mtDNA, or accumulation of multiple point mutations are likely possibilities, but more investigation is needed to prove these hypotheses.

Balance and control of the mitochondrial deoxynucleotide pools are essential for the maintenance of mtDNA copy number. Perturbation of this homeostatic control, as determined by defects of *dGK* and *TK2*, and possibly of *TYMP*, *RRM2B*, can lead to mtDNA depletion or multiple deletions. These enzymes are involved in the salvage pathways of mitochondrial deoxynucleotides, which constitute the major source of mtDNA precursors in stable tissues such as liver, brain, and muscle [62].

Elpeleg et al. [14] postulated that also mutations in the SUCLA2 gene could determine a defect in the last step of the mitochondrial deoxyribonucleoside triphosphate (dNTP) salvage pathway, since SCS-A copurifies with and is tightly associated in a complex with nucleoside diphosphate kinase (NDPK) [32, 34, 36]. NDPKs are ubiquitous protein kinases that also catalyze the exchange of terminal phosphates between tri- and di-phosphoribonucleosides and are crucial for maintaining the homeostasis of ribonucleotides and deoxyribonucleotides [56].

Very recently it has been reported that dominant mitochondrial damages due to *ANT1* pathogenic mutations are not caused by aberrant nucleotide transport, but by uncoupled mitochondrial respiration [76].

Finally, the function of *MPV17* and its role in the pathogenesis of MDS are still unknown [64] but studies on its yeast ortholog, *SYM1* suggest for this protein a role in the cellular response to metabolic stress [68]. The availability of Mpv17 KO mouse, provides an invaluable tool to elucidate the function of this gene in mitochondrial homeostasis and to investigate its role in the pathogenesis of this disease.

## 6.3 Diagnostic Considerations

Specific diagnosis is based on examination of muscle and, in the hepatopathic form of mtDNA depletion syndrome, liver biopsies. In all syndromes associated with multiple deletions, it is essential to have definite evidence of the mtDNA abnormalities in muscle. Southern-blot analysis of "linearized" muscle mtDNA from symptomatic and presymptomatic individuals shows the presence of numerous hybridization bands of different size, including a major 16.5 kb band, corresponding to wild-type mtDNA, as well as several smaller bands, corresponding to deleted mtDNAs. In mitochondrial DNA depletion syndromes, the residual mtDNA amount in affected tissues, which can be measured by either quantitative Southern-blot or real-time PCR analyses, must be less than 30–35% of control, and multiple deletions must be absent. In addition, there must be histological evidence of ragged-red and COX-negative fibers in multiple deletions syndromes. The latter are also present in mtDNA depletion, whereas the former are usually absent. Specific diagnosis is possible for mitochondrial neurogastrointestinal encephalomyopathy via the identification of mutations in the gene encoding thymidine phosphorylase or the measurement of its activity in peripheral leukocytes or the measurement of thymidine and deoxyuridine in plasma and urine [43, 63]. Likewise, screening for mutations in the genes encoding the adenine nucleotide translocator, *Twinkle* and *POLG*, should also be performed in all cases of PEO with multiple mtDNA deletions. If PEO is associated with optic atrophy *OPA1* should be analyzed. When Alpers-Huttenlocher syndrome, sensory-ataxia neuropathy, dysarthria and ophthalmoplegia, or ataxia-epilepsy syndromes are suspected, analysis of the *POLG1* gene is mandatory. In children with an Alpers-like phenotype with normal *POLG1*, analysis of *Twinkle* is recommended. When liver involvement seems to be the most most prominent feature, *DGUOK* and *MPV17* are mandatory.

**Acknowledgements** This work was supported by the Pierfranco and Luisa Mariani Foundation Italy, Fondazione Telethon-Italy grant number GGP07019, the Italian Ministry of University and Research (FIRB 2003 – project RBLA038RMA), MITOCIRCLE and EUMITOCOMBAT network grants from the European Union framework program 6.

## References

1. Agostino A, Valletta L, Chinnery PF, et al. Mutations of ANT1, Twinkle, and POLG1 in sporadic progressive external ophthalmoplegia (PEO). Neurology 2003;60: 1354–1356.
2. Alexander C, Votruba M, Pesch UE, et al. OPA1, encoding a dynamin-related GTPase, is mutated in autosomal dominant optic atrophy linked to chromosome 3q28. Nat Genet 2000; 26:211–215.
3. Amati-Bonneau P, Valentino ML, Reynier P, et al. OPA1 mutations induce mitochondrial DNA instability and optic atrophy 'plus' phenotypes. Brain 2008;131:338–351.
4. Bornstein B, Area E, Flanigan KM, Ganesh J, et al. Mitochondrial DNA depletion syndrome due to mutations in the RRM2B gene. Neuromuscul Disord 2008;18:453–459.
5. Bourdon A, Minai L, Serre V, et al. Mutation of RRM2B, encoding p53-controlled ribonucleotide reductase (p53R2), causes severe mitochondrial DNA depletion. Nat Genet 2007;39:776–780.
6. Calvo S, Jain M, Xie X, Sheth SA, et al. Systematic identification of human mitochondrial disease genes through integrative genomics. Nat Genet 2006;38:576–582.
7. Carrozzo R, Dionisi-Vici C, Steuerwald U, Lucioli S, et al. SUCLA2 mutations are associated with mild methylmalonic aciduria, Leigh-like encephalomyopathy, dystonia and deafness Brain 2007;130:862–874.
8. Clayton DA. Vertebrate mitochondrial DNA-a circle of surprises. Exp Cell Res 2000; 255:4–9.
9. Contamine V, Picard M. Maintenance and integrity of the mitochondrial genome: a plethora of nuclear genes in the budding yeast. Microbiol Mol Biol Rev 2000; 64:281–315.
10. Davidzon G, Mancuso M, Ferraris S, et al. POLG mutations and Alpers syndrome. Ann Neurol 2005;57:921–923.
11. Delettre C, Lenaers G, Griffoin JM, Gigarel N, et al. Nuclear gene OPA1, encoding a mitochondrial dynamin-related protein, is mutated in dominant optic atrophy. Nat Genet 2000; 26:207–210.
12. Deschauer M, Kiefer R, Blakely EL, et al. A novel Twinkle gene mutation in autosomal dominant progressive external ophthalmoplegia. Neuromuscul Disord 2003;13:568–572.
13. Duvezin-Caubet S, Jagasia R, Wagener J, Hofmann S, et al. Proteolytic processing of OPA1 links mitochondrial dysfunction to alterations in mitochondrial morphology. J Biol Chem 2006; 281:37972–37979.
14. Elpeleg O, Miller C, Hershkovitz E et al. Deficiency of the ADP-forming succinyl-CoA synthase activity is associated with encephalomyopathy and mitochondrial DNA depletion.Am J Hum Genet 2005;76:1081–1086.
15. Ferrari G, Lamantea E, Donati A, et al. Infantile hepatocerebral syndromes associated with mutations in the mitochondrial DNA polymerase-gammaA. Brain 2005;128:723–731.
16. Ferraro P, Pontarin G, Crocco L, et al. Mitochondrial deoxynucleotide pools in quiescent fibroblasts: a possible model for mitochondrial neurogastrointestinal encephalomyopathy (MNGIE). J Biol Chem 2005;280:24472–244780.
17. Filosto M, Mancuso M, Nishigaki Y, et al. Clinical and genetic heterogeneity in progressive external ophthalmoplegia due to mutations in polymerase gamma. Arch Neurol 2003;60:1279–1284.
18. Galassi G, Lamantea E, Invernizzi F, et al. Additive effects of POLG1 and ANT1 mutations in a complex encephalomyopathy. Neuromuscul Disord 2008;18:465–470.
19. Galbiati S, Bordoni A, Papadimitriou D, et al. New mutations in TK2 gene associated with mitochondrial DNA depletion. Pediatr Neurol 2006; 34:177–185.
20. Graziewicz MA, Longley MJ, et al. Structure-function defects of human mitochondrial DNA polymerase in autosomal dominant progressive external ophthalmoplegia. Nat Struct Mol Biol 2004;11:770–776.
21. Hakonen AH, Isohanni P, Paetau A, et al. Recessive Twinkle mutations in early onset encephalopathy with mtDNA depletion. Brain 2007;130:3032–3040.

22. Hakonen AH, Goffart S, Marjavaara S, Paetau A, et al. Infantile-onset spinocerebellar ataxia and mitochondrial recessive ataxia syndrome are associated with neuronal complex I defect and mtDNA depletion. Hum Mol Genet 2008 Sep 5. [Epub ahead of print].
23. Invernizzi F, Varanese S, Thomas A, et al. Two novel POLG1 mutations in a patient with progressive external ophthalmoplegia, levodopa-responsive pseudo-orthostatic tremor and parkinsonism. Neuromuscul Disord 2008;18:460–464.
24. Hirano M, Marti R, Ferreiro-Barros C, et al. Defects of intergenomic communication: autosomal disorders that cause multiple deletions and depletion of mitochondrial DNA. Semin Cell Dev Biol. 2001;12:417–427.
25. Hirano M, Silvestri G, Blake DM, et al. Mitochondrial neurogastrointestinal encephalomyopathy (MNGIE): clinical, biochemical, and genetic features of an autosomal recessive mitochondrial disorder. Neurology 1994;44:721–727.
26. Hirano M, Martí R, Casali C, Tadesse S, et al. Allogeneic stem cell transplantation corrects biochemical derangements in MNGIE. Neurology 2006;67:1458–1460.
27. Holt IJ, Lorimer HE, Jacobs HT. Coupled leading- and lagging-strand synthesis of mammalian mitochondrial DNA. Cell 2000;100:515–524.
28. Horvath R, Hudson G, Ferrari G, Futterer N et al. Phenotypic spectrum associated with mutations of the mitochondrial polymerase gamma gene. Brain 2006;129:1674–1684.
29. Hudson G, Amati-Bonneau P, Blakely EL et al. Mutation of OPA1 causes dominant optic atrophy with external ophthalmoplegia, ataxia, deafness and multiple mitochondrial DNA deletions: a novel disorder of mtDNA maintenance. Brain 2008;13:329–337.
30. Kajander OA, Rovio AT, Majamaa K, et al. Human mtDNA sublimons resemble rearranged mitochondrial genomes found in pathological states. Hum Mol Genet 2000; 9: 2821–2835.
31. Karadimas CL, Vu TH, Holve SA, et al. Navajo neurohepatopathy is caused by a mutation in the MPV17 gene. Am J Hum Genet 2006;79:544–548.
32. Kadrmas EF, Ray PD, Lambeth DO. Apparent ATP-linked succinate thiokinase activity and its relation to nucleoside diphosphate kinase in mitochondrial matrix preparations from rabbit. Biochim Biophys Acta 1991 Aug ;1074:339–346.
33. Kaukonen J, Juselius JK, Tiranti V, et al. Role of adenine nucleotide translocator 1 in mtDNA maintenance. Science 2000;289:782–785.
34. Kavanaugh-Black A, Connolly DM, et al. Characterization of nucleoside-diphosphate kinase from Pseudomonas aeruginosa: complex formation with succinyl-CoA synthetase. Proc Natl Acad Sci U S A 1994;91:5883–5887.
35. Kjer P. Infantile optic atrophy with dominant mode of inheritance: a clinical and genetic study of 19 Danish families. Acta Ophthalmol Scand 1959; 37:1–146.
36. Kowluru A, Tannous M, Chen HQ. Localization and characterization of the mitochondrial isoform of the nucleoside diphosphate kinase in the pancreatic beta cell: evidence for its complexation with mitochondrial succinyl-CoA synthetase. Arch Biochem Biophys 2002;398:160–169.
37. Lamantea E, Tiranti V, Bordoni A, et al. Mutations of mitochondrial DNA polymerase gammaA are a frequent cause of autosomal dominant or recessive progressive external ophthalmoplegia. Ann Neurol 2002;52:211–219.
38. Lara MC, Weiss B, Illa I, Madoz P, et al. Infusion of platelets transiently reduces nucleoside overload in MNGIE. Neurology 2006;67:1461–1463.
39. Lee YJ, Jeong SY, Karbowski M, et al. Roles of the mammalian mitochondrial fission and fusion mediators Fis1, Drp1, and Opa1 in apoptosis. Mol Biol Cell 2004; 15:5001–11.
40. Longley MJ, Clark S, Yu Wai Man C, et al. Mutant POLG2 disrupts DNA polymerase gamma subunits and causes progressive external ophthalmoplegia. Am J Hum Genet 2006;78: 1026–1034.
41. Luoma P, Melberg A, Rinne JO, et al. Parkinsonism, premature menopause, and mitochondrial DNA polymerase gamma mutations: clinical and molecular genetic study. Lancet 2004;364:875–882.

42. Mandel H, Szargel R, Labay V, et al. The deoxyguanosine kinase gene is mutated in individuals with depleted hepatocerebral mitochondrial DNA. Nat Genet 2001;29: 337–341.
43. Marti R, Spinazzola A, Tadesse S, Nishino I, Nishigaki Y, Hirano M. Definitive diagnosis of mitochondrial neurogastrointestinal encephalomyopathy by biochemical assays. Clin Chem 2004; 50:120–124.
44. Mizusawa H, Watanabe M, Kanazawa I, et al. Familial mitochondrial myopathy associated with peripheral neuropathy: partial deficiencies of complex 1 and complex 4. J Neurol Sci 1988;86:171–184.
45. Moraes CT, Shanske S, Tritschler HJ, et al. mtDNA depletion with variable tissue expression: a novel genetic abnormality in mitochondrial diseases. Am J Hum Genet 1991;48:492–501.
46. Moslemi AR, Melberg A, Holme E, Oldfors A. Clonal expansion of mitochondrial DNA with multiple deletions in autosomal dominant progressive external ophthalmoplegia. Ann Neurol 1996;40:707–713.
47. Naviaux RK, Nguyen KV. POLG mutations associated with Alpers′ syndrome and mitochondrial DNA depletion. Ann Neurol 2004;55:706–712.
48. Nikali K, Suomalainen A, Saharinen J, Kuokkanen M, et al. Infantile onset spinocerebellar ataxia is caused by recessive mutations in mitochondrial proteins Twinkle and Twinky. Hum Mol Genet. 2005;14:2981–2990.
49. Nishigaki Y, Marti R, Copeland WC, Hirano M. Site-specific somatic mitochondrial DNA point mutations in patients with thymidine phosphorylase deficiency. J Clin Invest 2003;111:1913–1921.
50. Nishino I, Spinazzola A, Hirano M. Thymidine phosphorylase gene mutations in MNGIE, a human mitochondrial disorder. Science 1999;283:689–692.
51. Nishino I, Spinazzola A, Papadimitriou A, Hammans S, et al. Mitochondrial neurogastrointestinal encephalomyopathy: an autosomal recessive disorder due to thymidine phosphorylase mutations. Ann Neurol 2000;47:792–800.
52. Oskoui M, Davidzon G, Pascual J, Erazo R et al. Clinical spectrum of mitochondrial DNA depletion due to mutations in the thymidine kinase 2 gene. Arch Neurol 2006;63:1122–1126.
53. Ostergaard E, Hansen FJ, Sorensen N, Duno M, et al. Mitochondrial encephalomyopathy with elevated methylmalonic acid is caused by SUCLA2 mutations. Brain 2007;130:853–861.
54. Ostergaard E, Christensen E, Kristensen E, Mogensen B, et al. Deficiency of the alpha subunit of succinate-coenzyme A ligase causes fatal infantile lactic acidosis with mitochondrial DNA depletion. Am J Hum Genet 2007;81:383–387.
55. Palmieri L, Alberio S, Pisano I, Lodi T et al. Complete loss-of-function of the heart/muscle-specific adenine nucleotide translocator is associated with mitochondrial myopathy and cardiomyopathy. Hum Mol Genet 2005;14:3079–3088.
56. Parks Jr, RE, Brown PR., Cheng YC, et al. Purine metabolism in primitive erythrocytes. Comp Biochem Physiol B 1973;45:355–364.
57. Saada A, Shaag A,Mandel H, et al. Mutant mitochondrial thymidine kinase in mitochondrial DNA depletion myopathy. Nat Genet 2001;29:342–344.
58. Salviati L., Sacconi S, Mancuso M, Otaegui D, et al. Mitochondrial DNA depletion and dGK gene mutations. Ann. Neurol 2002;52:311–316.
59. Sarzi E, Goffart S, Serre V, Chretien D, et al. Twinkle helicase (PEO1) gene mutation causes mitochondrial DNA depletion. Ann Neurol 2007;62:579–587.
60. Servidei S, Zeviani M, Manfredi G, et al. Dominantly inherited mitochondrial myopathy with multiple deletions of mitochondrial DNA: clinical, morphologic, and biochemical studies. Neurology 1991;41:1053–1059.
61. Song S, Wheeler LJ, Mathews CK. Deoxyribonucleotide pool imbalance stimulates deletions in HeLa cell mitochondrial DNA. J Biol Chem 2003;278:43893–43896.
62. Spelbrink JN, Li FY, Tiranti V, et al. Human mitochondrial DNA deletions associated with mutations in the gene encoding Twinkle, a phage T7 gene 4-like protein localized in mitochondria. Nat Genet 2001; 28:223–231.

63. Spinazzola A, Marti R, Nishino I et al. Altered thymidine metabolism due to defects of thymidine phosphorylase. J Biol Chem 2002;277:4128–4133.
64. Spinazzola A, Viscomi C, Fernandez-Vizarra E, et al. MPV17 encodes an inner mitochondrial membrane protein and is mutated in infantile hepatic mitochondrial DNA depletion. Nat Genet 2006;38:570–575.
65. Spinazzola A, Zeviani M. Disorders of nuclear-mitochondrial intergenomic signaling. Gene 2005;354:162–168.
66. Spinazzola A, Massa V, Hirano M, Zeviani M. Lack of founder effect for an identical mtDNA depletion syndrome (MDS)-associated MPV17 mutation shared by Navajos and Italians. Neuromuscul Disord 2008;18:315–318.
67. Spinazzola A, Santer R, Akman OH, et al. Hepatocerebral form of mitochondrial DNA depletion syndrome: novel MPV17 mutations. Arch Neurol 2008;65:1108–1113.
68. Tadiboyina VT, Rupar A, Atkison P, et al. Novel mutation in DGUOK in hepatocerebral mitochondrial DNA depletion syndrome associated with cystathioninuria. Am J Med Genet 2005;135:289–291.
69. Trifunovic A, Wredenberg A, Falkenberg M, et al. Premature ageing in mice expressing defective mitochondrial DNA polymerase. Nature 2004;429:417–423.
70. Tritschler HJ, Andreetta F, Moraes CT, et al. Mitochondrial myopathy of childhood associated with depletion of mitochondrial DNA. Neurology 1992;42:209–217.
71. Van Goethem G, Dermaut B, Lofgren A, Martin JJ, Van Broeckhoven C. Mutation of POLG is associated with progressive external ophthalmoplegia characterized by mtDNA deletions. Nat Genet 2001;28:211–212.
72. Van Goethem G, Luoma P, Rantamaki M, et al. POLG mutations in neurodegenerative disorders with ataxia but no muscle involvement. Neurology 2004;63:1251–1257.
73. Van Goethem G, Martin JJ, Dermaut B, et al. POLG mutations presenting with sensory and ataxic neuropathy in compound heterozygote patients with progressive external ophthalmoplegia. Neuromuscul Disord 2003;13:133–142.
74. Vu TH, Tanji K, Holve SA, Bonilla E, et al. Navajo neurohepatopathy: a mitochondrial DNA depletion syndrome? Hepatology 2001;34:116–120.
75. Winterthun S, Ferrari G, He L, et al. Autosomal recessive mitochondrial ataxic syndrome due to mitochondrial polymerase gamma mutations. Neurology 2005;64:1204–1208.
76. Wang X, Salinas K, Zuo X, Kucejova B, Chen XJ. Dominant membrane uncoupling by mutant adenine nucleotide translocase in mitochondrial diseases. Hum Molec Genet 2008 [Epub ahead of print].
77. Wong LJ, Brunetti-Pierri N, Zhang Q, Yazigi N, et al. Mutations in the MPV17 gene are responsible for rapidly progressive liver failure in infancy. Hepatology 2007;46:1218–1227.
78. Yavuz H, Ozel A, Christensen M, Christensen E, et al. Treatment of mitochondrial neurogastrointestinal encephalomyopathy with dialysis.Arch Neurol 2007;64:435–438.
79. Yuzaki M, Ohkoshi N, Kanazawa I, et al. Multiple deletions in mitochondrial DNA at direct repeats of non-D-loop regions in cases of familial mitochondrial myopathy. Biochem Biophys Res Commun 1989;164:1352–1357.
80. Zeviani M, Servidei S, Gellera C, et al. An autosomal dominant disorder with multiple deletions of mitochondrial DNA starting at the D-loop region. Nature 1989;339:309–311.

# Chapter 7
# Mitochondrial Disorders Due to Nuclear OXPHOS Gene Defects

**Cristina Ugalde, María Morán, Alberto Blázquez, Joaquín Arenas, and Miguel A. Martín**

**Abstract** Because of the bi-genomic origin of the OXPHOS system, mitochondrial disease-associated mutations have been found in both mtDNA and nuclear structural genes. In the last years, interest has shifted toward mendelian genetics in mitochondrial disease, not only because the majority of the OXPHOS system subunits are encoded by the nuclear genome, but also because a large number of yet unknown nuclear proteins, such as regulatory proteins and assembly factors, are likely involved in its biogenesis and function. A clinical-genetic classification can be proposed for nuclear defects that affect the biogenesis of the OXPHOS system, as follows: (i) disorders due to nuclear gene defects encoding structural components or assembly factors of the OXPHOS complexes, (ii) disorders due to gene defects in the biogenesis of protein constituents of the OXPHOS system, (iii) disorders due to defects in the biosynthesis of non-protein constituents of the respiratory chain, and (iv) disorders due to gene defects encoding proteins involved in mitochondrial dynamics.

**Keywords:** Mitochondrial biogenesis, Mitochondrial respiratory chain, Nuclear gene defect, OXPHOS disorders, Respiratory subunits assembly

## 7.1 Introduction

Adenosine 5'-triphosphate (ATP), the key energy source of the cell, is produced and regulated via a highly complicated cascade of cellular events. The final biochemical step involved in the production of ATP is the oxidative phosphorylation (OXPHOS) system. The OXPHOS system is composed of five multiprotein complexes (complexes I to V) and two mobile electron carriers (ubiquinone and cytochrome c) embedded in the lipid bilayer of the mitochondrial inner membrane. The subunits of these complexes are encoded either by the mitochondrial chromosome (mtDNA)

C. Ugalde (✉)
Mitochondrial diseases laboratory, Research Center, "12 de Octubre", University Hospital; U723, CIBER de Enfermedades Raras (CIBERER), Instituto de Salud Carlos III, Madrid, Spain

C. Espinós et al. (eds.), *Inherited Neuromuscular Diseases*, Advances in Experimental Medicine and Biology 652, DOI 10.1007/978-90-481-2813-6_7,

or the nuclear genome, except for complex II that is exclusively encoded by nuclear DNA (nDNA). The nuclear-encoded subunits are synthesized on free ribosomes in the cytosol and imported into mitochondria, where they assemble together with the mitochondrial-encoded subunits and prosthetic groups to form the OXPHOS complexes. Because of the bi-genomic origin of the OXPHOS system, mitochondrial disease-associated mutations have been found in both mtDNA and nuclear structural genes [163]. In the last years, interest has shifted toward mendelian genetics in mitochondrial disease, not only because the majority of the OXPHOS system subunits are encoded by the nuclear genome, but also because a large number of yet unknown nuclear proteins, such as regulatory proteins and assembly factors, are likely involved in its biogenesis and function. Mutations in these genes, as increasingly demonstrated, may also cause mitochondrial disorders [69].

A clinical-genetic classification can be proposed for nuclear defects that affect the biogenesis of the OXPHOS system, as follows: (i) disorders due to nuclear gene defects encoding structural components or assembly factors of the OXPHOS complexes, (ii) disorders due to gene defects in the biogenesis of protein constituents of the OXPHOS system, (iii) disorders due to defects in the biosynthesis of non-protein constituents of the respiratory chain, and (iv) disorders due to gene defects encoding proteins involved in mitochondrial dynamics.

The nuclear-encoded genes involved in impairment of the oxidative phosphorylation biogenesis, their recognized function and the clinical phenotypes associated with their defects are showed in Table 7.1.

## 7.2 Nuclear Gene Defects of Structural or Assembly Genes of the OXPHOS System

### *7.2.1 Complex I Deficiency*

Reduced nicotinamide adenine dinucleotide (NADH): ubiquinone oxidoreductase (complex I, EC1.6.5.3) is the principal entry point of electrons into the respiratory chain as it catalyzes the electron transfer from NADH to ubiquinone, accompanied by the translocation of protons across the inner mitochondrial membrane. Mammalian complex I, with a molecular weight of ~980 kDa, is the largest complex of the mitochondrial respiratory chain. It consists of 45 polypeptide subunits [45, 86], seven encoded by the mitochondrial DNA (mtDNA) and the remainder by the nuclear genome, a non-covalently bound flavomononucleotide (FMN) group and eight iron sulphur (Fe-S) clusters. It has an L-shape, with a hydrophobic membrane arm embedded in the inner mitochondrial membrane, and a hydrophilic peripheral arm protruding into the matrix [78]. Complex I can be functionally divided in three parts, the subunit composition of which has been (partially) determined by subfractionation experiments: (i) the dehydrogenase module, responsible for the binding and oxidation of NADH to NAD+, also contains the binding sites for the FMN and several Fe-S clusters, and includes at least the NDUFV2, NDUFV1, and

**Table 7.1** Nuclear-encoded genes involved in defects of the oxidative phosphorylation (OXPHOS) system biogenesis

| Gene | Pathophysiology | Clinical presentation | References |
|---|---|---|---|
| *Subunits of the OXPHOS system* | | | |
| *NDUFA1* | Complex I deficiency | Leigh Syndrome<br>Myoclonic epilepsy and developmental delay | [66] |
| *NDUFA2* | Complex I deficiency | Leigh Syndrome | [87] |
| *NDUFA11* | Complex I deficiency | Encephalocardiomyopathy, fatal infantile lactic academia | [31] |
| *NDUFS1* | Complex I deficiency | Leigh Syndrome | [27, 119] |
| *NDUFS2* | Complex I deficiency | Hypertrophic cardiomyopathy and encephalomyopathy | [115] |
| *NDUFS3* | Complex I deficiency | Leigh Syndrome | [30] |
| *NDUFS4* | Complex I deficiency | Leigh or Leigh-like Syndrome | [182, 38, 145, 28] |
| *NDUFS6* | Complex I deficiency | Encephalopathy and fatal neonatal lactic academia | [101] |
| *NDUFS7* | Complex I deficiency | Leigh Syndrome | [111, 162] |
| *NDUFS8* | Complex I deficiency | Leigh Syndrome | [113, 146] |
| *NDUFV1* | Complex I deficiency | Leukodystrophy and myoclonic epilepsy<br>Early-onset ophthalmoplegia, Leigh-like Syndrome | [161, 27] |
| *NDUFV2* | Complex I deficiency | Hypertrophic cardiomyopathy and encephalopathy<br>Association with bipolar disorder and schizophrenia. | [29, 82] |
| *SDHA* | Complex II deficiency | Leigh Syndrome<br>Late-onset optic atrophy, ataxia and myopathy | [36, 140,181] |
| *SDHB* | Complex II deficiency | Hereditary paraganglioma/pheochromocytoma (PGL4)<br>Cowden syndrome | [12] |
| *SDHC* | Complex II deficiency | Hereditary paraganglioma (PGL3) | [130] |
| *SDHD* | Complex II deficiency | Hereditary paraganglioma/pheochromocytoma (PGL1)<br>Cowden syndrome | [24] |
| *UQCRB* | Complex III deficiency | Hypoglycaemia and lactic acidemia | [83] |
| *UQCRQ* | Complex III deficiency | Leigh Syndrome | [17] |
| *COX6B1* | Complex IV deficiency | Encephalomyopathy | [121] |

**Table 7.1** (continued)

| Gene | Pathophysiology | Clinical presentation | References |
|---|---|---|---|
| *Assembly factors of the OXPHOS complexes* | | | |
| *NDUFA12L* | Complex I deficiency | Leukoencephalopathy with vanishing white matter | [134] |
| *NDUFAF1* | Complex I deficiency | Cardioencephalomyopathy | [64] |
| *C6ORF66* | Complex I deficiency | Infantile mitochondrial encephalomyopathy<br>Antenatal cardiomyopathy | [156] |
| *C8ORF38* | Complex I deficiency | Leigh Syndrome | [138] |
| *BCS1L* | Complex III deficiency | GRACILE or Björnstad Syndromes<br>Encephalopathy, and/or hepatopathy and tubulopathy | [184, 59,84,60] |
| *COX10* | Complex IV deficiency | Leigh or Leigh-like Syndrome<br>Tubulopathy and leukodystrophy<br>Hypertrophic cardiomyopathy and sensorineural deafness | [180, 8, 51] |
| *COX15* | Complex IV deficiency | Early-onset fatal hypertrophic cardiomyopathy<br>Leigh syndrome | [9, 135] |
| *LRPPRC* | Complex IV deficiency | Leigh syndrome, French-Canadian type (LSFC) | [128] |
| *SCO1* | Complex IV deficiency | Neonatal-onset hepatic failure and encephalopathy | [179] |
| *SCO2* | Complex IV deficiency | Fatal infantile cardioencephalomyopathy | [139] |
| *SURF1* | Complex IV deficiency | Leigh syndrome or leukodystrophy<br>Subacute encephalopathy<br>Villous atrophy, hypertricosis, mild neurological involvement | [200, 171, 149, 143] |
| *ATP12* | Complex V deficiency | Fatal neonatal lactic acidemia and cardiomyopathy | [61] |
| *Biogenesis of protein constituents of the OXPHOS system* | | | |
| *Mitochondrial protein translation* | | | |
| *DARS2* | Aspartyl-tRNA synthetase deficiency | Leukoencephalopathy with brain stem and spinal cord involvement and lactic acidemia | [159] |
| *EFG1* | Mitochondrial translation elongation | Leigh syndrome, severe hepatopathy | [52, 178] |
| *MRPS16* | Ribosome assembly | Fatal neonatal lactic acidosis, corpus callosum agenesis dysmorphis, dysmorphism | [124] |
| *MRPS22* | Ribosome assembly | Antenatal skin edema, hypotonia, cardiomyopathy and tubulopathy | [155] |
| *PUS1* | Mitochondrial tRNA pseudouridylation | Myopathy, lactic acidosis and sideroblastic anaemia (MLASA) | [41, 68] |

**Table 7.1** (continued)

| Gene | Pathophysiology | Clinical presentation | References |
|---|---|---|---|
| *TSFM* | Mitochondrial translation elongation | Encephalomyopathy or hypertrophic cardiomyopathy | [164] |
| *TUFM* | Mitochondrial translation elongation | Macrocystic leukodystrophy with micropolygyria | [178] |
| *Mitochondrial protein import* | | | |
| *DDP1* | Mitochondrial protein import | Mohr-Tranebjaerg syndrome | [98] |
| *DNAJC19* | Mitochondrial protein import | Dilated cardiomyopathy and ataxia (DCMA syndrome) | [54] |
| *HSPD1* | Mitochondrial chaperone | Autosomal dominant form of spastic paraplegia Pelizaeus-Merzbacher-like disease (PMLD) | [81, 118] |
| *FeS cluster biogenesis* | | | |
| *ABCB7* | Mitochondrial iron homeostasis | X-linked sideroblastic anemia with ataxia (XLSA/A) | [3] |
| *FRDA* | Iron-sulfur cluster biosynthesis | Friedrichs ́ ataxia | [55] |
| *GLRX5* | Iron-sulfur cluster biosynthesis | Sideroblastic-like microcytic anemia and iron overload | [42] |
| *ISCU* | Iron-sulfur cluster biosynthesis | Myopathy with exercise intolerance | [125] |
| *Mitochondrial protein metabolism* | | | |
| *SPG7*/Paraplegin | Protein quality control and ribosome assembly | Hereditary spastic paraplegia (HSP) | [46] |
| *Biogenesis of non-protein constituents of the OXPHOS system* | | | |
| *Coenzyme $Q_{10}$ biogenesis* | | | |
| *ADCK3/CABC1 COQ2* | Coenzyme $Q_{10}$deficiency | Childhood-onset cerebellar ataxia, and seizures | [127, 105] |
| *COQ2* | Coenzyme $Q_{10}$ deficiency | Infantile glomerular nephropathy and encephalomyopathy | [148] |
| *PDSS1* | Coenzyme $Q_{10}$ deficiency | Deafness, optic atrophy, mental retardation | [126] |
| *PDSS2* | Coenzyme $Q_{10}$ deficiency | Leigh syndrome with infantile nephropathy | [116] |
| *Cardiolipin synthesis* | | | |
| *TAZ* | Cardiolipin metabolism | Barth syndrome | [32] |

**Table 7.1** (continued)

| Gene | Pathophysiology | Clinical presentation | References |
|---|---|---|---|
| *Mitochondrial dynamics* | | | |
| *DLP1* | Mitochondrial and peroxisomal fission | Microcephaly, abnormal brain development, optic atrophy and hypoplasia, persistent lactic acidemia | [161] |
| *LETM1* | Mitochondrial cristae morphology | Wolf-Hirschhorn syndrome | [80] |
| *MFN2* | Mitochondrial fusion | Charcot-Marie-Tooth neuropathy type 2A | [198] |
| *OPA1* | Mitochondrial fusion | Dominant optic atrophy (DOA), plus mitochondrial myopathy with deafness, ataxia, ophthalmoplegia | [4, 57, 141] |

NDUFS1 subunits; (ii) the hydrogenase module, responsible for the Fe-S cluster-mediated electron transfer to ubiquinone, includes, at a minimum, the NDUFS2, NDUFS3, NDUFS7, and NDUFS8 subunits; and (iii) the proton translocation module, which constitutes the bulk of the membrane arm, includes at least the seven mtDNA-encoded ND subunits [45, 49, 75, 158, 185].

Isolated complex I deficiency [OMIM 252010] constitutes the most frequently diagnosed mitochondrial disorder, which represents largely one-third of all cases of respiratory chain enzyme defect [97, 100, 163]). It often involves severe infantile disease with a premature death and comprises disorders showing a great variety of clinical presentations [39, 97, 100, 114]). Most of the infantile cases fall into six main clinical phenotypes: Leigh syndrome, fatal infantile lactic acidosis (FILA), neonatal cardiomyopathy, leukoencephalopathy, pure myopathy, and combined hepatopathy and tubulopathy. Several presentations cannot be assigned to any of these groups, and are generically defined as mitochondrial encephalomyopathy with complex I deficiency [39, 97]. Given the complexity of the enzyme, most complex I defects remain undefined at the genetic and molecular levels, and pathogenic mutations have only been detected in approximately 20–40% of the cases [163].

#### 7.2.1.1 Mutations in Complex I Nuclear Structural Genes

The first patient with isolated complex I deficiency and Leigh Syndrome, due to a missense mutation in the *NDUFS8* gene, was reported in 1998 [113]. Since then, other pathogenic mutations have been identified in 12 out of the 38 nuclear complex I structural genes. The most typical clinical presentations include Leigh or Leigh-like syndrome, leukodystrophy and cardiomyopathy. These disease-causing genes (*NDUFA1* [OMIM 300078], *NDUFA2* [OMIM 602137], *NDUFA11* [OMIM 612698], *NDUFS1* [OMIM 157655], *NDUFS2* [OMIM 602985], *NDUFS3* [OMIM 603846], *NDUFS4* [OMIM 602694], *NDUFS6* [OMIM 603848], *NDUFS7* [OMIM 601825], *NDUFS8* [OMIM 602141], *NDUFV1* [OMIM 161015], and *NDUFV2*

[OMIM 600532]) [27, 28, 29, 30, 31, 66, 87, 101, 113, 115, 119, 161, 174, 182], are highly conserved across species and play a major role in the catalytic activity of the complex. Although mutations in these genes are all associated with autosomal recessive inheritance, the gene encoding the NDUFA1 subunit is on Xq24, and the corresponding mutations are X-linked recessive [66]. It is remarkable that all mutations, except for NDUFA1, are located in nuclear subunits of the peripheral arm. Seven of them, NDUFV1, NDUFV2, NDUFS1, NDUFS2, NDUFS3, NDUFS7 and NDUFS8, encode the so-called essential or core subunits and are evolutionarily conserved from bacteria to mammals. Mutations in all these subunits (except NDUFS3) are thought to affect intramolecular electron transfer. Abnormalities in structural subunits are usually associated with an accumulation of assembly intermediates, and their investigation has contributed to define the first models of complex I biogenesis [7, 69, 107, 176, 177, 182]. However, mutations in these genes only account for a minority of the complex I deficiency cases, suggesting that the molecular cause of the disease should be found in other subunits or factors involved in the catalytic regulation, assembly or maintenance of complex I.

#### 7.2.1.2 Mutations in Complex I Assembly Factors

This group comprises proteins that are involved in the assembly of complex I but are not part of the final structure of the enzyme. Pathogenic mutations are restricted to only four human genes: *NDUFA12L* [OMIM 609653] [18, 134], *NDUFAF1* [OMIM 606934] [64], *C6ORF66* [OMIM 611776] [156], and *C8ORF38* [138]. The first mutation was detected in a patient with progressive encephalopathy and impaired complex I assembly who harboured a null mutation in *NDUFA12L*, an ortholog to the human NDUFA12 structural subunit [134]. NDUFA12L is associated to an ~830-kDa complex I subassembly intermediate that is commonly present in NDUFV1-, NDUFS4-, and NDUFS6-deficient patients but not in controls [134, 187]. The complete absence of NDUFA12L does not entirely prevent the formation of fully-assembled complex I [18, 134], suggesting a role for this protein in the stabilization of assembly intermediates and possibly, in the incorporation of specific subunits into the nascent complex. A second mutation was reported in NDUFAF1, the human homolog of the *Neurospora Crassa* CIA30 protein, in a patient with cardiomyoencephalopathy and complex I deficiency [64]. RNA interference (RNAi) assays had previously shown that NDUFAF1 was necessary for complex I assembly [186] and immunoprecipitation studies demonstrated a specific interaction of NDUFAF1 with complex I subunits [64]. However, the presence of NDUFAF1 in two high-molecular-weight complexes of ~850 and ~500 kDa that did not contain complex I subunits, indicated that NDUFAF1 transiently interacts with complex I assembly intermediates but not with the fully-assembled complex [187]. These two complexes exhibited alterations in their relative proportion and size in patients with mutations in different nuclear structural subunits [186]. Another homozygous mutation in *C6ORF66* was associated with severe complex I deficiency in five consanguineous patients who presented with infantile mitochondrial encephalopathy or antenatal cardiomyopathy [156]. Mutant muscle mitochondria revealed decreased levels of fully-assembled complex I and an accumulation of assembly intermediates that were absent in controls. By using a combination of phylogenetic profiling, which uses shared evolutionary history to highlight functionally related proteins, together with homozygosity mapping, a new mutation in the *C8ORF38* gene was recently found in two Lebanese siblings presenting with Leigh Syndrome and severe isolated complex I deficiency [138]. Knockdown of *C8ORF38* showed a strong reduction of both complex I abundance and activity, suggesting that this protein is crucial for activity and/or assembly of endogenous complex I.

Given the intricacy of the complex I assembly pathway, it is conceivable that many other unknown assembly factors that await elucidation may be involved in mitochondrial disease. For example, the predicted human homologue of the assembly factor CIA84, found in the aerobic fungus *Neurospora crassa*, has been identified by means of comparative genomics (GenBank accession no. NM 015545), but still awaits functional characterization [74]. A recently proposed inventory of 19 mitochondrial proteins that share ancestry with a large subset of complex I proteins, to which the C8ORF38 protein belongs, also represent strong candidates for genes underlying clinical complex I deficiency [138].

### 7.2.2 Complex II Deficiency

Complex II is also named succinate dehydrogenase (SDH) or succinnate-coenzyme Q reductase (E.C. 1.3.5.1.). It is the only enzyme that participates in both the citric acid cycle and the mitochondrial electron transport chain. SDH is made up of four subunits two soluble proteins, the flavoprotein (SDHA) and the Fe-S protein (SDHB) which are anchored to the inner membrane by the subunits SDHC and SDHD which contain a b-type heme.

Isolated complex II deficiency is a very rare disorder (2–4% of all respiratory chain enzyme deficiencies). *SDHA* gene mutations [OMIM 600857] have been reported in only six families. Four of these presented with autosomal recessive Leigh syndrome [36, 77, 89, 140] and one child died suddenly at the age of 5 months from a severe deterioration of neuromuscular, cardiac, and hepatic symptoms after an intermittent infection [181]. In the sixth family, two siblings suffered from late onset optic atrophy, ataxia, and myopathy, and a heterozygous mutation was identified, suggesting a possible role of *SDHA* mutations in autosomal dominant neurodegenerative disorders [33]. Further studies showed that several patients with complex II deficiency do not have mutations in *SDHA* [89]. This fact suggests a role of additional nuclear genes involved in synthesis, assembly, or maintenance of SDH [140].

Dominant mutations of the three other SDH subunit genes (*SDHB, SDHC*, and *SDHD*) were only described in association with tumors, such as hereditary paraganglioma and abdominal phaeochromocytoma [13].

### *7.2.3 Complex III Deficiency*

Mitochondrial complex III (ubiquinol-cytochrome c reductase or cytochrome bc1 complex, E.C.1.10.2.2) is a multiprotein enzyme complex that catalyzes the transfer of electrons from reduced coenzyme Q to cytochrome c, with a concomitant translocation of protons across the inner mitochondrial membrane [19]. The purified bovine complex is a symmetric homodimer with a combined molecular mass of ~450 kDa [94, 194]. Each monomer is composed of 11 subunits, ten encoded in the nucleus and one (cytochrome b) in the mitochondrial genome. Three subunits contain the metal centres responsible for electron transfer: cytochrome b (MTCYB), cytochrome c1 (CYC1), and the Rieske Fe-S protein (UQCRFS1 or RISP). The function of the other eight subunits remains to be elucidated.

Mitochondrial respiratory chain complex III deficiency [OMIM 124000] is a relatively uncommon defect of the OXPHOS system, associated with a wide variety of visceral, muscular and neurological disorders [35]. Mitochondrial DNA (mtDNA) mutations in the *cytochrome b (MT-CYB)* gene [OMIM 516020] constitute a major cause of complex III deficiency [6], but they only account for a small percentage of the total number of complex III-deficient patients.

#### 7.2.3.1 Mutations in Complex III Nuclear Structural Genes

Mutations in nuclear structural genes are rare, and only the genes *UQCRQ* [OMIM 609653] and *UQCRB* [OMIM 609653] have been associated to complex III deficiency [17, 83]. A homozygous 4-bp deletion was first found in the *UQCRB* gene, which encodes the ubiquinone-binding protein (subunit 7 or QP-C subunit), in a young girl who exhibited complex III deficiency associated with hypoglycemia and liver dysfunction [83]. Recently, the first missense mutation in the *UQCRQ* gene, encoding subunit 8, was detected in an Israeli family presenting with severe psychomotor retardation and extrapyramidal signs [17].

#### 7.2.3.2 Mutations in BCS1L

More than 20 different pathogenic mutations have been reported in the gene encoding the complex III assembly factor BCS1L [OMIM 603647]. BCS1L is a mitochondrial inner membrane chaperone that belongs to the AAA+ family of proteins and plays a role in the insertion of the Rieske Fe/S subunit into complex III [53, 73]. These mutations underlie several phenotypes that range from Björnstad syndrome (autosomal recessive disorder characterized by sensorineural hearing loss and *pili torti*, OMIM 262000) [84] or GRACILE syndrome (fetal growth retardation, aminoaciduria, cholestasis, iron overload, lactic acidosis, and early death, OMIM 603358) [65, 184]), to fatal complex III deficiency in neonates presenting with encephalopathy, alone or in combination with congenital metabolic acidosis, neonatal proximal tubulopathy, and liver failure [59, 60, 67]. It has been hypothesized that the impact of the *BCS1L* mutations might account for the clinical spectrum of the disease [67, 84]. For example, GRACILE segregates with the Finnish disease heritage mutation S78G, which does not cause complex III deficiency but is associated with the most severe clinical phenotype (50% die before 12 days and 50% before 4 months), complex III deficiency-associated mutations are relatively less severe (children live from 3 months to 5 years), and Bjornstad syndrome mutations are the mildest (normal life span). In this regard, complex III deficiency and GRACILE syndrome are only subtly different. Although BCS1L seems to be ubiquitously expressed, observations during mice embryonal development support tissue-specific differences in its expression that might account for the organ specific manifestations of disease [104].

In yeast, the protein was demonstrated to interact with the cytochrome bc1 complex, but as in humans, a substantial fraction of the protein was present in a high molecular-weight complex in which no complex III subunits were found [53, 69, 196]. Although these findings suggest a transient interaction between BCS1L and complex III, this physical association has not been confirmed in human samples [69].

### 7.2.4 Complex IV Deficiency

Cytochrome *c* oxidase (COX, complex IV, E.C. 1.9.3.1) is the terminal enzyme of the mitochondrial respiratory chain. It catalyses the oxidation of cytochrome *c* and the consequent reduction of oxygen to water. In this reaction four protons taken from the matrix are transferred to oxygen to form water and four protons are translocated across the inner mitochondrial membrane. The resulting electrochemical gradient is used to drive the synthesis of ATP.

The three dimensional structure of bovine heart cytochrome *c* oxidase is known [175]. Mammalian COX is composed of 13 subunits. The catalytic core of the enzyme is composed of the three mtDNA-encoded subunits. MTCO1 and MTCO2 contain the two heme A moieties (a and a3) and the two copper centers (CuA and CuB) responsible for the electron transfer. MTCO3 is part of the structural core and may play a role in proton pumping. The remaining ten subunits (COX4, COX5A, COX5B, COX6A, COX6B, COX6C, COX7A, COX7B, COX7C and COX8) are encoded in the nuclear DNA and must be coordinately translated and guided into the mitochondrial inner membrane to get assembled, along with the mtDNA-encoded subunits, heme groups and metal ions, into a functional complex. The function of these nuclear-encoded subunits remains unknown, although they may play regulatory and stabilization roles [11].

Isolated COX deficiency [OMIM 220110] constitutes, together with complex I deficiency, the most frequently observed abnormality of the OXPHOS system. In a study of 157 patients with respiratory chain defects [188] were found that the deficiency resided in complex I in 33%, in complex IV in 28%, and in a combined complex I and IV in 28%. Interestingly, a high rate of parental consanguinity was observed in complex IV (20%) and complex I+IV (28%) deficiencies.

#### 7.2.4.1 Mutations in Complex IV Nuclear Structural Subunits

Until very recently, no disease mutations had been reported in nuclear-encoded complex IV subunits. After thorough searches, the first pathogenic mutation has been finally detected in the ubiquitously expressed COX6B1 subunit [OMIM 124089] in two cases of mitochondrial encephalomyopathy and isolated COX deficiency [121].

#### 7.2.4.2 Mutations in Complex IV Assembly Factors

Most syndromes with isolated complex IV deficiency are caused by mutations in genes encoding COX assembly factors. The COX assembly pathway is believed to be a sequential process in which pools of unassembled subunits exist, and the findings of assembly intermediates led to the proposal of the first COX assembly model [131]. In yeast, around 20 proteins regulate this process to ensure the proper assembly and functioning of the enzyme. They include proteins involved in all stages of the assembly process, such as processing and translation of mitochondrially encoded mRNAs, insertion of newly synthesised polypeptides into the inner membrane and the addition of cofactors [72].

In humans, mutations have been found in nuclear genes encoding COX assembly factors that affect the stability and incorporation of COX subunits into the assembled complex. These genes correspond to *SURF1* [OMIM 185620] [143, 171, 200], *SCO1* [OMIM 603644] [179], *SCO2* [OMIM 604272] [96, 139], *COX10* [OMIM 602125] [8, 51, 180], *COX 15* [OMIM 603646] [9, 40, 135] and *LRPPRC* [OMIM 607544] [128, 195]. They are associated with different clinical presentations of COX deficiency, including encephalopathies such as Leigh Syndrome, fatal cardiomyopathy, hepatic failure and leukodystrophy [69]. COX-deficient patients carrying mutations in different assembly factors, such as SURF1, SCO1, COX10 or COX15 show distinct subunit assembly profiles, suggesting that the loss of enzyme activity in these cases is caused by an improper assembly of COX [69, 168, 172, 192]. When the assembly or maintenance of COX fails, reduced levels of fully-assembled COX occurs in the mitochondrial inner membrane, together with an aggregation of potentially toxic, unassembled, hydrophobic mitochondrial translation products [131].

The discovery of the first mutation in a nuclear OXPHOS assembly gene resulted from chromosomal transfer experiments, and led to the identification of *SURF1* as a complex IV deficiency-responsible gene. Since then, more than 40 different pathogenic mutations have been reported in *SURF1* [142, 143, 178, 200], which constitutes the most frequent cause of COX deficiency. All mutations lead to an almost complete loss of the SURF1 protein, and almost uniformly cause Leigh syndrome with isolated complex IV deficiency. As rare exceptions, patients presenting other clinical presentations such as with villous atrophy, hypertrichosis and mild neurological involvement or leukodystrophy have been reported [149, 189]. SURF1 patients usually show decreased levels of fully assembled COX and an accumulation of the specific assembly intermediates S1 and S2, suggesting a role for this protein in the early stages of complex IV assembly [131, 168, 172, 192]. Several SURF1 knockout animal models have been studied [1, 16, 56, 197]. Constitutive knockdown of Surf1 expression in *Drosophila melanogaster* [197] and in *Dario rerio* [16] produced COX deficiency associated with developmental impairment and nervous system involvement. However, pan-neuronal knockdown in *D. melanogaster* was associated with increased lifetime. Curiously, the mouse model did not display neurological symptoms, but it also exhibited increased longevity and resistance to calcium-dependent excitotoxic brain damage, suggesting that a moderate impairment of respiratory chain activity may prolong life span [1, 56].

Two other COX assembly genes, SCO1 and SCO2, encode metallochaperone proteins that play a role in general copper homeostasis and facilitate, together with other factors like COX17, the insertion of copper into COX subunits I and II, and the subsequent incorporation of these subunits into the holoenzyme [108, 109]. Mutations in SCO1 have been associated with neonatal hepatopathy and encephalopathy [179], while mutations in SCO2 have been mainly found in patients with infantile cardiomyopathy [96, 139], in both cases accompanied by a profound complex IV defect.

Additionally, mutations have been found in the *COX10* and *COX15* genes, which encode two enzymes involved in the biosynthesis of an essential COX prostethic group, heme A [20, 21, 22]. Mutations in *COX10* have been associated with leukodystrophy and renal tubulopathy [180], Leigh syndrome, or fatal infantile hypertrophic cardiomyopathy [8, 51]. Mutations in *COX15* may cause fatal infantile hypertrophic cardiomyopathy [9] or Leigh syndrome [40, 135]. In both cases, the total heme A content is reduced [8, 9].

Finally, mutations in LRPPRC, a factor involved in the stabilization of *MTCO1* and *MTCO3* mRNAs, cause COX-defective French-Canadian Leigh syndrome (LSFC) [128, 194].

### 7.2.5 Complex V Deficiency

Mitochondrial ATP synthase (FoF$_1$-ATPase, E.C. 3.6.3.14) is composed of a catalytic F$_1$ part connected by two stalks with a membrane-embedded Fo part. The mammalian enzyme is made of at least 16 different subunits (F$_1$: $\alpha3\beta3\gamma\delta\epsilon$+IF1, Fo: a, b, c10, d, e, f, g, F6, A6L, OSCP, (factor B)) [169] of which two Fo subunits – the subunit a (subunit 6) and subunit A6L are encoded by mtDNA. ATP synthase complex is formed with the assistance of several assembly factors, of which in mammalian cells only the orthologs of yeast ATP11 (ATPAF1) and ATP12 (ATPAF2) but no Fo-specific assembly factors have been found [190].

At present, more than dozen patients with ATP synthase deficiency (others than those with mutations in mtDNA encoded ATP6 and ATP8 subunits) have been diagnosed [90, 91, 167]. Most of them display a remarkably uniform phenotype with an early onset in newborns, severe and often fatal hyperlactacidemia, hypertrophic cardiomyopathy and elevated levels of 3-methylglutaconic acid in urine. About half of the patients die within the first days or months of life; however, long survival even for years, is also possible. Degenerative necrotic changes in the brain stem, typical of mtDNA ATP6 mutations, are rare. Studies of $F_1$ assembly factors have shown in one case with brain atrophy a homozygous missense mutation in the *ATP12* gene [OMIM 604273] encoding the ATP12 assembly factor [61]. However, in others, clinically distinct case mutations in *ATP11* and *ATP12* were excluded, and normal expression of the F1-assembly genes was found.

## 7.3 Nuclear Gene Defects in the Biogenesis of Protein Constituents of the OXPHOS System

### *7.3.1 Defects in the Mitochondrial Translation Machinery*

The genetic code, tRNAs and rRNAs of mitochondria are different from those of the cytoplasm. Translation of mtDNA protein-encoding genes (analogous to the yeast mit genes) is carried out in the mitochondrial matrix by a specific protein-synthesis machinery, which is composed of tRNAs and rRNAs synthesized *in situ* from the corresponding mitochondrial genes (analogous to the yeast syn genes) and a number of proteins encoded by nuclear DNA. These include mitoribosomal proteins (29 for the small subunit and 48 for the large subunit), several tRNA maturation enzymes (e.g. several pseudouridylate synthases), the aminoacyl-tRNA synthetases, the translation initiation, elongation and termination factors, and a large number of unidentified factors such as ribosome assembly factors [95]. Seven proteins have been associated with mitochondrial diseases: three translation elongation factors [52, 164, 178], two ribosomal proteins [124, 155] and two enzymes of tRNA maturation [41, 159]. Most patients have deficiencies of several complex of mitochondrial respiratory chain. Clinical pictures are heterogeneous presenting with lethal neonatal acidosis, infantile encephalomyopathy, hypertrophic cardiomyopathy with encephalomyopathy, and leukoencephalopathy with brain stem and spinal cord involvement and lactic acidosis.

Miller and coworkers [124] observed a defect in the assembly and/or stability of mitoribosomal small subunits, and an OXPHOS defect in fibroblasts from a patient with a fatal neonatal encephalopathy including lactic acidosis and agenesis of corpus callosum, born to consanguineous parents who were from a Bedouin family. They found a nonsense homozygous mutation in *MRPS16* [OMIM 609204], a nuclear gene encoding a protein of the mitoribosomal small subunit. In two female infants with edema on prenatal ultrasounds, and neonatal hypertrophic cardiomyopathy, renal tubulopathy and fatal severe lactic acidosis, were found a mutation in another nuclear-encoded protein of the mitochondrial ribosome, MRPS22 [OMIM 605810] [155].

Coenen and coworkers [52] identified by means of chromosome transfer, microsatellite and sequence analysis, and complementation in cell culture, a missense mutation in *EFG1* [OMIM 606639], one of two genes encoding different isoforms of the mitochondrial translational elongation factor G. This mutation affects a conserved residue of the GTP-binding site of *EFG1*, which predicts a effect on nucleotide binding and hence enzymatic function. EFG1 is a GTPase which catalyzes the translocation of peptidyl-tRNA from the ribosomal-acceptor site to the peptidyl site after peptide-bond formation. Concomitant movement of mRNA exposes the next codon in the acceptor site. The mutation was encountered in two siblings presenting with postnatal liver failure and a generalized deficiency of mitochondrial protein synthesis in patient-derived fibroblasts. Another patient with mutations in EFG1 had also hepatic involvement [10], however, a clearly different phenotype has been described in one infant [178] characterized by severe neonatal lactic acidosis and early-onset Leigh syndrome. The same homozygous mutation in mitochondrial TSFM [OMIM 604723], a gene coding for the mitochondrial translation elongation factor EFTs, has been identified in individuals with distinct syndromes, an encephalomyopathy and hypertrophic cardiomyopathy [164]. Authors suggest that an adaptive mechanism may act differently in different patients, thereby contributing to the clinical variability.

In a patient with diffuse cystic leukodystrophy and micropolygyria and mild liver involvement was identified a missense mutation in the elongation factor EFTu (TUFM) [OMIM 602389] [178]. EFTu is a GTPase that loads an aminoacyl-tRNA from its specific transferase, escorts it to the ribosomal A site, and is then released from the ribosome.

Base modification at the wobble base of tRNAs seems to be important for efficient and accurate decoding. It is the molecular 'target' of pathological tRNA gene mutations in mtDNA [102]. Several types of base modifications (e.g. pseudouridylation) might be relevant for maintaining the tertiary structure. In yeast, a family of pseudourine synthases with multiple specificities is involved in carrying out this isomerization at diffetent positions of cytosolic and mitochondrial tRNAs and rRNAs. The human gene pseudouridylate synthase 1 (PUS1) [OMIM 608109], which encodes one such isoenzyme, has been shown to harbour a mutation in a patient with mitochondrial myopathy and sideroblastic anemia (MLASA) [41], associated with defective OXPHOS in bone marrow and skeletal muscle. Although a direct effect on mitochondrial protein synthesis has not been demonstrated, deficient pseudouridylation of one or more mitochondrial tRNAs seems to be the disease mechanism. More recently, a homozygous stop mutation was identified in two Italian brothers affected by MLASA [68].

Mutations in the nuclear gene *DARS* [OMIM 603084] that encodes for the mitochondrial aspartyl-tRNA synthetase cause progressive pyramidal, cerebellar and posterior column signs, possible mild cognitive impairment, a diagnostic magnetic resonance imaging pattern involving cerebral, brain stem and spinal white matter and increased lactate in white matter on magnetic resonance spectroscopy [159].

### 7.3.2 Defects in Mitochondrial Protein Import

The targeting and import of mitochondrial proteins involve a series of receptors, import pores, and molecular machines located in the four intra-mitochondrial locations; the outer and inner mitochondrial membranes, the intermembrane space (IMS), and the matrix [88]. Most mitochondrial proteins are synthesized by free cytosolic ribosomes and are imported posttranslationally. Abnormalities in these pathways can lead to a protein not reaching its final destination, finally causing disease.

The X-linked neurodegenerative disorder Human Deafness Dystonia Syndrome (DDS), also known as Mohr–Tranebjaerg syndrome, is characterized by postlingual progressive sensorineural deafness, dystonia, spasticity, dysphagia, mental deterioration, paranoia, and cortical blindness [173]. DDS is caused by mutations in the IMS protein deafness dystonia peptide 1 (DDP1/Tim8) [OMIM 300356]. It has been suggested that the mechanism underlying the severity of DDS is the impaired biogenesis of the TIM23 complex as a result of a dysfunctional Tim8. This would lead to severe mitochondrial dysfunction because the TIM23 complex is used for the import of all matrix proteins and some IMS and inner membrane proteins. The $Ca^{2+}$-responsive aspartate/glutamate carriers, aralar1 and citrin, are also chaperoned across the IMS by DDP1/Tim8–Tim13 to TIM22 for insertion into the inner membrane, suggesting that Tim8 defects could lead to decreased NADH levels due to a decrease in the aspartate/malate NADH shuttle [151].

By homozygosity mapping in 11 patients with dilated autosomal recessive cardiomyopathy with ataxia (DCMA) in the Canadian Dariusleut Hutterite population, a novel gene was identified, *DNAJC19* [OMIM 608977], that shares similarity with fungal Pam18 [54]. The condition is characterized by early onset dilated cardiomyopathy, nonprogressive cerebral ataxia, growth failure, and in some patients, testicular dysgenesis and 3-methylglutaconic aciduria. Pam18 is an inner membrane protein in all eukaryotes that participates in the import of matrix proteins. All DCMA patients were homozygous for a mutation in the acceptor site of intron 3 of *DNAJC19* predicting the synthesis of a truncated protein. If DNAJC19 is confirmed as the human ortholog of Pam18 the DCMA phenotype might be the result of dysfunctional import of matrix proteins through the TIM23 complex [117].

*HSPD1* gene encodes heat shock 60 kDa protein 1 (HSP60), a mitochondrial matrix chaperone involved in the folding and assembly of mitochondrial proteins. Patients with an atypical mitochondria disease involving multisystem failure have been linked to a deficiency in HSP60 [OMIM 118190] [2, 37]. Analysis of their mitochondria indicated that there was a decreased number of mitochondria per cell, ultrastructural abnormalities and the organelles were mainly localized around the nucleus. Enzyme activities of oxidative phosphorylation, pyruvate metabolism, β-oxidation, and other mitochondrial metabolic pathways were markedly decreased in cultured fibroblasts, and the reduced levels were associated with decreased amounts of HSP60 [92]. In contrast, the activities of enzymes localized in the outer membrane or in other cellular organelles were normal [2]. Although the authors suggested that the molecular basis of the disease was a primary defect in HSP60 synthesis and/or processing, there has been no evidence presented to date to support this hypothesis. HSP60 has also been implicated in causing the autosomal dominant spastic paraplegia-13 (SPG13) [OMIM 605280] [71, 81]. A missense mutation in HSP60 was found in a family with SPG13 [81]. Herediatry spastic paraplegias are group of neurodegenerative disorders characterized by progressive weakness and spasticity of the lower limbs due to degeneration of corticospinal axons. Recently, a homozygous missense mutation in *HPSD1* has been identified in a consanguineous kindred with autosomal-recessive Pelizaeus-Merzbacher-like disease (PMLD) [OMIM 612233] [118], a lethal early-onset neurodegenerative disorder associated with brain hypomyelination and leukodystrophy. HSP60 defects can cause neurodegerative pathologies of varying severity.

### *7.3.3 Defects in Mitochondrial Fe-S Cluster Biosynthesis*

Iron–sulfur (Fe–S) clusters are prosthetic groups that are essential for respiration. The most common Fe–S clusters in eukaryotes, the [2Fe–2S] and [4Fe–4S] clusters, are able to accept or donate single electrons, catalyze enzymatic reactions or function as regulatory proteins. Fe–S clusters are essential components of respiratory electron transfer complexes as well as the tricarboxylic acid cycle (TCA cycle) enzymes, aconitase and succinate dehydrogenase. Moreover, Fe–S cluster assembly and disassembly in mammalian iron regulatory protein-1 (IRP1) determines whether IRP1 binds its mRNA targets in response to oxidative stress and intracellular iron levels [153]. Defects in this pathway are the cause of several human diseases.

Over the last decade, studies of the autosomal recessive neurodegenerative disorder Friedreich ataxia revealed that deficiency of the protein frataxin (FXN) [OMIM 606829], commonest caused by the expansion of the unstable GAA triplet repeat in the intronic sequence of the gene, results in the loss of Fe–S protein activities, mitochondrial iron overload, oxidative damage and ultimately mitochondrial failure [15]. FXN deficiency is a spinocerebellar ataxia associated with disartria, muscle weakness, cardiomyopathy and diabetes. Several evidences indicate that frataxin is an iron-binding chaperone protein for Fe-S cluster assembly which is enable to bind to the scaffold protein, ISCU (Iron-sulfur cluster scaffold homolog), or the heme biosynthetic enzyme, Ferrochelatase [26].

In the last 2 years, two other defects in Fe–S cluster assembly genes have been recognized: a splicing defect in glutaredoxin 5 (GLRX5) [OMIM 609588] was associated as a cause of sideroblastic anemia [43], and very recently, a splicing defect in the ISCU protein [OMIM 611911] in skeletal muscle leading to a myopathy with severe exercise intolerance, and deficiency in the TCA succinate dehydrogenase and aconitase activities [125, 137]. GLRX5 function is still unclear, although studies in yeast indicated that GLRX5 might be involved in the transfer of Fe-S clusters from the scaffold to target proteins [153]. Studies in GLRX5 deficient zebrafish [193] suggest that anemia is presumably caused by the decrease of heme synthesis as a result of the repression translational IRP1 (a mammalian RNA-binding protein that play a role in regulating intracellular iron metabolism)-mediated of the erythroide isoform of aminolevulinate synthase (eALAS), which catalizes the first step in heme biosynthesis.

In the 1960s, five families were described presenting with a hereditary myopathy characterized by exercise intolerance and elevated amounts of lactate and pyruvate, having muscle weakness, myoglobinuria in the severe cases. Biochemical and histochemical analysis suggested a defect in complex II of the mitochondrial respiratory chain. Muscle biopsies showed electronic microscopy abnormalities typical of mitochondrial myopathy. After exercise, fibres, were markedly depleted of glycogen and contained large amounts of lipid droplets. Low levels of the 30-kD iron–sulphur protein and the 13.5-kD protein of complex II, but close to normal levels of the 70-kD protein were found. Furthermore, reduced levels of mitochondrial aconitase and abnormalities in the respiratory chain subunits with iron–sulphur centres in complexes I and III were also observed. These results suggested that the patients suffered from a dysfunction in the synthesis, import, processing or assembly of iron–sulphur clusters. All but one family were genealogically related. In one family has been recently identified a *ISCU* gene mutation [137]. Furthermore, the skeletal muscle mitochondria in the ISCU patients were overloaded with ferric ion [125].

The mitochondrial transporter cassette *ABCB7* is the mutated gene for the X-linked sideroblastica anemia with ataxia (XLSA/A) [25]. ABCB7, although not directly involve in Fe-S cluster byosinthesys, is implicate in the export of Fe-S clusters to the cytosol. RNAi experiments has shown reduced activities in two cytosolic Fe-S enzymes (xanthine oxidase and cytosolic aconitase) but succinate dehydrogenase activity has not been affected, suggesting ABCB7 is not involved in the process of mitochondrial Fe-S cluster biogenesis [42].

### *7.3.4 Defects in Mitochondrial Protein Metabolism*

Mutations in the spastic-paraplegia-7 gene (*SPG7*) have been associated with a complicated form of autosomal recessive hereditary spastic paraplegia [OMIM 602783] [46]. Besides spastic paraparesis, some of these patients have cerebral and cerebellar atrophy and optic atrophy. *SPG7* encodes a putative mitochondrial protease that is known as paraplegin. Paraplegin is highly homologous to mitochondrial metalloproteases that belong to the AAA family of P-loop ATPases, and form large proteolytic complexes in the inner membrane of mitochondria [154]. Paraplegin is part of the m-AAA protease which controls protein quality and regulates ribosome assembly [132]. *SPG7* mutations may alter the substrate specificity of the m-AAA protease [103], leading to loss of function. The consequences of a paraplegin deficiency in mice are axonal swellings caused by massive accumulation of organelles and neurofilaments, reminiscent of those observed in spastin-deficient mice [70], which suggests that altered axonal trafficking and mitochondrial deficiencies may be intimately related.

## 7.4 Nuclear Gene Defects in the Biogenesis of Non-Protein Constituents of the OXPHOS System

### *7.4.1 Coenzyme Q10 Deficiency*

CoQ funnels electrons from complexes I and II to complex III. CoQ deficiency, primary or secondary, is the a mitochondrial disorder that may respond to replacement therapy. The first patients with muscle coenzyme Q10 (CoQ10) deficiency were reported in 1989 [133]. It has been associated with different clinical phenotypes: (i) myopathic presentation, including exercise intolerance and recurrent episodes of exercise-related myoglobinuria, was described in five patients, who also had CNS involvement, manifested by seizures, ataxia, or mental retardation [62, 133, 166]. Their muscle biopsies had RRF and lipid storage. All patients improved after oral CoQ10 supplementation, (ii) childhood-onset cerebellar ataxia as the dominant symptom, which was associated with cerebellar atrophy [129]. Frequent accompanying features included seizures, weakness, mental retardation, pyramidal signs, and peripheral neuropathy [106, 129]. Other more rare are late-onset ataxia and hypergonadotropic hypogonadism [76], psychomotor delay, deafness, ataxia, mild spasticity, and neuroradiological Leigh syndrome [183]. The muscle biopsy is normal. Ataxic patients also improve with CoQ10 supplementation. Linkage analysis of one led to the identification of a mutation in the aprataxin *(APTX)* gene [OMIM 606350], causing ataxiaoculomotor- apraxia 1 *(AOA1)* [110, 147]. Mutations in the *ADCK3* gene [OMIM 606980] (also called *COQ8* or *CABC1*) [105, 127]) were also recently found in patients with this phenotype. ADCK3 belongs to the family of atypical kinases, which includes phosphoinositide and choline kinases, suggesting that ADCK3 plays an indirect regulatory role in ubiquinone biosynthesis possibly as part of a feedback loop that regulates ATP production [105], (iii) severe infantil encephalomyopathy associated with renal disease [150, 152]. CoQ10 supplementation in these patients appears to be lifesaving. Mutations in three genes of CoQ biosynthesis pathway, *COQ2*, para-hydroxybenzoate-polyprenyl transferase [OMIM 609825] [126, 148], decaprenyl diphosphate synthases, subunit 2, *PDSS2* [OMIM 610564] [116] and subunit 1, *PDSS1* [OMIM 607429] [127] were reported in patients presenting with this phenotype, (iv) Two cases have reported to present with the myopathic and ataxic phenotypes: a 4-year-old boy with delayed motor development, mitochondrial myopathy, bilateral ptosis, generalized seizures, pigmentary retinopathy, ataxia, and cerebellar atrophy [34], and an 11-year-old girl, who presented at 3 years of age with exercise intolerance, weakness, and mitochondrial myopathy, and developed ataxia and cerebellar atrophy at age 7 [14].

### *7.4.2 Barth Syndrome*

In patients with Barth syndrome have been identified mutations in the Taffazine gene *(TAZ)* which is located in X-chromosome [OMIM 300394] [32]. Patients present with heart failure, myopathy, neutropenia, and growth retardation. They have reduced concentration and altered composition of cardiolipin, the specific mitochondrial phospholipid, and additionally have elevated urinary excretion of 3-methylglutaconic acid and hypocholesterolemia [23]. Tissue biopsies show abnormal ultrastructural changes in mitochondria and deficiencies of the respiratory function. Tafazzin, which is located in mitochondria, is a phospholipid acyltransferase and presumably involved in acyl-specific remodelling of cardiolipin, which promotes structural uniformity and molecular symmetry among the cardiolipin molecular species. Inhibition of this pathway would lead to changes in mitochondrial architecture and function [123]. A study of mitochondrial respiratory supercomplexes in Barth syndrome patient's lymphoblast showed that function of complexes I, III and IV were affected directly and that complex IV stability was reduced, which in turn destabilizes the CI/CIII2/CIV supercomplex resulting in lower levels of complex I, further reducing enzymatic activity [122].

Recently, a 12-year-old proband having left ventricular non-compaction and dilated cardiomyopathy was shown to be a compound harboring two novel mutations, one frame-shift of maternal origin in the *TAZ* gene and one missense of paternal origin in the *LDB3* (LIM domain-binding 3) [120].

## 7.5 Nuclear Gene Defects of Mitochondrial Dynamics

Mitochondrial fusion involves the coordinated fusion of both the outer and inner mitochondrial membranes. Mitofusins are GTPases localized to the outer mitochondrial membrane. Two mitofusin genes, *Mfn1* and *Mfn2* [47] are known in mammals. The dynamin family GTPase OPA1 is also essential for mitochondrial fusion [48, 50]. This mitochondrial intermembrane space protein is associated with the inner membrane and may also play a role in control of inner membrane structure [79]. The opposing process of mitochondrial fission depends on the dynamin-related protein DRP1 [165]. This GTPase localizes to discrete spots on mitochondria, and a subset of such spots mark sites of future fission. Mutations in MFN2 [OMIM 608507] cause Charcot-Marie-Tooth (CMT) subtype 2A [198]. CMT is a group of diseases characterized by pathology in the longest motor and sensory nerves, which enervate the hands and feet [199]. The most common phenotypes of CMT result from demyelination of peripheral nerves. CMT2A is an axonal form, thought to be caused by defects in the neurons themselves. CMT2A has an autosomal-dominant pattern of inheritance. Due to the extraordinary length of motor and sensory neurons that enervate the distal limbs, it is plausible that quite subtle perturbations in mitochondrial dynamics can, over time, lead to severe physiological defects. Ganglioside induced differentiation associated protein 1 (GDAP1) [OMIM 606598], a mitochondrial outer membrane protein with tandem glutathione transferase domains, is mutated in CMT4A [199]. CMT4A is a recessive disease with clinical heterogeneity; patients can have features of myelin or axonal defects, or a combination of both. Reduction of GDAP1 expression leads to elongation of mitochondrial tubules, suggesting that GDAP1 is involved in mitochondrial fission.

Autosomal dominant optic atrophy (DOA), also termed Kjer's disease, results in loss of visual acuity and is caused by degeneration of retinal ganglion cells [58]. The major form of DOA is caused by mutations in OPA1 [OMIM 605290], a mitochondrial intermembrane space protein that is essential for mitochondrial fusion. Over 100 OPA1 mutations (http://lbbma.univ-angers.fr/ eOPA1/) have been identified; some patients have reduced copies of mtDNA and diminished oxidative phosphorylation in skeletal muscle [99, 112], which may be related to the severe reduction of mitochondrial membrane potential and respiration that occur in OPA1-deficient cells [48]. Loss of OPA1 also leads to apoptosis [136] and severe lesions mitochondrial cristae [79]. LHON, one of the most common mtDNA diseases, results in optic nerve atrophy and has clinical features similar to DOA [44]. These suggest that perturbations in mitochondrial dynamics can affect some of the same tissues as mtDNA diseases. Recently [5, 93], it has been shown that some dominant missense mutations in *OPA1* gene cause the accumulation of multiple mtDNA deletions in skeletal muscle. The syndrome reported was complex, consisting of a combination of DOA with PEO, peripheral neuropathy, ataxia and deafness. Most of these patients have ragged-red and cytochrome-c-oxidase negative muscle fibres, with paracrystalline inclusions in abnormally shaped mitochondria. These features are typical of mitochondrial encephalomyopathies associated with multiple mtDNA deletions, suggesting that the instability of mtDNA acts as a common mechanism in all forms of autosomal-dominant PEO. Patients with the 'OPA1-plus' syndrome all harbor heterozygous missense mutations which suggests that a 'toxic' effect exerted by an abnormal OPA1 species, rather than the simple decrease of protein content, causes mtDNA damage.

For mitochondrial fission, another GTPase, called dynamin-like protein 1 (DLP1) or dynamin-related protein 1 (DRP1), is required. DLP1 is recruited from the cytoplasm to the surface of the mitochondrion, where it is thought to assemble into large oligomeric complexes that use GTP hydrolysis to constrict mitochondrial tubules during fission. It is established that FIS1, an outer-membrane protein, is critical to DLP1-mediated fission. Both DLP1 and FIS1 are also required for the fission of peroxisomes, organelles involved in oxidative reactions that are important for the catabolism of fatty acids and hydrogen peroxide. Recently, a *DLP1* mutation [OMIM 603850] [191] was reported in an infant with microcephaly and metabolic aberrations who died at 37 days of age, affecting both mitochondrial and peroxisomal fission.

LETM1 encodes for the human homologue of yeast Mdm38p, a mitochondria-shaping protein of unclear function. LETM1 is located in the inner membrane, exposed to the matrix and oligomerized in higher molecular weight complexes. Wolf-Hirschhorn syndrome (WHS) is a complex congenital syndrome caused by a monoallelic deletion of the short arm of chromosome 4 that includes *LETM1* gene [OMIM 604407] [63, 160]. It seems to be that BCS1L, an assembly protein of mitochondrial complex III, stimulates the assembly of the LETM1 complex. Furhtermore, *BCS1L* knockdown caused disassembly of the respiratory chains as well as LETM1 [170].

## References


1. Agostino A, Invernizzi F, Tiveron C, Fagiolari G, Prelle A, Lamantea E, Giavazzi A, Battaglia G, Tatangelo L, Tiranti V, Zeviani M (2003). Constitutive knockout of *Surf1* is associated with high embryonic lethality, mitochondrial disease and cytochrome c oxidase deficiency in mice. Hum Mol Genet 12:399–413.
2. Agsteribbe E, Huckriede A., Veenhuis M, Ruiters MH, Niezen-Koning KE, Skjeldal OH, Skullerud K, Gupta RS, Hallberg R, van Diggelen OP (1993). A fatal, systemic mitochondrial disease with decreased mitochondrial enzyme activities, abnormal ultrastructure of the mitochondria and deficiency of heat shock protein 60. Biochem Biophys Res Commun 193:146–154.
3. Allikmets R, Raskind WH, Hutchinson A, Schueck ND, Dean M, Koeller DM (1999). Mutation of a putative mitochondrial iron transporter gene *(ABC7)* in X-linked sideroblastic anemia and ataxia (XLSA/A). Hum Molec Genet 8:743–749.
4. Amati-Bonneau P, Guichet A, Olichon A, Chevrollier A, Viala F, Miot S, Ayuso C, Odent S, Arrouet C, Verny C, Calmels M-N, Simard G, Belenguer P, Wang J, Puel J-L, Hamel C, Malthiery Y, Bonneau D, Lenaers G, Reynier P (2005). *OPA1* R445H mutation in optic atrophy associated with sensorineural deafness. Ann Neurol 58:958–963.
5. Amati-Bonneau P, Valentino ML, Reynier P, Gallardo ME, Bornstein B, Boissière A, Campos Y, Rivera H, de la Aleja JG, Carroccia R, Iommarini L, Labauge P, Figarella-Branger D, Marcorelles P, Furby A, Beauvais K, Letournel F, Liguori R, La Morgia C, Montagna P, Liguori M, Zanna C, Rugolo M, Cossarizza A, Wissinger B, Verny C, Schwarzenbacher R, Martín MA, Arenas J, Ayuso C, Garesse R, Lenaers G, Bonneau D, Carelli V (2008). *OPA1* mutations induce mitochondrial DNA instability and optic atrophy 'plus' phenotypes. Brain 131:338–351.
6. Andreu AL, Hanna MG, Reichmann H, Bruno C, Penn, AS, Tanji K, Pallotti F, Iwata S, Bonilla E, Lach B, Morgan-Hughes J, DiMauro S (1999). Exercise intolerance due to mutations in the *cytochrome b* gene of mitochondrial DNA. N Engl J Med 341: 1037–1044.
7. Antonicka H, Ogilvie I, Taivassalo T, Anitori RP, Haller RG, Vissing J, Kennaway NG, Shoubridge EA (2003). Identification and characterization of a common set of complex I assembly intermediates in mitochondria from patients with complex I deficiency. J Biol Chem 278:43081–43088.
8. Antonicka H, Leary SC, Guercin GH, Agar, Horvath R, Kennaway NG, Harding CO, Jaksch M, Shoubridge EA (2003). Mutations in *COX10* result in a defect in mitochondrial heme A biosynthesis and account for multiple, early-onset clinical phenotypes associated with isolated COX deficiency. Hum Mol Genet 12:2693–2702.
9. Antonicka H, Mattman A, Carlson CG, Glerum DM, Hoffbuhr KC, Leary SC, Kennaway NG, Shoubridge EA (2003). Mutations in *COX15* produce a defect in the mitochondrial heme biosynthetic pathway, causing early-onset fatal hypertrophic cardiomyopathy. Am J Hum Genet 72:101–114.
10. Antonicka H, Sasarman F, Kennaway NG, Shoubridge EA (2006). The molecular basis for tissue specificity of the oxidative phosphorylation deficiencies in patients with mutations in the mitochondrial translation factor EFG1. Hum Mol Genet 15:1835–1846.
11. Arnold S, Kadenbach B (1997). Cell respiration is controlled by ATP, an allosteric inhibitor of cytochrome-c oxidase. Eur J Biochem 249:350–354.
12. Astuti D, Latif F, Dallol A, Dahia PLM, Douglas F, George E, Skoldberg F, Husebye ES, Eng C, Maher ER (2001). Gene mutations in the succinate dehydrogenase subunit SDHB cause susceptibility to familial pheochromocytoma and to familial paraganglioma. Am J Hum Genet 69:49–54.
13. Astuti D, Morris M, Krona C, Abel F, Gentle D, Martinsson T, Kogner P, Neumann HP, Voutilainen R, Eng C, Rustin P, Latif F, Maher ER (2004). Investigation of the role

of SDHB inactivation in sporadic phaeochromocytoma and neuroblastoma. Br J Cancer 91:1835–1841.
14. Auré K, Benoist JF, Ogier de Baulny H, Romero NB, Rigal O, Lombès A (2004). Progression despite replacement of a myopathic form of coenzyme Q10 defect. Neurology 63:727–729.
15. Babady NE, Carelle N, Wells RD, Rouault TA, Hirano M, Lynch DR, Delatycki MB, Wilson RB, Isaya G, Puccio H (2007). Advancements in the pathophysiology of Friedreich's Ataxia and new prospects for treatments. Mol Genet Metab 92:23–35.
16. Baden KN, Murray J, Capaldi RA, Guillemin K (2007). Early developmental pathology due to cytochrome c oxidase deficiency is revealed by a new zebrafish model. J Biol Chem 282:34839–34849.
17. Barel O, Shorer Z, Flusser H, Ofir R, Narkis G, Finer G, Shalev H, Nasasra A, Saada A, Birk OS (2008). Mitochondrial complex III deficiency associated with a homozygous mutation in *UQCRQ*. Am J Hum Genet 82:1211–1216.
18. Barghuti F, Elian K, Gomori JM, Shaag A, Edvardson S, Saada A, Elpeleg O (2008). The unique neuroradiology of complex I deficiency due to NDUFA12L defect. Mol Genet Metab. 94:78–82.
19. Baum H, Rieske JS, Silman HI, Lipton SH (1967). On the mechanism of electron transfer in complex III of the electron transfer chain. Proc Natl Acad Sci U S A 57:798–805.
20. Barros MH, Carlson CG, Glerum DM, Tzagaloff A (2001). Involvement of mitochondrial ferredoxin and Cox15p in hydroxylation of heme O. FEBS Lett 492:133–138.
21. Barros MH, Nobrega FG, Tzagoloff A (2002). Mitochondrial ferredoxin is required for heme A synthesys in *Saccharomyces cervisiae*. J Biol chem. 277:9997–10002.
22. Barros MH, Tzagoloff A (2002). regulation of heme A biosynthetic pathway in *Saccharomyces cerevisiae*. FEBS Lett 516: 119–123.
23. Barth PG, Valianpour F, Bowen VM, Lam J, Duran M, Vaz FM, Wanders RJ (2004). X-linked cardioskeletal myopathy and neutropenia (Barth syndrome): an update. Am J Med Genet A 126A:349–354.
24. Baysal BE, Ferrell RE, Willett-Brozick JE, Lawrence EC, Myssiorek D, Bosch A, van der May A, Taschner PEM, Rubinstein WS, Myers EN, Richard CWIII, Cornelisse CJ, Devilee P, Devlin B (2000). Mutations in *SDHD*, a mitochondrial complex II gene, in hereditary paraganglioma. Science 287:848–851.
25. Bekri S, Kispal G, Lange H, Fitzsimons E, Tolmie J, Lill R, Bishop DF (2000). Human ABC7 transporter: gene structure and mutation causing X-linked sideroblastic anemia with ataxia with disruption of cytosolic iron-sulfur protein maturation. Blood 96: 3256–3264.
26. Bencze KZ, Kondapalli KC, Cook JD, McMahon S, Millán-Pacheco C, Pastor N, Stemmler TL (2006). The structure and function of frataxin. Crit Rev Biochem Mol Biol 41:269–291.
27. Benit P, Chretien D, Kadhom N, De Lonlay-Debeney P, Cormier-Daire V, Cabral A, Peudenier S, Rustin P, Munnich A, Rotig A (2001). Large-scale deletion and point mutations of the nuclear *NDUFV1* and *NDUFS1* genes in mitochondrial complex I deficiency. Am J Hum Genet 68:1344–1352.
28. Benit P, Steffann J, Lebon S, Chretien D, Kadhom N, de Lonlay P, Goldenberg A, Dumez Y, Dommergues M, Rustin P, Munnich A, Rotig A (2003). Genotyping microsatellite DNA markers at putative disease loci in inbred/multiplex families with respiratory chain complex I deficiency allows rapid identification of a novel nonsense mutation (IVS1nt -1) in the *NDUFS4* gene in Leigh syndrome. Hum Genet 112:563–566.
29. Benit P, Beugnot R, Chretien D, Giurgea I, De Lonlay-Debeney P, Issartel JP, Corral-Debrinski M, Kerscher S, Rustin P, Rotig A, Munnich A (2003). Mutant NDUFV2 subunit of mitochondrial complex I causes early onset hypertrophic cardiomyopathy and encephalopathy. Hum Mutat 21:582–586.
30. Benit P, Slama A, Cartault F, Giurgea I, Chretien D, Lebon S, Marsac C, Munnich A, Rotig A, Rustin P (2004). Mutant NDUFS3 subunit of mitochondrial complex I causes Leigh syndrome. J Med Genet 41:14–17.

31. Berger I, Hershkovitz E, Shaag A, Edvardson S, Saada A, Elpeleg O (2008). Mitochondrial complex I deficiency caused by a deleterious *NDUFA11* mutation. Ann Neurol 63:405–408.
32. Bione S, D'Adamo P, Maestrini E, Gedeon AK, Bolhuis PA, Toniolo D (1996). A novel X-linked gene, G4.5. (sic) is responsible for Barth syndrome. Nature Genet 12:385–389.
33. Birch-Machin MA, Taylor RW, Cochran B, Ackrell BA, Turnbull DM (2000). Late-onset optic atrophy, ataxia, and myopathy associated with a mutation of a complex II gene. Ann Neurol 48:330–335.
34. Boitier E, Degoul F, Desguerre I, Charpentier C, François D, Ponsot G, Diry M, Rustin P, Marsac C (1998). A case of mitochondrial encephalomyopathy associated with a muscle coenzyme Q10 deficiency. J Neurol Sci 156:41–46.
35. Borisov VB (2002). Defects in mitochondrial respiratory complexes III and IV, and human pathologies. Mol Aspects Med 23:385–412.
36. Bourgeron T, Rustin P, Chretien D, Birch-Machin M, Bourgeois M, Viegas-Péquignot E, Munnich A, Rötig A (1995). Mutation of a nuclear succinate dehydrogenase gene results in mitochondrial respiratory chain deficiency. Nat Genet 11:144–149.
37. Briones P, Vilaseca MA, Ribes A, Vernet A, Lluch M, Cusi V, Huckriede A, Agsteribbe E (1997). A new case of multiple mitochondrial enzyme deficiencies with decreased amount of heat shock protein 60. J Inherit Metab Dis 20:569–577.
38. Budde SMS, van den Heuvel LPWJ, Janssen AJ, Smeets RJP, Buskens CAF, DeMeirleir L, Van Coster R, Baethmann M, Voit T, Trijbels JMF, Smeitink JAM (2000). Combined enzymatic complex I and III deficiency associated with mutations in the nuclear encoded *NDUFS4* gene. Biochem Biophys Res Commun 275:63–68.
39. Bugiani M, Invernizzi F, Alberio S, Briem E, Lamantea E, Carrara F, Moroni I, Farina L, Spada M, Donati MA, Uziel G, Zeviani M (2004). Clinical and molecular findings in children with complex I deficiency. Biochim Biophys Acta 1659:136–147.
40. Bugiani M, Tiranti V, Farina L, Uziel G, Zeviani M (2005). Novel mutations in *COX15* in a long surviving Leigh syndrome patient with cytochrome c oxidase deficiency. J Med Genet 42:e28.
41. Bykhovskaya Y, Casas K, Mengesha E, Inbal A, Fischel-Ghodsian N (2004). Missense mutation in pseudouridine synthase 1 *(PUS1)* causes mitochondrial myopathy and sideroblastic anemia (MLASA). Am J Hum Genet 74:1303–1308.
42. Camaschella C (2008). Recent advances in the understanding of inherited sideroblastic anaemia. Br J Hematol 143:27–38.
43. Camaschella C, Campanella A, De Falco L, Boschetto L, Merlini R, Silvestri L, Levi S, Iolascon A (2007). The human counterpart of zebrafish shiraz shows sideroblastic-like microcytic anemia and iron overload. Blood 110: 1353–1358.
44. Carelli V, La Morgia C, Iommarini L, Carroccia R, Mattiazzi M, Sangiorgi S, Farne' S, Maresca A, Foscarini B, Lanzi L, Amadori M, Bellan M, Valentino ML (2007). Mitochondrial optic neuropathies: how two genomes may kill the same cell type? Biosci Rep 27:173–184.
45. Carroll J, Fearnley IM, Shannon RJ, Hirst J, Walker JE (2003). Analysis of the subunit composition of complex I from bovine heart mitochondria. Mol Cell Proteomics 2:117–126.
46. Casari G, De Fusco M, Ciarmatori S, Zeviani M, Mora M, Fernandez P, De Michele G, Filla A, Cocozza S, Marconi R, Dürr A, Fontaine B, Ballabio A (1998). Spastic paraplegia and OXPHOS impairmentcaused by mutations in paraplegin, a nuclear-encoded mitochondrial metalloprotease. Cell 93:973–983.
47. Chen H, Detmer SA, Ewald AJ, Griffin EE, Fraser SE, Chan DC (2003). Mitofusins Mfn1 and Mfn2 coordinately regulate mitochondrial fusion and are essential for embryonic development. J Cell Biol 160:189–200.
48. Chen H, Chan DC (2004). Mitochondrial dynamics in mammals. Curr Top Dev Biol 59: 119–144
49. Chomyn A, Patel SD, Cleeter MW, Ragan CI, Attardi G (1988). The site of synthesis of the iron-sulfur subunits of the flavoprotein and iron-protein fractions of human NADH dehydrogenase. J Biol Chem 263:16395–16400.

50. Cipolat S, Martins de Brito O, Dal Zilio B, Scorrano L (2004). OPA1 requires mitofusin 1 to promote mitochondrial fusion. Proc Natl Acad Sci U S A 101:15927–15932.
51. Coenen MJ, van den Heuvel LP, Ugalde C, Ten Brinke M, Nijtmans LG, Trijbels FJ, Beblo S, Maier EM, Muntau AC, Smeitink JA (2004). Cytochrome c oxidase biogenesis in a patient with a mutation in *COX10* gene. Ann Neurol 56: 560–564.
52. Coenen MJ, Antonicka H, Ugalde C, Sasarman F, Rossi R, Heister JG, Newbold RF, Trijbels FJ, van den Heuvel LP, Shoubridge EA, Smeitink JA (2004). Mutant mitochondrial elongation factor G1 and combined oxidative phosphorylation deficiency. N Engl J Med 351:2080–2086.
53. Cruciat CM, Hell K, Folsch H, Neupert W, Stuart RA (1999). Bcs1p, an AAA-family member, is a chaperone for the assembly of the cytochrome bc(1) complex. EMBO J 18:5226–5233.
54. Davey KM, Parboosingh JS, McLeod DR, Chan A, Casey R, Ferreira P, Snyder FF, Bridge PJ, Bernier FP (2006). Mutation of *DNAJC19*, a human homologue of yeast inner mitochondrial membrane co-chaperones, causes DCMA syndrome, a novel autosomal recessive Barth syndrome-like condition. J Med Genet 43:385–393.
55. Delatycki MB, Williamson R, Forrest SM (2000). Friedreich ataxia: an overview. J Med Genet 37:1–8.
56. Dell'agnello C, Leo S, Agostino A, Szabadkai G, Tiveron C, Zulian a, Prelle A, Roubertoux P, Rizzuto R, Zeviani M (2007). Increased longevity and refractoriness to Ca2+-dependent neurodegeneration in *Surf1* knockout mice. Hum Mol Genet 16:431–444.
57. Delettre C, Lenaers G, Griffoin J-M, Gigarel N, Lorenzo C, Belenguer P, Pelloquin L, Grosgeorge J, Turc-Carel C, Perret E, Astarie-Dequeker C, Lasquellec L, Arnaud B, Ducommun B, Kaplan J, Hamel CP (2000). Nuclear gene *OPA1*, encoding a mitochondrial dynamin-related protein, is mutated in dominant optic atrophy. Nature Genet 26:207–210.
58. Delettre C, Lenaers G, Pelloquin L, Belenguer P, Hamel CP (2002). *OPA1* (Kjer type) dominant optic atrophy: a novel mitochondrial disease. Mol Genet Metab 75:97–107.
59. De Lonlay P, Valnot I, Barrientos A, Gorbatyuk M, Tzagoloff A, Taanman JW, Benayoun E, Chretien D, Kadhom N, Lombes A, de Baulny HO, Niaudet P, Munnich, A, Rustin, P, Rotig A (2001). A mutant mitochondrial respiratory chain assembly protein causes complex III deficiency in patients with tubulopathy, encephalopathy and liver failure. Nat Genet 29: 57–60.
60. De Meirleir L, Seneca S, Damis E, Sepulchre B, Hoorens A, Gerlo E, Garcia Silva MT, Hernandez EM, Lissens W, Van Coster R (2003). Clinical and diagnostic characteristics of complex III deficiency due to mutations in the *BCS1L* gene. Am J Med Genet 121: 126–131.
61. De Meirleir L, Seneca S, Lissens W, De Clercq I, Eyskens F, Gerlo E, Smet J, Van Coster R (2004). Respiratory chain complex V deficiency due to a mutation in the assembly gene *ATP12*. J Med Genet 41:120–124.
62. Di Giovanni S, Mirabella M, Spinazzola A, Crociani P, Silvestri G, Broccolini A, Tonali P, Di Mauro S, Servidei S (2001). Coenzyme Q10 reverses pathological phenotype and reduces apoptosis in familial CoQ10 deficiency. Neurology 57:515–518.
63. Dimmer KS, Navoni F, Casarin A, Trevisson E, Endele S, Winterpacht A, Salviati L, Scorrano L (2008). *LETM1*, deleted in Wolf-Hirschhorn syndrome is required for normal mitochondrial morphology and cellular viability. Hum Mol Genet 17:201–214.
64. Dunning CJ, McKenzie M, Sugiana C, Lazarou M, Silke J, Connelly A, Fletcher JM, Kirby DM, Thorburn DR, Ryan MT (2007). Human CIA30 is involved in the early assembly of mitochondrial complex I and mutations in its gene cause disease, EMBO J 26:3227–3237.
65. Fellman V (2002). The GRACILE syndrome, a neonatal lethal metabolic disorder with iron overload. Blood Cells Mol Dis 29:4444–4450.
66. Fernandez-Moreira D, Ugalde C, Smeets R, Rodenburg RJ, Lopez-Laso E, Ruiz-Falco ML, Briones P, Martin MA, Smeitink JA, Arenas J (2007). X-linked *NDUFA1* gene mutations associated with mitochondrial encephalomyopathy. Ann Neurol 61:73–83.

67. Fernandez-Vizarra E, Bugiani M., Goffrini P, Carrara F, Farina L, Procopio E, Donati A, Uziel G, Ferrero I, Zeviani M (2007). Impaired complex III assembly associated with *BCS1L* gene mutations in isolated mitochondrial encephalopathy. Hum Mol Genet 16: 1241–1252.
68. Fernandez-Vizarra E, Berardinelli A, Valente L, Tiranti V, Zeviani M (2007). Nonsense mutation in pseudouridylate synthase 1 *(PUS1)* in two brothers affected by myopathy, lactic acidosis and sideroblastic anemia (MLASA). J Med Genet 44:173–180.
69. Fernandez-Vizarra E, Tiranti V, Zeviani M (2009). Assembly of the oxidative phosphorylation system in humans: What we have learned by studying its defects. Biochim Biophys Acta. 1993: 200–211.
70. Ferreirinha F, Quattrini A, Pirozzi M, Valsecchi V, Dina G, Broccoli V, Auricchio A, Piemonte F, Tozzi G, Gaeta L, Casari G, Ballabio A, Rugarli EI (2004). Axonal degeneration in paraplegin deficient mice is associated with abnormal mitochondria and impairment of axonal transport. J Clin Invest 113:231–242.
71. Fontaine B, Davoine CS, Dürr A, Paternotte C, Feki I, Weissenbach J, Hazan J, Brice A (2000). A new locus for autosomal dominant pure spastic paraplegia, on chromosome 2q24–q34. Am J Hum Genet 66:702–707.
72. Fontanesi F, Soto IC, Horn D, Barrientos A (2006). Assembly of mitochondrial cytochrome c-oxidase, a complicated and highly regulated cellular process. Am J Physiol Cell Physiol 291:C1129–C1147.
73. Frickey T, Lupas AN (2004). Phylogenetic analysis of AAA proteins. J Struct Biol 146:2–10.
74. Gabaldon T, Rainey D, Huynen MA (2005). Tracing the evolution of a large protein complex in the eukaryotes, NADH:ubiquinone oxidoreductase (Complex I). J Mol Biol 348: 857–870.
75. Galante YM, Hatefi Y (1978). Resolution of complex I and isolation of NADH dehydrogenase and an iron-sulfur protein. Methods Enzymol 53:15–21.
76. Gironi M, Lamperti C, Nemni R, Moggio M, Comi G, Guerini FR, Ferrante P, Canal N, Naini A, Bresolin N, DiMauro S (2004). Late-onset cerebellar ataxia with hypogonadism and muscle coenzyme Q10 deficiency. Neurology 62:818–820.
77. Goto Y, Murayama K, Hirawake H, et al. (1998). Molecular genetics of 5 patients with succinate dehydrogenase deficiency [abstract]. Am J Hum Genet (suppl 63): A362.
78. Grigorieff N (1999). Structure of the respiratory NADH: ubiquinone oxidoreductase (complex I). Curr Opin Struct Biol 9:476–483.
79. Griparic L, van der Wel NN, Orozco IJ, Peters PJ, van der Bliek AM (2004). Loss of the intermembrane space protein Mgm1/OPA1 induces swelling and localized constrictions along the lengths of mitochondria. J Biol Chem 279:18792–18798.
80. Gusella JF, Tanzi RE, Bader PI, Phelan MC, Stevenson R, Hayden MR, Hofman KJ, Faryniarz AG, Gibbons K (1985). Deletion of Huntington's disease-linked G8 (D4S10) locus in Wolf-Hirschhorn syndrome. Nature 318:75–78.
81. Hansen JJ, Durr A, Cournu-Rebeix I, Georgopoulos C, Ang D, Nielsen MN, Davoine CS, Brice A, Fontaine B, Gregersen N, Bross P (2002). Hereditary spastic paraplegia SPG13 is associated with a mutation in the gene encoding the mitochondrial chaperonin Hsp60. Am J Hum Genet 70:1328–1332.
82. Hattori N, Yoshino H, Tanaka M, Suzuki H, Mizuno Y (1998). Genotype in the 24-kDa subunit gene *(NDUFV2)* of mitochondrial complex I and susceptibility to Parkinson disease. Genomics 49:52–58.
83. Haut S, Brivet M, Touati G, Rustin P, Lebon S, Garcia-Cazorla A, Saudubray J M, Boutron A, Legrand A, Slama A (2003). A deletion in the human *QP-C* gene causes a complex III deficiency resulting in hypoglycaemia and lactic acidosis. Hum Genet 113:118–122.
84. Hinson JT, Fantin VR, Schonberger J, Breivik N, Siem G, McDonough B, Sharma P, Keogh I, Godinho R, Santos F, Esparza A, Nicolau Y, Selvaag E, Cohen BH, Hoppel CL, Tranebjaerg L, Eavey RD, Seidman JG, Seidman CE (2007). Missense mutations in the *BCS1L* gene as a cause of the Bjornstad syndrome. N Engl J Med 356:809–819.

85. Hinttala R, Uusimaa J, Remes AM, Rantala H, Hassinen IE, Majamaa K (2005). Sequence analysis of nuclear genes encoding functionally important complex I subunits in children with encephalomyopathy. J Mol Med 83:786–794.
86. Hirst J, Carroll J, Fearnley IM, Shannon RJ, Walker JE (2003). The nuclear encoded subunits of complex I from bovine heart mitochondria. Biochim Biophys Acta 1604:135–150.
87. Hoefs SJ, Dieteren CE, Distelmaier F, Janssen RJ, Epplen A, Swarts HG, Forkink M, Rodenburg RJ, Nijtmans LG, Willems PH, Smeitink JA, van den Heuvel LP (2008). *NDUFA2* complex I mutation leads to Leigh disease. Am J Hum Genet 82:1306–1315.
88. Hoogenraad N, Ward L, Ryan M (2002). Import and assembly of proteins into mitochondria of mammalian cells. Biochim Biophys Acta 1592:97–105.
89. Horvath R, Abicht A, Holinski-Feder E, Laner A, Gempel K, Prokisch H, Lochmuller H, Klopstock T, Jaksch M (2006). Leigh syndrome caused by mutations in the flavoprotein (Fp) subunit of succinate dehydrogenase (SDHA). J Neurol Neurosurg Psychiatry 77:74–76.
90. Houstek J, Klement P, Floryk D, Antonická H, Hermanská J, Kalous M, Hansíková H, Hout'ková H, Chowdhury SK, Rosipal T, Kmoch S, Stratilová L, Zeman J (1999). A novel deficiency of mitochondrial ATPase of nuclear origin. Hum Mol Genet 8:1967–1974.
91. Houstek J, Pícková A, Vojtísková A, Mrácek T, Pecina P, Jesina P (2006). Mitochondrial diseases and genetic defects of ATP synthase. Biochim Biophys Acta 1757:1400–1405.
92. Huckriede A, Heikema A, Sjollema K, Briones P, Agsteribbe E (1995). Morphology of the mitochondria in heat shock protein 60 deficient fibroblasts from mitochondrial myopathy patients. Effects of stress conditions. Virchows Arch 427:159–165.
93. Hudson G, Amati-Bonneau P, Blakely EL, Stewart JD, He L, Schaefer AM, Griffiths PG, Ahlqvist K, Suomalainen A, Reynier P, McFarland R, Turnbull DM, Chinnery PF, Taylor RW (2008). Mutation of *OPA1* causes dominant optic atrophy with external ophthalmoplegia, ataxia, deafness and multiple mitochondrial DNAdeletions: a novel disorder of mtDNA maintenance. Brain 131:329–337.
94. Iwata S, Lee JW, Okada K, Lee JK, Iwata M, Rasmussen B, Link TA, Ramaswamy S, Jap BK (1998). Complete structure of the 11-subunit bovine mitochondrial cytochrome bc1 complex. Science 281:64–71.
95. Jacobs HT, Turnbull DM (2005). Nuclear genes and mitochondrial translation: a new class of genetic disease. Trends Genet 21:312–314.
96. Jaksch M, Ogilvie I, Yao J, Kortenhaus G, Bresser HG, Gerbitz KD, Shoubridge EA (2000). Mutations in *SCO2* are associated with a distinct form of hypertrophic cardiomyopathy and cytochrome c oxidase deficiency. Hum Mol Genet 9:795–801.
97. Janssen RJ, Nijtmans LG, van den Heuvel LP, Smeitink JA (2006). Mitochondrial complex I: structure, function and pathology. J Inherit Metab Dis 29:499–515.
98. Jin H, May M, Tranebjaerg L, Kendall E, Fontan G, Jackson J, Subramony SH, Arena F, Lubs H, Smith S, Stevenson R, Schwartz C, Vetrie D (1996). A novel X-linked gene, *DDP*, shows mutations in families with deafness (DFN-1), dystonia, mental deficiency and blindness. Nature Genet 14:177–180.
99. Kim JY, Hwang JM, Ko HS, Seong MW, Park BJ, Park SS (2005). Mitochondrial DNA content is decreased in autosomal dominant optic atrophy. Neurology 64:966–972.
100. Kirby DM, Crawford M, Cleary MA, Dahl HH, Dennett X, Thorburn DR (1999). Respiratory chain complex I deficiency: an underdiagnosed energy generation disorder. Neurology 52:1255–1264.
101. Kirby DM, Salemi R, Sugiana C, Ohtake A, Parry L, Bell KM, Kirk EP, Boneh A, Taylor RW, Dahl HH, Ryan MT, Thorburn DR (2004). *NDUFS6* mutations are a novel cause of lethal neonatal mitochondrial complex I deficiency. J Clin Invest 114:837–845.
102. Kirino Y, Yasukawa T, Ohta S, Akira S, Ishihara K, Watanabe K, Suzuki T (2004). Codon-specific translational defect caused by a wobble modification deficiency in mutant tRNA from a human mitochondrial disease. Proc Natl Acad Sci U S A 101:15070–15075.
103. Koppen M, Metodiev MD, Casari G, Rugarli EI, Langer T (2007). Variable and tissue-specific subunit composition of mitochondrial m-AAA protease complexes linked to hereditary spastic paraplegia. Mol Cell Biol 27:758–767.

104. Kotarsky H, Tabasum I, Mannisto S, Heikinheimo M, Hansson S, Fellman V (2007). BCS1L is expressed in critical regions for neural development during ontogenesis in mice. Gene Expr Patterns 7:266–273.
105. Lagier-Tourenne C, Tazir M, López LC, Quinzii CM, Assoum M, Drouot N, Busso C, Makri S, Ali-Pacha L, Benhassine T, Anheim M, Lynch DR, Thibault C, Plewniak F, Bianchetti L, Tranchant C, Poch O, DiMauro S, Mandel JL, Barros MH, Hirano M, Koenig M (2008). ADCK3, an ancestral kinase, is mutated in a form of recessive ataxia associated with coenzyme Q10 deficiency. Am J Hum Genet 82:661–672.
106. Lamperti C, Naini A, Hirano M, De Vivo DC, Bertini E, Servidei S, Valeriani M, Lynch D, Banwell B, Berg M, Dubrovsky T, Chiriboga C, Angelini C, Pegoraro E, DiMauro S (2003). Cerebellar ataxia and coenzyme Q10 deficiency. Neurology 60:1206–1208.
107. Lazarou M, Thorburn DR, Ryan MT, McKenzie M (2009). Assembly of mitochondrial complex I and defects in disease. Biochim Biophys Acta. 1793: 77–88
108. Leary SC, Kaufman BA, Pellecchia G, Guercin GH, Mattman A, Jaksch M, Shoubridge EA (2004). Human SCO1 and SCO2 have independent, cooperative functions in copper delivery to cytochrome c oxidase. Hum Mol Genet 13:1839–1848.
109. Leary SC, Cobine PA, Kaufman BA, Guercin GH, Mattmanm A, Palaty J, Lockitch G, Winge DR, Rustin P, Horvath P, Shoubridge EA (2007). The human cytochrome c oxidase assembly factors SCO1 and SCO2 have regulatory roles in the maintenance of cellular copper homeostasis. Cell Metab 5:6909–6920.
110. Le Ber I, Dubourg O, Benoist JF, Jardel C, Mochel F, Koenig M, Brice A, Lombès A, Dürr A (2007). Muscle coenzyme Q10 deficiencies in ataxia with oculomotor apraxia 1. Neurology 68:295–297.
111. Lebon S, Rodriguez D, Bridoux D, Zerrad A, Rotig A, Munnich A, Legrand A, Slama A (2007). A novel mutation in the human complex I *NDUFS7* subunit associated with Leigh syndrome. Molec Genet Metab 90:379–382.
112. Lodi R, Tonon C, Valentino ML, Iotti S, Clementi V, Malucelli E, Barboni P, Longanesi L, Schimpf S, Wissinger B, Baruzzi A, Barbiroli B, Carelli V (2004). Deficit of in vivo mitochondrial ATP production in OPA1-related dominant optic atrophy. Ann Neurol 56:719–723.
113. Loeffen J, Smeitink J, Triepels R, Smeets R, Schuelke M, Sengers R, Trijbels F, Hamel B, Mullaart R, van den Heuvel L (1998). The first nuclear-encoded complex I mutation in a patient with Leigh syndrome. Am J Hum Genet 63:1598–1608.
114. Loeffen JL, Smeitink JA, Trijbels JM, Janssen AJ, Triepels RH, Sengers RC, van den Heuvel LP (2000). Isolated complex I deficiency in children: clinical, biochemical and genetic aspects. Hum Mutat 15:123–134.
115. Loeffen J, Elpeleg O, Smeitink J, Smeets R, Stockler-Ipsiroglu S, Mandel H, Sengers R, Trijbels F, van den Heuvel L (2001). Mutations in the complex I *NDUFS2* gene of patients with cardiomyopathy and encephalomyopathy. Ann Neurol 49:195–201.
116. López LC, Schuelke M, Quinzii CM, Kanki T, Rodenburg RJ, Naini A, Dimauro S, Hirano M (2006). Leigh syndrome with nephropathy and CoQ10 deficiency due to decaprenyl diphosphate synthase subunit 2 *(PDSS2)* mutations. Am J Hum Genet 79:1125–1129.
117. MacKenzie JA, Payne RM (2007). Mitochondrial protein import and human health and disease. Biochimica et Biophysica Acta 1772:509–523.
118. Magen D, Georgopoulus C, Bross P, Ang D, Segev Y, Goldsher D, Nemirovski A, Shahar E, Ravid S, Luder A, Heno B, Gershoni-Baruch R, Skorecki K, Mandel H (2008). Mitochondrial hsp60 chaperonopahty causes an autosomal-recessive neurodegenerative disorder linked to brain hypomyelination and leukodystrophy. Am J Hum Genet 83:30–42
119. Martin MA, Blazquez A, Gutierrez-Solana LG, Fernandez-Moreira D, Briones P, Andreu AL, Garesse R, Campos Y, Arenas J (2005). Leigh syndrome associated with mitochondrial complex I deficiency due to a novel mutation in the *NDUFS1* gene. Arch Neurol 62: 659–661.
120. Marziliano N, Mannarino S, Nespoli L, Diegoli M, Pasotti M, Malattia C, Grasso M, Pilotto A, Porcu E, Raisaro A, Raineri C, Dore R, Maggio PP, Brega A, Arbustini E (2007). Barth

syndrome associated with compound hemizygosity and heterozygosity of the *TAZ* and *LDB3* genes. Am J Med Genet A 143A:907–915.

121. Massa V, Fernandez-Vizarra E, Alshahwan S, Bakhsh E, Goffrini P, Ferrero I, Mereghetti P, D'Adamo P, Gasparini P, Zeviani M (2008). Severe infantile encephalomyopathy caused by a mutation in *COX6B1*, a nucleus-encoded subunit of cytochrome c oxidase. Am J Hum Genet 82:1281–1289.
122. McKenzie M, Lazarou M, Thorburn DR, Ryan MT (2006). Mitochondrial respiratory chain supercomplexes are destabilized in Barth syndrome patients. J Mol Biol 361:462–469.
123. Michael S, Mindong R (2006). Barth syndrome, a human disorder of cardiolipin metabolism. FEBS Lett 580:5450–5455.
124. Miller C, Saada A, Shaul N, Shabtai N, Ben-Shalom E, Shaag A, Hershkovitz E, Elpeleg O (2004). Defective mitochondrial translation caused by a ribosomal protein (MRPS16) mutation. Ann Neurol 56:734–738.
125. Mochel F, Knight MA, Tong WH, Hernandez D, Ayyad K, Taivassalo T, Andersen PM, Singleton A, Rouault TA, Fischbeck KH, Haller RG (2008). Splice mutation in the iron-sulfur cluster scaffold protein ISCU causes myopathy with exercise intolerance. Am J Hum Genet 82:652–660.
126. Mollet J, Giurgea I, Schlemmer D, Dallner G, Chretien D, Delahodde A, Bacq D, de Lonlay P, Munnich A, Rötig A (2007). Prenyldiphosphate synthase, subunit 1 (PDSS1) and OH-benzoate polyprenyltransferase (COQ2) mutations in ubiquinone deficiency and oxidative phosphorylation disorders. J Clin Invest 117:765–772.
127. Mollet J, Delahodde A, Serre V, Chretien D, Schlemmer D, Lombes A, Boddaert N, Desguerre I, de Lonlay P, de Baulny HO, Munnich A, Rötig A (2008). *CABC1* gene mutations cause ubiquinone deficiency with cerebellar ataxia and seizures. Am J Hum Genet 82:623–630.
128. Mootha VK, Lepage P, Miller K, Bunkenborg J, Reich M, Hjerrild M, Delmonte T, Villeneuve A, Sladek R, Xu F, Mitchell GA, Morin C, Mann M, Hudson TJ, Robinson B, Rioux JD, Lander ES (2003). Identification of a gene causing human cytochrome c oxidase deficiency by integrative genomics. Proc Natl Acad Sci U S A 100:605–610.
129. Musumeci O, Naini A, Slonim AE, Skavin N, Hadjigeorgiou GL, Krawiecki N, Weissman BM, Tsao CY, Mendell JR, Shanske S, De Vivo DC, Hirano M, DiMauro S (2001). Familial cerebellar ataxia with muscle coenzyme Q10 deficiency. Neurology 56:849–855.
130. Niemann S, Muller U (2000). Mutations in *SDHC* cause autosomal dominant paraganglioma, type 3. Nature Genet 26:268–270.
131. Nijtmans LG, Taanman JW, Muijsers AO, Speijer D, Van den Bogert C (1998). Assembly of cytochrome-c oxidase in cultured human cells. Eur J Biochem 254:389–394.
132. Nolden M, Ehses S, Koppen M, Bernacchia A, Rugarli EI, Langer T (2005). The m-AAA protease defective in hereditary spastic paraplegia controls ribosome assembly in mitochondria. Cell 123:277–289.
133. Ogasahara S, Engel AG, Frens D, Mack D (1989). Muscle coenzyme Q deficiency in familial mitochondrial encephalomyopathy. Proc Natl Acad Sci U S A 86:2379–2382.
134. Ogilvie I, Kennaway NG, Shoubridge EA (2005). A molecular chaperone for mitochondrial complex I assembly is mutated in a progressive encephalopathy. J Clin Invest 115: 2784–2792.
135. Oquendo CE, Antonicka H, Shoubridge EA, Reardon W, Brown GK (2004). Functional and genetic studies demonstrate that mutation in the *COX15* gene can cause Leigh syndrome. J Med Genet 41:540–544.
136. Olichon A, Baricault L, Gas N, Guillou E, Valette A, Belenguer P, Lenaers G (2003). Loss of OPA1 perturbates the mitochondrial inner membrane structure and integrity, leading to cytochrome c release and apoptosis. J Biol Chem 278:7743–7746.
137. Olsson A, Lind L, Thornell LE, Holmberg M (2008). Myopathy with lactic acidosis is linked to chromosome 12q23.3-24.11 and caused by an intron mutation in the *ISCU* gene resulting in a splicing defect. Hum Mol Genet 17:1666–1672.

138. Pagliarini DJ, Calvo SE, Chang B, Sheth SA, Vafai SB, Ong SE, Walford GA, Sugiana C, Boneh A, Chen WK, Hill DE, Vidal M, Evans JG, Thorburn DR, Carr SA, Mootha VK (2008). A mitochondrial protein compendium elucidates complex I disease biology. Cell 134:112–123.
139. Papadopoulou LC, Sue CM, Davidson MM, Tanji K, Nishino I, Sadlock JE, Krishna S, Walker W, Selby J, Glerum DM, Coster RV, Lyon G, Scalais E, Lebel R, Kaplan P, Shanske S, De VD, Bonilla E, Hirano M, DiMauro S, Schon EA. (1999). Fatal infantile cardioencephalomyopathy with COX deficiency and mutations in *SCO2*, a COX assembly gene. Nat Genet 23:333–337.
140. Parfait B, Chretien D, Rötig A, Marsac C, Munnich A, Rustin P (2000). Compound heterozygous mutations in the flavoprotein gene of the respiratory chain complex II in a patient with Leigh syndrome. Hum Genet 106:236–243.
141. Payne M, Yang Z, Katz BJ, Warner JEA, Weight CJ, Zhao Y, Pearson ED, Treft RL, Hillman T, Kennedy RJ, Meire FM, Zhang K (2004). Dominant optic atrophy, sensorineural hearing loss, ptosis, and ophthalmoplegia: a syndrome caused by a missense mutation in *OPA1*. Am J Ophthal 138:749–755.
142. Pecina P, Houstkova H, Hansikova H, Zeman J, Houstek J (2004). Genetic defects of cytochrome c oxidase assembly. Physiol Res 53 (Suppl 1):S213–223.
143. Pequignot MO, Dey R, Zeviani M, Tiranti V, Godinot C, Poyau A, Sue C, DiMauro S, Abitbol M, Marsac C (2001). Mutations in the *SURF1* gene associated with Leigh syndrome and cytochrome c oxidase deficiency. Hum Mutat 17:374–381.
144. Petruzzella V, Tiranti V, Fernandez P, Ianna P, Carrozzo R, Zeviani M (1998). Identification and characterization of human cDNAs specific to *BCS1*, *PET112*, *SCO1*, *COX15*, and *COX11*, five genes involved in the formation and function of the mitochondrial respiratory chain. Genomics 54:494-504.
145. Petruzzella V, Vergari R, Puzziferri I, Boffoli D, Lamantea E, Zeviani M, Papa S (2001). A nonsense mutation in the *NDUFS4* gene encoding the 18 kDa (AQDQ) subunit of complex I abolishes assembly and activity of the complex in a patient with Leigh-like syndrome. Hum Molec Genet 10:529–535.
146. Procaccio V, Wallace DC (2004). Late-onset Leigh syndrome in a patient with mitochondrial complex I *NDUFS8* mutations. Neurology 62:1899–1901.
147. Quinzii CM, Kattah AG, Naini A, Akman HO, Mootha VK, DiMauro S, Hirano M (2005). Coenzyme Q deficiency and cerebellar ataxia associated with an aprataxin mutation. Neurology 64:534–541.
148. Quinzii C, Naini A, Salviati L, Trevisson E, Navas P, DiMauro S, Hirano MA (2006). Mutation in para-hydroxybenzoatepolyprenyl transferase *(COQ2)* causes primary coenzyme Q10 deficiency. Am J Hum Genet 78:345–349.
149. Rahman S, Brown RM, Chong WK, Wilson CJ, Brown GK (2001). A *SURF1* gene mutation presenting as isolated leukodystrophy. Ann Neurol 49:797–800.
150. Rahman S, Hargreaves I, Clayton P, Heales S (2001). Neonatal presentation of coenzyme Q10 deficiency. J Pediatr139:456–458.
151. Roesch K, Hynds PJ, Varga R, Tranebjaerg L, Koehler CM (2004). The calcium-binding aspartate/glutamate carriers, citrin and aralar1, are new substrates for the DDP1/TIMM8a–TIMM13 complex. Hum Mol Genet 13:2101–2111.
152. Rötig A, Appelkvist EL, Geromel V, Chretien D, Kadhom N, Edery P, Lebideau M, Dallner G, Munnich A, Ernster L, Rustin P (2000). Quinone-responsive multiple respiratory-chain dysfunction due to widespread coenzyme Q10 deficiency. Lancet 356:391–395.
153. Rouault TA, Tong WH (2008). Iron-sulfur cluster biogenesis and human disease. Trends in Genet 24:398–407.
154. Rugarli EI, Langer T (2006). Translating m-AAA protease function in mitochondria to hereditary spastic paraplegia. Trends Mol Med12:262–269.
155. Saada A, Shaag A, Arnon S, Dolfin T, Miller C, Fuchs-Telem D, Lombes A, Elpeleg O (2007). Antenatal mitochondrial disease caused by mitochondrial ribosomal protein (MRPS22) mutation. J Med Genet 44:784–786.

156. Saada A, Edvardson S, Rapoport M, Shaag A, Amry K, Miller C, Lorberboum-Galski H, Elpeleg O (2008). *C6ORF66* is an assembly factor of mitochondrial complex I. Am J Hum Genet 82:32–38.
157. Salviati L, Freehauf C, Sacconi S, DiMauro S, Thoma J, Tsai AC-H (2004). Novel *SURF1* mutation in a child with subacute encephalopathy and without the radiological features of Leigh syndrome. Am J Med Genet 128A:195–198.
158. Sazanov LA, Peak-Chew SY, Fearnley IM, Walker JE (2000). Resolution of the membrane domain of bovine complex I into subcomplexes: Implications for the structural organization of the enzyme. Biochemistry 39:7229–7235.
159. Scheper GC, van der Klok T, van Andel RJ, van Berkel CG, Sissler M, Smet J, Muravina TI, Serkov SV, Uziel G, Bugiani M, Schiffmann R, Krägeloh-Mann I, Smeitink JA, Florentz C, Van Coster R, Pronk JC, van der Knaap MS (2007). Mitochondrial aspartyl-tRNA synthetase deficiency causes leukoencephalopathy with brain stem and spinal cord involvement and lactate elevation. Nat Genet 39:534–539.
160. Schlickum S, Moghekar A, Simpson JC, Steglich C, O'Brien RJ, Winterpacht A, Endele SU (2004). *LETM1*, a gene deleted in Wolf-Hirschhorn syndrome, encodes an evolutionarily conserved mitochondrial protein. Genomics 83:254–261.
161. Schuelke M, Smeitink J, Mariman E, Loeffen J, Plecko B, Trijbels F, Stockler-Ipsiroglu S, van den Heuvel L (1999). Mutant NDUFV1 subunit of mitochondrial complex I causes leukodystrophy and myoclonic epilepsy. Nat Genet 21:260–261.
162. Smeitink J, van den Heuvel L (1999). Human mitochondrial complex I in health and disease. Am J Hum Genet 64:1505–1510.
163. Smeitink J, van den Heuvel L, DiMauro S (2001). The genetics and pathology of oxidative phosphorylation. Nat Rev Genet 2:342–352.
164. Smeitink JA, Elpeleg O, Antonicka H, Diepstra H, Saada A, Smits P, Sasarman F, Vriend G, Jacob-Hirsch J, Shaag A, Rechavi G, Welling B, Horst J, Rodenburg RJ, van den Heuvel B, Shoubridge EA (2006). Distinct clinical phenotypes associated with a mutation in the mitochondrial translation elongation factor EFTs. Am J Hum Genet 79: 869–877.
165. Smirnova E, Griparic L, Shurland DL, van Der Bliek AM (2001). Dynamin-related protein drp1 is required for mitochondrial division in mammalian cells. Mol Biol Cell 12: 2245–2256.
166. Sobreira C, Hirano M, Shanske S, Keller RK, Haller RG, Davidson E, Santorelli FM, Miranda AF, Bonilla E, Mojon DS, Barreira AA, King MP, DiMauro S (1997). Mitochondrial encephalomyopathy with coenzyme Q10 deficiency. Neurology 48: 1238–1243.
167. Sperl W, Jesina P, Zeman J, Mayr JA, Demeirleir L, VanCoster R, Píckova A, Hansíkovå H, Houstkovå H, Krejcík Z, Koch J, Smet J, Muss W, Holme E, Houstek J (2006). Deficiency of mitochondrial ATP synthase of nuclear genetic origin. Neuromuscul Disord 16:821–829.
168. Stiburek L, Vesela K, Hansikova H, Pecina P, Tesarova M, Cerna L, Houstek J, Zeman J (2005). Tissue-specific cytochrome c oxidase assembly defects due to mutations in *SCO2* and *SURF1*. Biochem J 392:625–632.
169. Stock D, Leslie AG, Walker JE (1999). Molecular architecture of the rotary motor in ATP synthase. Science 286:1700–1705.
170. Tamai S, Iida H, Yokota S, Sayano T, Kiguchiya S, Ishihara N, Hayashi J, Mihara K, Oka T (2008). Characterization of the mitochondrial protein LETM1, which maintains the mitochondrial tubular shapes and interacts with the AAA-ATPase BCS1L. J Cell Sci 121:2588–2600.
171. Tiranti V, Hoertnagel K, Carrozzo R, Galimberti C, Munaro M, Granatiero M, Zelante L, Gasparini P, Marzella R, Rocchi M, Bayona-Bafaluy MP, Enriquez JA, Uziel G, Bertini E, Dionisi-Vici C, Franco B, Meitinger T, Zeviani M (1998), Mutations of *SURF-1* in Leigh disease associated with cytochrome c oxidase deficiency, Am J Hum Genet 63:1609–1621.

172. Tiranti V, Galimberti C, Nijtmans L, Bovolenta S, Perini MP, Zeviani M (1999). Characterization of *SURF-1* expression and Surf-1p function in normal and disease conditions. Hum Mol Genet 8:2533–2540, 1043.
173. Tranebjaerg L, Schwartz C, Eriksen H, Andreasson S, Ponjavic V, Dahl A, Stevenson RE, May M, Arena F, Barker D (1995). A new X linked recessive deafness syndrome with blindness, dystonia, fractures, and mental deficiency is linked to Xq22. J Med Genet 32:257–263.
174. Triepels RH, van den Heuvel LP, Loeffen JL, Buskens CA, Smeets RJ, Rubio GM, Budde SM, Mariman EC, Wijburg FA, Barth PG, Trijbels JM, Smeitink JA (1999). Leigh syndrome associated with a mutation in the NDUFS7 (PSST) nuclear encoded subunit of complex I. Ann Neurol 45:787–790.
175. Tsukihara T, Aoyama H, Yamashita E, Tomizaki T, Yamaguchi H, Shinzawa-Itoh K, Nakashima R, Yaono R, Yoshikawa S (1996). The whole structure of the 13-subunit oxidized cytochrome c oxidase at 2.8 A. Science 272:1136–1144.
176. Ugalde C, Janssen RJRJ, Smeitink JAM, Nijtmans LGJ (2004). Differences in assembly or stability of complex I and other mitochondrial OXPHOS complexes in inherited complex I deficiency. Hum Mol Genet 13:659–667.
177. Ugalde C, Vogel R, Huijbens R, van den Heuvel B, Smeitink J, Nijtmans L (2004). Human mitochondrial complex I assembles through the combination of evolutionary conserved modules: a framework to interpret complex I deficiencies. Hum Mol Genet 13:2461–2472.
178. Valente L, Tiranti V, Marsano RM, Malfatti E, Fernandez-Vizarra E, Donnini C, Mereghetti P, De Gioia L, Burlina A, Castellan C, Comi GP, Savasta S, Ferrero I, Zeviani M (2007). Infantile encephalopathy and defective mitochondrial DNA translation in patients with mutations of mitochondrial elongation factors EFG1 and EFTu. Am J Hum Genet 80:44–58.
179. Valnot I, Osmond S, Gigarel N, Mehaye B, Amiel J, Cormier-Daire V, Munnich A, Bonnefont JP, Rustin P, Rotig A (2000). Mutations of the *SCO1* gene in mitochondrial cytochrome c oxidase deficiency with neonatal-onset hepatic failure and encephalopathy. Am J Hum Genet 67:1104–1109.
180. Valnot I, von Kleist-Retzow JC, Barrientos A, Gorbatyuk M, Taanman JW, Mehaye B, Rustin P, Tzagoloff A, Munnich A, Rotig A (2000). A mutation in the human heme A:farnesyltransferase gene *(COX10)* causes cytochrome c oxidase deficiency. Hum Mol Genet 9:1245–1249.
181. Van Coster R, Seneca S, Smet J, Van Hecke R, Gerlo E, Devreese B, Van Beeumen J, Leroy JG, De Meirleir L, Lissens W (2003). Homozygous Gly555Glu mutation in the nuclear-encoded 70 kDa flavoprotein gene causes instability of the respiratory chain complex II. Am J Med Genet 120A:13–18.
182. van den Heuvel L, Ruitenbeek W, Smeets R, Gelman-Kohan Z, Elpeleg O, Loeffen J, Trijbels F, Mariman E, de Bruijn D, Smeitink J (1998). Demonstration of a new pathogenic mutation in human complex I deficiency: a 5-bp duplication in the nuclear gene encoding the 18-kD (AQDQ) subunit. Am J Hum Genet 62:262–268.
183. Van Maldergem L, Trijbels F, DiMauro S, Sindelar PJ, Musumeci O, Janssen A, Delberghe X, Martin JJ, Gillerot Y (2002). Coenzyme Q responsive Leigh's encephalopathy in two sisters. Ann Neurol 52:750–754.
184. Visapää I, Fellman V, Vesa J, Dasvarma A, Hutton JL, Kumar V, Payne GS, Makarow M, Van Coster R, Taylor RW, Turnbull DM, Suomalainen A, Peltonen L (2002). GRACILE syndrome, a lethal metabolic disorder with iron overload, is caused by a point mutation in *BCS1L*. Am J Hum Genet 71:863–876.
185. Vogel R, Nijtmans L, Ugalde C, van den Heuvel L, Smeitink J (2004). Complex I assembly: a puzzling problem. Curr Opin Neurol 17:179–186.
186. Vogel R, Janssen RJ, Ugalde C, Grovenstein M, Huijbens RJ, Visch HJ, van den Heuvel LP, Willems PH, Zeviani M,. Smeitink JA, Nijtmans LG (2005). Human mitochondrial complex I assembly is mediated by NDUFAF1. FEBS J 272:5317–5326.
187. Vogel R, Smeitink JA, Nijtmans LG (2007). Human mitochondrial complex I assembly: a dynamic and versatile process. Biochim Biophys Acta 1767:1215–1227.

188. von Kleist-Retzow JC, Cormier-Daire V, de Lonlay P, Parfait B, Chretien D, Rustin P, Feingold J, Rötig A, and Munnich A (1998). A high rate (20%-30%) of parental consanguinity in cytochrome-oxidase deficiency. Am J Hum Genet 6:428–435.
189. von Kleist-Retzow JC, Yao J, Taanman JW, Chantrel K, Chretien D, Cormier-Daire V, Rotig A, Munnich A, Rustin P, Shoubridge EA (2001). Mutations in *SURF1* are not specifically associated with Leigh syndrome. J Med Genet 38:109–113.
190. Wang WG, White PS, Ackerman SH (2001). Atp11p and Atp12p are assembly factors for the F(1)-ATPase in human mitochondria. J Biol Chem 2276:30773–30778.
191. Waterham HR, Koster J, van Roermund CWT, Mooyer PAW, Wanders RJA, Leonard JV (2007). A lethal defect of mitochondrial and peroxisomal fission. N Engl J Med 356: 1736–1741.
192. Williams SL, Valnot I, Rustin P, Taanman JW (2004). Cytochrome c oxidase subassemblies in fibroblast cultures from patients carrying mutations in *COX10*, *SCO1*, or *SURF1*. J Biol Chem 279:7462–7469.
193. Wingert RA, Galloway JL, Barut B, Foott H, Fraenkel P, Axe JL, Weber GJ, Dooley K, Davidson AJ, Schmid B, Paw BH, Shaw GC, Kingsley P, Palis J, Schubert H, Chen O, Kaplan J, Zon LI (2005). Tübingen 2000 Screen Consortium. Deficiency of glutaredoxin 5 reveals Fe-S clusters are required for vertebrate haem synthesis. Nature 436: 1035–1039.
194. Xia D, Yu CA, Kim H, Xia JZ, Kachurin AM, Zhang L, Yu L, Deisenhofer J (1997). Crystal structure of the cytochrome bc1 complex from bovine heart mitochondria. Science 277: 60–66.
195. Xu F, Morin C, Mitchell G, Ackerley C, Robinson BH (2004). The role of the *LRPPRC* (leucine-rich pentatricopeptide repeat cassette) gene in cytochrome oxidase assembly: mutation causes lowered levels of COX (cytochrome c oxidase) I and COX III mRNA. Biochem J 382:331–336.
196. Zara V, Conte L, Trumpower BL (2007). Identification and characterization of cytochrome bc(1) subcomplexes in mitochondria from yeast with single and double deletions of genes encoding cytochrome bc(1) subunits. FEBS J 274:4526–4539.
197. Zordan MA, Cisotto P, Benna C, Agostino A, Rizzo G, Piccin A, Pegoraro M, Sandrelli F, Perini G, Tognon G, De Caro R, Peron S, Kronnie TT, Megighian A, Reggiani C, Zeviani M, Costa R (2006). Post-transcriptional silencing and functional characterization of the *Drosophila melanogaster* homolog of human *Surf1*. Genetics 172:229–241.
198. Züchner S, Mersiyanova IV, Muglia M, Bissar-Tadmouri N, Rochelle J, Dadali EL, Zappia M, Nelis E, Patitucci A, Senderek J, Parman Y, Evgrafov O, Jonghe PD, Takahashi Y, Tsuji S, Pericak-Vance MA, Quattrone A, Battaloglu E, Polyakov AV, Timmerman V, Schröder JM, Vance JM. (2004). Mutations in the mitochondrial GTPase mitofusin 2 cause Charcot-Marie-Tooth neuropathy type 2A. Nat Genet 36:449–451.
199. Züchner S, Vance JM (2005). Emerging pathways for hereditary axonopathies. J Mol Med 83:935–943.
200. Zhu Z, Yao J, Johns T, Fu K, De BI, Macmillan C, Cuthbert AP, Newbold RF, Wang J, Chevrette M, Brown GK., Brown RM., Shoubridge EA (1998). *SURF1*, encoding a factor involved in the biogenesis of cytochrome c oxidase, is mutated in Leigh syndrome. Nat Genet 20:337–343.

# Chapter 8
# Coenzyme $Q_{10}$ Deficiencies in Neuromuscular Diseases

**Rafael Artuch, Leonardo Salviati, Sandra Jackson, Michio Hirano, and Plácido Navas**

**Abstract** Coenzyme Q (CoQ) is an essential component of the respiratory chain but also participates in other mitochondrial functions such as regulation of the transition pore and uncoupling proteins. Furthermore, this compound is a specific substrate for enzymes of the fatty acids β–oxidation pathway and pyrimidine nucleotide biosynthesis. Furthermore, CoQ is an antioxidant that acts in all cellular membranes and lipoproteins. A complex of at least ten nuclear (*COQ*) genes encoded proteins synthesizes CoQ but its regulation is unknown. Since 1989, a growing number of patients with multisystemic mitochondrial disorders and neuromuscular disorders showing deficiencies of CoQ have been identified. CoQ deficiency caused by mutation(s) in any of the *COQ* genes is designated primary deficiency. Other patients have displayed other genetic defects independent on the CoQ biosynthesis pathway, and are considered to have secondary deficiencies. This review updates the clinical and molecular aspects of both types of CoQ deficiencies and proposes new approaches to understanding their molecular bases.

**Keywords:** Mitochondria · Coenzyme Q deficiency · Neuromuscular diseases

## 8.1 Introduction

Coenzyme Q (CoQ) is an electron carrier in the mitochondrial respiratory chain (MRC) transferring electrons from complex I and II to complex III [13]. It also transfers protons to the mitochondrial inter-membrane space contributing to establish the membrane potential and then to the ATP biosynthesis. Furthermore, CoQ has a key role in the mitochondrial respiratory chain function through the stabilization of the complex III [53]. In addition, CoQ is an essential factor for uncoupling proteins activation [17] and in the opening of the permeability transition pore [19], and acts as antioxidant in the different cellular membranes and even in cholesterol transport lipoproteins [57]. The antioxidant function of CoQ directly prevents the

R. Artuch (✉)
Biochemistry Department, Hospital Sant Joan de Déu, Barcelona; CIBER de Enfermedades Raras(CIBERER), ISCII, Spain

C. Espinós et al. (eds.), *Inherited Neuromuscular Diseases*, Advances in Experimental Medicine and Biology 652, DOI 10.1007/978-90-481-2813-6_8,

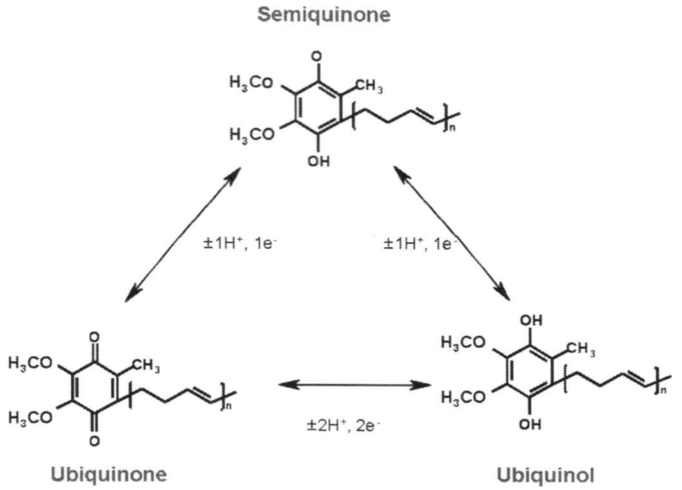


**Fig. 8.1** *Scheme of CoQ redox isoforms*. It includes the fully oxidized form ubiquinone, the fully reduced form ubiquinol, and the intermediate redox form semiubiquinone. The polyprenyl residue in carbon 2 contains isoprene units repeated several times (*n*). The number of isoprene residues *n* is specific in different organisms

progression of lipid peroxidation in membranes or by recycling other antioxidants such as vitamin E or ascorbate [26, 61]. CoQ of plasma membrane also regulates the ceramide-dependent apoptosis pathway by the non-competitive inhibition of the neutral sphingomyelinase activity [40]. Redox functions of CoQ are due to its ability to exchange two electrons in a redox cycle between the oxidized (ubiquinone) and the reduced (ubiquinol) forms. The redox reaction can be driven by one step of two electrons or by two steps of one electron through the semiquinone form (Fig. 8.1).

CoQ is composed of a benzoquinone ring, synthesized from tyrosine through 4. hydroxybenzoate, and a polyprenyl side-chain, generated from acetyl-coA through the mevalonate pathway [57], which is also required for the synthesis of other compounds, such as cholesterol and dolichol-phosphate. The polyprenyl side-chain is built with different isoprenoid units, which is species specific. Human CoQ contains predominantly ten isoprenoid units and is therefore designated CoQ10.

## 8.2 CoQ Biosynthesis Pathway

The biosynthesis of CoQ is a very complex process (Fig. 8.2), which involves at least ten genes as shown in Table 8.1.

Most of the proteins of these genes have not been yet purified and, although the regulation of this biosynthesis pathway is largely unknown [56, 57], it is up-regulated under oxidative stress in rats [15]. The yeast *coq-1* homologue encodes for a polyprenyl diphosphate synthase and is a branch-point enzyme of the mevalonate pathway [42].

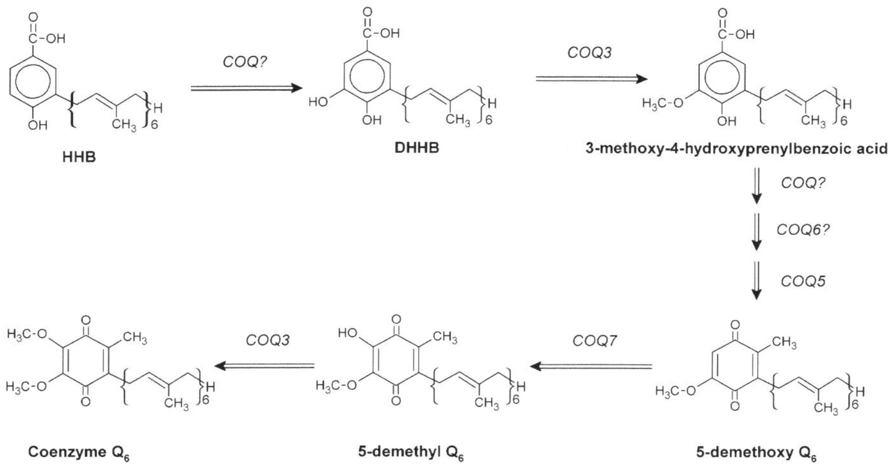


**Fig. 8.2** *A model for the biosynthesis pathway of CoQ in eukaryotic organisms.* The tentative gene participating in the different steps is indicated. 5-demethoxy-CoQ (*DMQ*) is accumulated in yeast and *C. elegans clk-1/coq-7* mutants. SAM=s-adenosyl-methionine; SAH=S-Adenosyl-homocysteine (modified from [43])

**Table 8.1** Homologue sequences of genes required for coenzyme Q biosynthesis in S. cerevisiae and H. sapiens. GeneBank(http://www.ncbi.nlm.nih.gov/Genbank) accession numbers for genes and proteins (in parenthesis) are listed

| Common name | *Chromosome location* | *H. sapiens* |
|---|---|---|
| *coq-1(PDSS1)* | 10p12.1 | NM_014317(NP_055132.1) |
| *coq-1(PDSS2)* | 6q21 | NM_020381.3(NP_065114.3) |
| *coq-2(COQ2)* | 4q21.3 | NM_015697(AAH20728.1) |
| *coq-3(COQ3)* | 6q16.3 | NM_017421(NP_059117) |
| *coq-4(COQ5)* | 9q34.13 | NM_016035(NP_057119.1) |
| *coq-5(COQ5)* | 12q24.31 | NM_032314(NP_115690.1) |
| *coq-6(COQ6)* | 14q24.1 | NM_182476,IsoA(NP_872282) NM_182480,IsoB(NP_872286) |
| *coq-7/clk-1(COQ7)* | 16p13.11-p12.3 | NM_016138(NP_057222) |
| *coq-8 (COQ8)* (ABC1,ADCK3) | 1q42.13 | NM_020247(NP_064632.1) |
| *coq-9(COQ9)* | 16q13 | NM_020312.1(NP_064708.1) |

Yeast *coq-2* and human homologue, *COQ2*, encodes a polyprenyltransferase that catalyses the prenylation of the 4-hydroxybenzoic acid intermediate [6, 20]. The gene products of yeast and rat *coq-3* homologues catalyze two O-methyltransferase steps in CoQ biosynthesis [44], essential for the final structure of quinone ring. Yeast *coq-5* encodes for a C-methyltransferase although a direct demonstration of its kinetic function is not available [24], it is required for the stability of other *coq*-encoded proteins [7]. Protein encoded by yeast and nematode *clk-1/coq-7* has been described as a membrane-bound hydroxylase of the benzoquinonic ring [25, 55], and it has been also demonstrated to participate in CoQ biosynthesis in mammal species such as mouse and human [59].

Yeast Coq-6 protein catalytic participation in CoQ biosynthesis is still unknown, but its sequence has a high homology with a large family of proteins that function as flavoprotein monoxygenases [22]. No catalytic function on CoQ biosynthesis has been shown for *coq-4*-encoded peptide although it has been demonstrated to be required for CoQ biosynthesis in eukaryotic species including humans [5, 9, 11]. It has been demonstrated that the gene *coq-8/ABC1* participates in CoQ biosynthesis in *S. cerevisiae* and *C. elegans* [5, 16]. There is no current demonstration of the exact role of *coq-8* in CoQ biosynthesis, but based on sequence similarity, *coq-8* belongs to a protein kinase family harboring an ATP-binding cassette similar to those found in regulatory kinases [30]. *C. elegans coq-8* knockout animals showed embryo development arrest that coincided with the triggering of expression of *coq-8* in embryo, and these animals are the only *coq* knockout strain that maintain a basal CoQ biosynthesis level [4]. This gene encodes probably a regulatory protein of the CoQ biosynthesis complex through a phosphorylation mechanism.

The gene *COQ10* encodes for a CoQ binding protein that is required for the activity of CoQ in respiration. *COQ10* null mutants of both *Saccharomyces cerevisiae* and *Schizosaccharomyces pombe* have normal levels of CoQ but partially lack respiration [8, 14]. A human *COQ10* ortholog is able to complement *S. pombe* null strain [14]. Although this gene is not required for CoQ biosynthesis, it must be considered in the study of mitochondrial disorders based on CoQ deficiency or dysfunction.

The CoQ biosynthetic intermediate 3-hexaprenyl-4-hydroxybenzoic acid (HHB) is the metabolite only accumulated in yeast *coq* null mutants suggesting that interactions among Coq proteins lead to the formation of a CoQ biosynthetic complex [24, 56]. These results also would indicate a regulation point at the Coq2 enzyme reaction [57]. Furthermore, mutations in *coq7/clk-1* induced the accumulation of 2-hexaprenyl-3-methyl-6-methoxy-1,4-benzoquinone (demethoxy-Q, DMQ) in yeast and nematodes indicating a second regulation point of CoQ biosynthesis pathway. This regulation is independent on Coq7p and modulated by either CoQ or Coq8p in yeast [43]. As a consequence, we would speculate that a mutation in any of the *coq* genes but also in those unknown that regulate the CoQ biosynthesis pathway would induce a CoQ deficiency as a primary event.

## 8.3 Primary CoQ Deficiencies

### *8.3.1 Clinical and Biochemical Diagnosis*

In 1989, Ogasahara et al. reported the first two siblings with a syndrome characterized by myoglobinuria, brain involvement, ragged-red-fibers and lipid storage. The patients had severe CoQ deficiency in muscle causing an impairment of the mitochondrial respiratory chain (MRC). Since then, more than 30 patients with deficiencies of CoQ have been reported and with diverse clinical phenotypes with proven or postulated autosomal recessive inheritance [45]. In addition to the original encephalomyopathic form, with myoglobinuria, epilepsy, mental retardation, and ataxia [41], other phenotypes include: a severe infantile multisystemic syndrome with prominent central nervous system involvement, including initially nystagmus, followed by bilateral visual loss, due to retinitis pigmentosa and optic nerve atrophy, sensorineural deafness, progressive ataxia, and a nephrotic syndrome causing terminal renal failure [50] and an ataxic form presenting with cerebellar ataxia and atrophy, nystagmus, seizures, absence of tendon reflexes, and in some cases, mental retardation [3, 28, 39]. This phenotype includes patients with early- or late-onset presentation and diagnosis and is the most frequent one associated with CoQ deficiency. Finally, other forms are expanding the clinical phenotype, since CoQ deficiency has been related with cases of neonatal presentation and fatal evolution, presenting poor feeding, hypothermia, hypotonia, renal tubulopathy, cardiac involvement, seizures with generalized cerebral and cerebellar atrophy [49], or failure of liver and pancreas and Leigh syndrome [31]. It has even been reported in adult patients with Leigh syndrome plus growth retardation, infantilism, ataxia, deafness and altered neuroimaging [60], or an adult form was diagnosed in a 48-year-old man who was completely normal up to 39 years of age and then presented late-onset ataxia with cerebellar atrophy and hypergonadotrophic hypogonadism [23]. Thus, considering this variety of clinical phenotypes, in any patient suspected of having a mitochondrial disorder CoQ deficiency should be considered, and therefore the next step in the investigation is the biochemical analysis of CoQ content in tissues. In all patients reported hitherto, metabolic investigations for the diagnosis of MRC defects, metabolite analysis in body fluids showed controversial results: some cases revealed increased blood or CSF lactate levels [10], while other patients did not show such metabolic alterations [23].

Therefore, peripheral markers of MRC disorders appear to have limited value in the initial diagnosis and selection of patients with CoQ deficiency syndrome.

The hallmark of CoQ deficiency syndrome is a decreased CoQ concentration in muscle and/or fibroblasts, which is usually measured by HPLC with electrochemical detection procedures. Patients showed a variable degree of CoQ deficiency in muscle and/or fibroblasts (with normal plasma CoQ concentration) causing decreased activities of NADH:cytochrome *c* oxidoreductase (complex I + CoQ + III) and succinate:cytochrome *c* oxidoreductase (complex II + CoQ + III) of the MRC. The degree of the deficiency is very broad, ranging from extremely low CoQ concentration in muscle [10, 50, 54] to milder deficiencies [3, 39], or even normal results in fibroblasts. Another biochemical sign commonly observed was a dramatic recovery of MRC enzyme activities in muscle and/or fibroblasts (complex I + III and II + III) after the addition of CoQ analogues in vitro [50]. Other very helpful laboratory tool for CoQ deficiency diagnosis is the quantification of CoQ biosynthesis in fibroblasts by means of the measurement of the incorporation of radiolabeled 4-hydroxy[U-$^{14}$C] benzoic acid to the quinone ring of CoQ, or the incorporation of radiolabeled $^{3}$H-mevalonate to the polyisoprenoid side chain [36, 47]. After incubation of fibroblasts, the CoQ synthesized is separated by reverse phase HPLC and quantified by radiometric detection.

## *8.3.2 Molecular Studies of CoQ Biosynthesis*

At present, 10 genes have directly been associated in CoQ biosynthesis in humans.

All of the genes participate in the terminal part of CoQ biosynthesis, including decaprenyl diphosphate synthesis, the incorporation of this polyprenyl side chain to the benzoquinone ring, and both the condensation and modification of the quinone moiety to its final structure as CoQ (Fig. 8.1). Among these genes, at present primary CoQ deficiency has been linked to 4 of them (*PDSS1, PDSS2, COQ2* and *COQ8*, also known as *CABC1* or *ADCK3*). We will briefly review these different molecular defects:

### 8.3.2.1 Mutations in *COQ2* Gene (MIM #609825)

Mutations in *COQ2* gene were the first genetic defects in the CoQ biosynthesis pathway in humans [47]. *COQ2* encodes for an enzyme called *para*hydroxybenzoate polyprenyl transferase, which catalyses the conjugation of the benzoquinone ring with the decaprenyl side chain. This paper describes a homozygous missense mutation (A→G transition in position 890) that changes a highly conserved tyrosine to cysteine at amino acid 297 in Coq2 peptide in 2 siblings presenting a severe CoQ deficiency (18% of mean control value in fibroblasts). This pathogenic mutation was associated with the infantile form of CoQ deficiency characterized by encephalopathy and nephropathy (one of the well-recognized clinical phenotypes of CoQ deficiency syndromes) [52]. Biochemically, both patients showed clear CoQ deficiency both in muscle and fibroblasts (more severe in the latter), affecting MRC enzymes dependent on CoQ (complexes I + III and II + III), and also the incorporation of both radiolabeled para-hydroxybenzoate and decaprenyl-PP into CoQ. Interestingly, further pathogenic mechanisms besides bioenergetics problems were demonstrated in these patients. Cultured fibroblasts from the patients revealed decreased ATP synthesis, increased production of reactive oxygen species and oxidative stress [48]. The decrease of CoQ biosynthesis impaired *de novo* pyrimidine nucleotide synthesis since ubiquinone is an essential cofactor for dihydro-orotate dehydrogenase, an enzyme located in the inner mitochondrial membrane [33]. Later, Mollet et al. [35] reported a new patient presenting pathogenic mutations in *COQ2* gene associated with CoQ deficiency (24% of mean control value in fibroblasts) In this case, the clinical phenotype was more severe than the previous description since the patient presented a fatal infantile multiorgan disease. The mutation was a homozygous single base pair frameshift deletion resulting in a premature stop codon (c.1198delT, N401fsX415). Biochemical data and pathogenic mechanism investigations consistently demonstrated that these mutations were the cause of the oxidative phosphorylation defect in this new case.

### 8.3.2.2 Mutations in *PDSS1* and *PDSS2* Subunits of Prenyldiphosphate Synthase

Prenyldiphosphate synthase, one of the rate-limiting enzymes in CoQ biosynthesis, is an enzyme that catalyses the elongation of geranyl-geranyl diphophate to decaprenyl diphosphate prior to conjugation of this polyprenyl side chain to the benzoquinone ring. It is therefore an essential enzyme for CoQ biosynthesis in humans, which encodes the human ortholog of the yeast (*Saccharomyces cerevisiae*) *COQ1* gene. Prenyldiphosphate synthase in humans is a heterotetramer composed by 2 copies of both subunit 1 and subunit 2 molecules encoded by different genes. Mutations in both subunits have been demonstrated as pathogenic in several patients.

In 2006, Lopez et al. [32] reported a single case with a severe infantile phenotype characterized by Leigh syndrome (a severe encephalopathy) with nephrotic syndrome associated with severe CoQ deficiency (13% of mean control value in fibroblasts) and MRC function impairment. Direct sequencing of eight genes involved in CoQ biosynthesis revealed the presence of 2 non-synonymous nucleotide changes in the *PDSS2* gene; a heterozygous C→T transition at nucleotide 964, changing amino acid 322 from glutamine to a stop codon, and a heterozygous C→T transition at nucleotide 1145 changing amino acid 382 from serine to leucine in the seventh conserved domain of the enzyme. These changes (MIM #610564) were not found in DNA from 210 chromosomes from controls (analyzed by RFLP). Pathogenic effect of this mutation was further confirmed by a clear reduction of the incorporation of radiolabeled substrate 14C-PHB to CoQ in fibroblasts. Interestingly, the incorporation of other substrate to CoQ, 3H-decaprenyl-PP, was normal, indicating that the biosynthetic pathway after decaprenyl diphosphate synthase was intact.

In 2007, Mollet et al. reported a family (2 siblings) presenting a multisystem disease with early-onset deafness, encephaloneuropathy, obesity, livedo reticularis and valvulopathy associated with a profound CoQ deficiency (3.3% of the mean control values in fibroblasts), and thus, expanding the clinical phenotype of these syndromes. A genome-wide search for homozygosity identified several candidate chromosome regions, including 10p12.2-p12.1, which corresponds to the *PDSS1* gene.

Direct sequencing of this gene revealed a homozygous T→G transversion at nucleotide 977, resulting in the change of a highly conserved aspartic acid into a glutamic acid (D308E). This mutation (MIM #607429) probably affects some of the catalytic properties of the *PDSS1* encoded enzyme. Pathogenic effect of this mutation was further confirmed by MRC analysis and CoQ biosynthesis in either in muscle or fibroblasts, which was clearly impaired, and by functional complementation studies in yeasts.

#### 8.3.2.3 Mutations in *COQ8* Gene (MIM #606980)

*COQ8* gene, also called *ADCK3* or *CABC1*, function is still unknown in CoQ biosynthesis pathway, although it has been proposed that could act as a protein kinase, since several kinase motifs have been found in its amino acid sequence [30], and it has been demonstrated to have regulatory functions in *C. elegans* [4]. In 2008, 2 groups reported independently and simultaneously mutations in this gene associated with CoQ deficiency and an ataxic form with seizures.

In one report [34], 3 unrelated families (4 patients) were reported presenting pathogenic mutations in different parts of the *COQ8* gene. Pathogenicity was consistently demonstrated by analysis of the amino acid residues conservation, by the absence of these mutations in 460 chromosomes in healthy controls, and by the study of functional complementation studies in yeasts. Furthermore, MRC function was impaired as well. CoQ values in muscle were clearly reduced in 3 patients, but this defect was not demonstrated in fibroblasts, because CoQ content was normal in 2 cases. These findings support the necessity of studying both muscle and fibroblast samples for CoQ deficiency syndrome diagnosis. Clinically, all these patients showed ataxia and cerebellar atrophy, with different degrees of affectation of other neurological functions (seizures) or extra-neurological signs, such as exercise intolerance.

In the second report, Lagier-Tourenne et al. [27] after a SNP-based genome widescan in a consanguineous family, found linkage of the ataxia to chromosome 1q41 that led to *COQ8* gene sequencing analysis. After this initial analysis, 7 additional patients presenting with childhood-onset ataxia andcerebellar atrophy were found to harbor several types of mutations in both alleles of *COQ8* gene. Pathogenicity of these mutations was demonstrated by different ways, including molecular studies of *COQ8* gene, complementation analysis in yeasts, and MRC function analysis. Direct quantification of CoQ in fibroblasts revealed that also one case did not present a CoQ deficiency when compared with controls, while the other 3 patients studied presented a moderate deficiency. In the index case however, a more severe reduction in CoQ content in muscle biopsy was demonstrated. The authors proposed for this new disorder the name ARCA2 (autosomal-recessive cerebellar ataxia 2).

## 8.4 Secondary CoQ Deficiency in Mitochondrial Disorders

The first studies concerning CoQ deficiency and mitochondrial encephalomyopathies were performed in Kearns-Sayre syndrome patients and other conditions caused by mutations in mitochondrial DNA (mtDNA) [18, 62]. Other nuclear and mitochondrial DNA defects has been described to be associated with CoQ deficiency, but no mechanism were described to be causing this deficiency. Some patients showing non-Friedreich ataxia showed a CoQ deficiency but it was not demonstrated any mutation in *COQ* gene sequences, indicating that either other *COQ* genes are unknown or it is a secondary defect [2]. In fact, some of the ataxic patients with CoQ deficiency harbor mutations in *aprataxin* that encodes for a DNA repair factor [29, 46].

Isolated myopathy has been also associated to CoQ deficiency but it is also a secondary event because it has been described a pathogenic mutation in the *ETFDH* gene [21]. This gene encodes an electron transport flavoprotein dehydrogenase that depends on CoQ as substrate and it is an essential step of β–oxidation of fatty acids. It has been also recently demonstrated a case of a child that presented a secondary CoQ deficiency associated to a depletion of mtDNA [37]. Finally, secondary CoQ deficiency has been reported in a patient with cardiofaciocutaneous syndrome due to a *BRAF* gene mutation [1].

In recent years, decreased plasma and tissue concentrations of CoQ have been associated with several diseases, such as cardiomyopathies, degenerative muscle and central nervous system diseases, cancer, phenylketonuria, and an impaired CoQ reduced status in plasma, free radical damage and lipid peroxidation in vascular diseases [58]. Thus, a significant impairment of the distribution of CoQ between plasma and blood cells has been demonstrated in fibromyalgia [12]. The mechanisms that may explain these deficiencies in this heterogeneous group of diseases is far to be understood but it could be explained by either a decreased endogenous CoQ biosynthesis or an increased catabolism, although the involvement of dietary CoQ supply can not be discarded.

The existence of the secondary CoQ deficiency indicates that CoQ would be a marker of mitochondria homeostasis and also that CoQ biosynthesis complex must have a connection to other complexes such as MRC or pathways inside mitochondria. The highly conserved biosynthesis pathway and structure of CoQ could guarantee its function as central sensor of mitochondria homeostasis and would coordinate the equilibrium between bioenergetics and nucleotides available for cellular physiology.

## 8.5 Therapy of CoQ Deficient Patients

Primary CoQ deficiency is unique among mitochondrial disease (and among genetic diseases in general) because an effective therapy is available, at least for some for patients. It has been known since the first report by Ogasahara et al. [41] that CoQ deficient patients benefit from oral CoQ10 supplementation. Subsequent reports have clearly shown that high-dose oral CoQ supplementation can stop the progression of the encephalopathy in patients with the multisystemic form of the disease [50, 52]. Moreover in 2008 two reports have shown that oral CoQ supplementation is effective also on renal manifestations, both in a patient with *COQ2* mutations [38], and in *PDSS2* knockout mice [51]. However, it is clear from these reports that therapy must be instituted as early as possible to be effective, because only limited recovery is seen after development of significant tissue damage.

CoQ10 appears to be less effective in patients with the ataxic form of the disease caused by *ADCK3* mutations. No effect was noted in the patients reported by Mollet et al. [34], while there were only mild improvements in the only patient treated by Lagier-Tourenne et al. [27]. It is not clear if the lack of significant clinical response is due to the fact that treatment was instituted too late in the course of the disease to be effective, or because *ADCK3* has also other biological functions unrelated to CoQ biosynthesis. The optimal dosage and formulation of this supplement are also still debated. Doses ranging from 5 mg/kg/day [50] to 30 mg/kg/day [38] have been employed without significant side effects in patients, while a dose of 200 mg/kg/day was shown to be effective in mice [51]. To date, however, there are no studies that correlate CoQ10 dosage with plasma or tissue levels in patients, and there is no practical way to monitor the efficacy of therapy. Moreover, CoQ10 is available in different formulations, and there are no data regarding their bioavailability in CoQ10-deficient patients of individual formulations.

## 8.6 Concluding Remarks

The importance of CoQ in human metabolism was increased in the last years after the description of patients with neuromuscular diseases associated to a deficiency of this compound. The primary deficiency implies not only the role of CoQ in bioenergetics but also the different functions that would have the COQ encoded proteins in mitochondria. The secondary deficiency opens a new view of mitochondria homeostasis through the coordination of the different pathways acting in the organelle. This coordination would involve mainly the MRC and CoQ biosynthesis complexes through either direct interactions or intermediate regulatory proteins.

**Acknowledgements** This work has been partially supported by the European Union contract LSHB-CT-2004-005151 and the NIH grant 1R01HD057543-01.

## References

1. Aeby, A., Sznajer, Y., Cave, H., Rebuffat, E., Van Coster, R., Rigal, O., and Van Bogaert, P. J. 2007, Inherit. Metab. Dis. 30, 827.
2. Artuch, R., Brea-Calvo, G., Briones, P., Aracil, A., Galván, M., Espinós, C., Corral, J., Volponi, V., Ribes, A., Andreu, A.L., Palau, F., Sánchez-Alcázar, J.A., Navas, P., and Pineda, M. 2006, J. Neurol. Sci. 246, 153.
3. Artuch, R., Briones, P., Aracil, A., Ribes, A., Pineda, J., Galván, M., Sánchez-Alcázar, J.A., Brea, G., Navas, P., and Pineda, M. 2004, J. Inherit. Metab. Dis., 27 (Suppl 1), 118.
4. Asencio, C. Rodríguez-Aguilera, J.C., Johnson, T.E., Cabello, J., Schnabel, R., and Navas, P. 2009, Mech Ageing Dev. 130:145.
5. Asencio, C., Rodriguez-Aguilera, J.C., Ruiz-Ferrer, M., Vela, J., and Navas, P. 2003. FASEB J, 17, 1135.
6. Ashby, M.N., Kutsunai, S.Y., Ackerman, S., Tzagoloff, A., and Edwards, P.A., 1992. J. Biol. Chem. 267, 4128.
7. Baba, S.W., Belogrudov, G.I., Lee, J.C., Lee, P.T., Strahan, J., Shepherd, J.N., and Clarke, C.F. 2004, J. Biol. Chem. 279, 10052.
8. Barros, M.H., Johnson, A., Gin, P., Marbois, B.N., Clarke, C.F., and Tzagoloff, A. 2005, J. Biol. Chem. 280, 42627.
9. Belogrudov, G.I., Lee, P.T., Jonassen, T., Hsu, A.Y., Gin, P., and Clarke, C.F. 2001, Arch. Biochim. Biophys. 392, 48.
10. Boitier, E., Degoul, F., Desguerre, I., Charpentier, C., Francois, D., Ponsot, G., Diry, M., Rustin, P., Marsac, C., and Desguerre, I. 1998, J. Neurol. Sci., 156, 41.
11. Casarin, A., Jimenez-Ortega, J.C., Trevisson, E., Pertegato, V., Doimo, M., Ferrero-Gomez, M.L., Abbadi, S., Artuch, R., Quinzii, C., Hirano, M., Basso, G., Santos-Ocaña, C., Navas, P., and Salviati, L. 2008, Biochem. Biophys. Res. Commun. 372, 35.
12. Cordero, M.D., Moreno-Fernández, A.M., de Miguel, M., Bonal, P., Campa, F., Jiménez-Jiménez, L.M., Ruiz-Losada, A., Sánchez-Domínguez, B., Sánchez-Alcázar JA, Salviati, L., and Navas P. 2009, Clin. Biochem. 42:732.
13. Crane, F.L., Hatefi, Y., Lester, R.L., and Widmer, C. 1957, Biochim. Biophys. Acta. 25, 220.
14. Cui, T.-Z., and Kawamukai, M. 2008, FEBS J. doi:10.111/j.1742-4658.2008.06821.x.

15. De Cabo R., Burgess, J.R., and Navas, P. 2006, J. Bioenerg. Biomembr. 38, 309.
16. Do, T.Q., Hsu, A.Y., Jonassen, T., Lee, P.T., and Clarke, C.F. 2001, J. Biol. Chem. 276, 18161.
17. Echtay, K.S., Bienengraeber, M., and Klingenberg, M., 2001, Biochemistry 40, 5243.
18. Fischer, J.C., Ruitenbeek, W., Gabreels, F.J., Janssen, A.J., Renier, W.O., Sengers, R.C., Stadhouders, A.M., ter Laak, H.J., Trijbels, J.M., and Veerkamp, J.H. 1986, Eur. J. Pediatr., 144, 441.
19. Fontaine, E., Ichas, F., and Bernardi, P. 1998, J. Biol. Chem. 273, 25734.
20. Forsgren, M., Attersand, A., Lake, S., Grunler, J., Swiezewska, E., Dallner, G., and Climent, I. 2004, Biochem. J. 382, 519.
21. Gempel, K., Topaloglu, H., Talim, B., Schneiderat, P., Schoser, B.G., Hans, V.H., Palmafy, B., Kale, G., Tokatli, A., Quinzii, C., Hirano, M., Naini, A., DiMauro, S., Prokisch, H., Lochmuller, H., and Horvath, R. 2007, Brain 130, 2037.
22. Gin, P., Hsu, A.Y., Rothman, S.C., Jonassen, T., Lee, P.T., Tzagoloff, A., and Clarke, C.F. 2003, J. Biol. Chem. 278, 25308.
23. Gironi, M., Lamperti, C., Nemni, R., Moggio, M., Comi, G., Guerini, F.R., Ferrante, P., Canal, N., Naini, A., Bresolin, N. and DiMauro, S. 2004, Neurology, 62, 818.
24. Hsu, A.Y., Do, T.Q., Lee, P.T., and Clarke, C.F. 2000, Biochim. Biophys. Acta 1484, 287.
25. Jonassen, T., Proft, M., Randez-Gil, F., Schultz, J.R., Marbois, B.N., Entian, K.D., and Clarke, C.F., 1998, J. Biol. Chem. 273, 3351.
26. Kagan, V.E., Arroyo, A., Tyurin, V.A., Tyurina, Y.Y., Villalba, J.M., and Navas, P. 1998, FEBS Lett. 428, 43.
27. Lagier-Tourenne, C., Tazir, M., López, L.C., Quinzii, C.M., Assoum, M., Drouot, N., Busso, C., Makri, S., Ali-Pacha, L., Benhassine, T., Anheim, M., Lynch, D.R., Thibault, C., Plewniak, F., Bianchetti, L., Tranchant, C., Poch, O., DiMauro, S., Mandel, J.L., Barros, M.H., Hirano, M., and Koenig, M. 2008, Am. J. Hum. Genet. 82, 661.
28. Lamperti, C., Naini, A., Hirano, M., De Vivo, D.C., Bertini, E., Servidei, S., Valeriani, M., Lynch, D., Banwell, B., Berg, M., Dubrovsky, T., Chiriboga, C., Angelini, C., Pegoraro, E., and DiMauro, S. 2003, Neurology, 60, 1206.
29. Le Ber, I., Dubourg, O., Benoist, J.F., Jardel, C., Mochel, F., Koenig, M., Brice, M., Lombes, A., and Durr, A. 2007, Neurology, 68, 295.
30. Leonard, C.J., Aravind, L., and Koonin, E.V. 1998, Genome Res 8, 1038.
31. Leshinsky-Silver, E., Levine, A., Nissenkorn, A., Barash, V., Perach, M., Buzhaker, E., Shahmurov, M., Polak-Charcon, S., Lev, D., and Lerman-Sagie, T. 2003, Mol. Gen. Metab., 79, 288.
32. Lopez, L.C., Schuelke, M., Quinzii, C.M., Kanki, T., Rodenburg, R.J., Naini, A., Dimauro, S., and Hirano, M. 2006, Am J Hum Genet 79, 1125.
33. López-Martín, J.M., Salviati, L., DiMauro, S., Hirano, M., Rodriguez-Hernandez, A., Cordero, M., Sanchez Alcazar, J.A., Santos-Ocaña, C. and Navas, P. 2007, Hum. Mol. Genet. 16, 1091.
34. Mollet, J., Delahodde, A., Serre, V., Chretien, D., Schlemmer, D., Lombes, A., Boddaert, N., Desguerre, I., de Lonlay, P., de Baulny, H.O., Munnich, A., and Rötig, A. 2008, Am J Hum Genet. 82, 623.
35. Mollet, J., Giurgea, I., Schlemmer, D., Dallner, G., Chretien, D., Delahodde, A., Bacq, D., de Lonlay, P., Munnich, A., and Rotig, A. 2007, J Clin Invest 117, 765.
36. Montero, R., Sáchez-Alcázar, J.A., Briones, P., Rodríguez-Hernández, M.A., Cordero, M.D., Trevisson, E., Salviati, L., Pineda, M., Garcia-Cazorla, A., Navas, P., and Artuch, R. 2008, Clin Biochem. 41, 697.
37. Montero, R., Sánchez-Alcázar, J.A., Briones, P., Navarro-Sastre, A., Gallardo, E., Bornstein, B., Herrero-Martín, D., Rivera, H., Martin, M.A., Marti, R., García-Cazorla, A., Montoya, J., Navas, P., and Artuch, R. 2009, Clin Biochem. 42:742.
38. Montini, G., Malaventura, C., and Salviati, L. 2008, N Engl J Med. 358, 2849.

39. Musumeci, O., Naini, A., Slonim, A.E., Skavin, N., Hadjigeorgiou, G.L., Krawiecki, N., Weissman, B.M., Tsao, C.Y., Mendell, J.R., Shanske, S., De Vivo, D.C., Hirano, M., and DiMauro, S. 2001, Neurology, 56, 849.
40. Navas, P., Fernández-Ayala, D.M., Martín, S.F., López-Lluch, G., De Cabo, R., Rodríguez-Aguilera, J.C., and Villalba, J.M. 2002, Free Radic. Res. 36, 369.
41. Ogasahara, S., Engel, A.G., Frens, D., and Mack, D. 1989, Proc. Natl. Acad. Sci. USA, 86, 2379.
42. Okada, K., Minehira, M., Zhu, X., Suzuki, K., Nakagawa, T., Matsuda, H., and Kawamukai, M., 1997, J. Bacteriol. 179, 3058.
43. Padilla, S., Santos-Ocaña, J.C., Jiménez-Hidalgo, M., López-Martín, J.M., Martín-Montalvo, A., and Navas, P. 2008, Cell. Mol. Life Sci. doi: 10.1007/s00018-008-8547-7.
44. Poon, W.W., Barkovich, R.J., Hsu, A.Y., Frankel, A., Lee, P.T., Shepherd, J.N., Myles, D.C., and Clarke, C.F. 1999, J. Biol. Chem. 274, 21665.
45. Quinzii, C.M., DiMauro, S., and Hirano, M. 2007, Neurochem. Res. 32, 723.
46. Quinzii, C.M., Kattah, A.G., Naini, A., Akman, H.O., Mootha, V.K., DiMauro, S., and Hirano, M. 2005, Neurology 64, 539.
47. Quinzii, C., Naini, A., Salviati, L., Trevisson, E., Navas, P., DiMauro, S., and Hirano, M. 2006, Am J Hum Genet 78, 345.
48. Quinzii, C., López, L.C., Von-Moltke, J., Naini, A., Krishna, S., Schuelke, M., Salviati, L., Navas, P. DiMauro, S., and Hirano, M. 2008, FASEB J 22, 1874.
49. Rahman, S., Hargreaves, I., Clayton, P., and Heales, S. 2001, J. Pediatr., 139, 456.
50. Rötig, A., Appelkvist, E.L., Geromel, V., Chretien, D., Kadhom, N., Edery, P., Lebideau, M., Dallner, G., Munnich, A., Ernster, L., and Rustin, P. 2000, Lancet, 356, 391.
51. Saiki, R., Lunceford, A.L., Shi, Y., Marbois, B., King, R., Pachuski, J., Kawamukai, M., Gasser, D.L., and Clarke. C,F. 2008, Am. J. Physiol. Renal. Physiol. 295, F1535.
52. Salviati, L., Sacconi, S., Murer, L., Zacchello, G., Franceschini; L., Laverda, A.M., Basso, G., Quinzii; C., Angelini, C., Hirano, M., Naini, A. B., Navas, P., DiMauro, S., and Montini, G. 2005, Neurology 65, 606.
53. Santos-Ocaña, C., Do, T.Q., Padilla, S., Navas, P., and Clarke, C.F, 2002, J. Biol. Chem. 277, 10973.
54. Sobreira, C., Hirano, M., Shanske, S., Keller, R.K., Haller, R.G., Davidson, E., Santorelli, F.M., Miranda, A.F., Bonilla, E., Mojon, D.S., Barreira, A.A., King, M.P., and DiMauro, S. 1997, Neurology, 48, 1238.
55. Stenmark, P., Grunler, J., Mattsson, J., Sindelar, P.J., Nordlund, P., and Berthold, D.A., 2001, J. Biol. Chem. 276, 33297.
56. Tran, U.C., and Clarke, C.F. 2007, Mitochondrion, 7 Suppl, S62.
57. Turunen, M., Olsson, J., and Dallner, G. 2004, Biochim. Biophys. Acta 1660, 171.
58. Turunen, M., Swiezewska, E., Chojnacki, T., Sindelar, P., and Dallner, G. 2002, Free Rad. Res., 36, 437.
59. Vajo, Z., King, L.M., Jonassen, T., Wilkin, D.J., Ho, N., Munnich, A., Clarke, C.F., and Francomano, C.A., 1999, Mamm. Genome 10, 1000.
60. Van Maldergem, L., Trijbels, F., DiMauro, S., Sindelar, P.J., Musumeci, O., Janssen, A., Delberghe, X., Martin, J.J., and Gillerot, Y. 2002, Ann. Neurol., 52, 750.
61. Villalba, J.M., Navarro, F., Córdoba, F., Serrano, A., Arroyo, A., Crane, F.L., and Navas, P. 1995, Proc. Natl. Acad. Sci. USA 92, 4887.
62. Zierz, S., Jahns, G., and Jerusalem, F. 1989, J. Neurol., 236, 97.

# Chapter 9
# The Role of Mitochondrial Network Dynamics in the Pathogenesis of Charcot-Marie-Tooth Disease

**Francesc Palau, Anna Estela, David Pla-Martín, and Maribel Sánchez-Piris**

**Abstract** Mitochondrial dysfunction plays a relevant role in the pathogenesis of neurological and neuromuscular diseases. Mitochondria may be involved as a primary defect of either the mtDNA or nuclear genome encoded subunits of the respiratory chain. These organelles have also been directly involved in the pathogenesis of Mendelian neurodegenerative disorders caused by mutations in nuclear-encoded proteins targeted to mitochondria, such as Friedreich ataxia, hereditary spastic paraplegia, or some monogenic forms of Parkinson disease. In addition, mitochondria also participate in the pathogenic mechanisms affecting neurodegenerative disorders such Huntington disease or amyotrophic lateral sclerosis. Cell death in neurodegeneration associated with neurological diseases usually occurs by apoptosis being the most common route the intrinsic mitochondria pathway. Along with regulation of apoptosis, mitochondria also modulate cell pathogenesis by means of energy production, reactive oxygen species (ROS) generation, and calcium buffering. Mitochondria form dynamic tubular networks that continually change their shape and move throughout the cell. Here we review the critical role of mitochondria in monogenic neuromuscular disorders, especially inherited peripheral neuropathies caused by abnormal mitochondrial network dynamics. In yeast, at least three proteins are required for mitochondrial fusion, Fzo1, Ugo1 and Mgm1. The human counterparts of Fzo1p and Mgm1p, MFN1/MFN2 and OPA1 respectively, are related to human disease. Mutations in the *MFN2* gene cause the most frequent form of autosomal dominant axonal Charcot-Marie-Tooth disease, CMT2A. Mutations in *OPA1* cause autosomal dominant optic atrophy (ADOA). For the opposite process of mitochondrial fission, four proteins are at least involved in yeast. Very recently a mutation in the *DRP1* gene (the human homologue of yeast *Dnm1*) has been reported in an infant with a syndrome with encephalopathy, optic atrophy and lactic acidosis. GDAP1 has been recently related to the mitochondrial fission in mammalian cells and, interestingly, mutations in the *GDAP1* gene are the cause of

F. Palau (✉)
Laboratory of Genetics and Molecular Medicine, Instituto de Biomedicina de Valencia, CSIC and CIBER de Enfermedades Raras (CIBERER), C/ Jaume Roig 11, 46010 Valencia, Spain
e-mail: fpalau@ibv.csic.es

C. Espinós et al. (eds.), *Inherited Neuromuscular Diseases*, Advances in Experimental Medicine and Biology 652, DOI 10.1007/978-90-481-2813-6_9,

the most common form of autosomal recessive CMT, either axonal or demyelinating. These and other disorders are the most recent instances of disease related with mitochondrial abnormal motility, fusion and fission. We propose that the pathomechanisms underlying these disorders also include a complex relationship between mitochondrial dynamics and transport across the axon.

**Keywords:** Mitochondrial dynamics · Mitochondrial fusion and fission · Charcot-Marie-Tooth disease · Mitofusins · Mitofusin 2 · GDAP1

## 9.1 Introduction

Charcot-Marie-Tooth (CMT) disease is a clinical and genetic heterogeneous group of inherited peripheral neuropathies that affect 1 in 2500 people worldwide [1, 2]. Mendelian segregation in families may follow either autosomal dominant, autosomal recessive or X-linked patterns. Autosomal recessive forms are described more frequently in specific populations and geographical areas such as the Mediterranean basin. Molecular genetic studies and positional cloning approaches have unraveled a wide number of genes involved in the etiology of CMT disease and related peripheral neuropathies. As an immediate consequence, molecular diagnosis has become an important tool in clinical practice, and patients and families have been beneficiated of more specific genetic counseling. Moreover, it has been proved that a large number of molecules, cell structures and pathways are involved in the pathogenesis of CMT neuropathies.

Involvement of mitochondria in the pathogenesis of neurological disease was traditionally associated with defects of the electron transfer chain (ETC) and oxidative phosphorylation (OXPHOS) pathway, referred to as encephalomyopathies. These disorders were originally caused by point or large mutations in the mitochondrial genome (mtDNA). In the last years, research in the field of mitochondria and neurology has undergone a deep change. In a recent review, DiMauro and Schon [3] distinguished four main groups of disorders for which mitochondria are an important factor in the pathogenesis: (i) diseases of the mitochondrial respiratory chain (or OXPHOS disorders), caused by mutations either in the mtDNA or in the nuclear DNA, (ii) diseases caused by impaired mitochondrial protein import, (iii) diseases caused by aberrant mitochondrial dynamics, and (iv) neurodegenerative diseases caused by mutations in nuclear-encoded proteins targeted to mitochondria.

The role of mitochondria in neurodegeneration is an important piece in understanding how neurons are affected in neurological diseases [4]. Mitochondria participate in the cell pathophysiology and pathogenesis of common neurodegenerative disorders such as Parkinson disease (PD) and Alzheimer disease (AD), and less frequent disorders that include Huntington disease (HD) and amyotrophic lateral sclerosis (ALS). Indicators of mitochondrial dysfunction observed in neurodegenerative disorders include ultrastructural changes, mtDNA depletion, involvement of ETC and decreased ATP production, abnormal oxidative stress with increased in reactive oxidative species, impaired $Ca^{2+}$ buffering, and loss of mitochondrial membrane potential [5]. Only very recently dysfunction of the fusion and fission pathways of the mitochondrial network dynamics has been recognized as a mechanism of neurodegeneration. Recent examples are the monogenic forms of PD caused by mutations in *PINK1* (PARK6) or *PARKIN* (PARK2). Genetic experiments in *Drosophila* suggest that *pink1* and *parkin* regulates mitochondrial dynamics in the fly by promoting mitochondrial fission and/or inhibiting fusion [6, 7].

## 9.2 Fusion and Fission in Mitochondrial Dynamics

Mitochondrial fusion and fission have been described in yeast and mammalian cells [8–10]. Fusion and fission processes generate a dynamic morphology and network that are essential for the physiology of mitochondrial functions in cells. There is nowhere more evident than in large neurons where mitochondria need to move along axons from the cell bodies to the synapses. Motility is particularly relevant in the axons of peripheral nerves. Fusion and fission allow mixing of mtDNA and metabolites, and the proper distribution of mitochondria in the cell so as to supply changing energy requirements. Both fusion and fission are coupled, but each process is finely regulated with specific molecules. Both routes have in common the presence of molecules belonging to the family of GTPases. These pathways were originally described in *Saccharomyces cerevisiae*. The proteins that participate in yeast mitochondrial fusion are Fzo1p (*Drosophila* fuzzy onion protein), Mgm1p (mitochondrial genome maintenance protein 1) and Ugo1p [8]. The mammalian orthologues of Fzo1p are mitofusin 1 (MFN1) and mitofusin 2 (MFN2), which are derived from two different genes, *MFN1* and *MFN2*. OPA1 is the mammalian orthologue of Mgm1p. Mutations in *MFN2* and *OPA1* have been associated with human disease, and to more specifically neuropathies. Mutations in *MFN2* cause autosomal dominant axonal Charcot-Marie-Tooth disease (CMT2A) [11, 12], whereas mutations in OPA1 cause the autosomal dominant optic atrophy (ADOA) [13, 14], the most frequent form of optic atrophy. MFN1 and MFN2 are essential GTPases localized in the outer mitochondrial membrane and it has been proposed that they mediate mitochondrial fusion by tethering outer membranes together through interactions of their coiled-coil domains in *trans* [15, 16]. However, MFN2 appears as a multifunctional protein whose biologic function is not restricted to mitochondrial shape regulation. Different roles such as regulating mitochondrial metabolism, apoptosis, shape of other organelles and even progression through cell cycle has been proposed for MFN2 [17]. Very recently, a new function for MFN2 has been reported. A fraction of MFN2 present in the endoplasmic reticulum (ER) assembles with MFN1 or MFN2 at the outer mitochondrial membrane and tethers ER to mitochondria, which facilitates the efficient $Ca^{2+}$ uptake into mitochondria [18].

OPA1 is a dynamin-related GTPase and localizes in the intermembrane space of mitochondria, where it is tightly associated to the inner mitochondrial membrane [9, 19]. OPA1 shows genetic interaction with MFN1, which suggests its role in coordinating the fusion mechanisms of outer and inner membranes [4, 20].

In yeast at least four proteins have been related with the fission pathway: Dnm1p or Dnm1lp (dynamin 1-like protein), Fis1p (fission-related protein 1), Mvd1p (mitochondrial division protein 1), and Caf4p (carbon catabolite repression-associated factor). Whereas Fis1p localizes in the outer membrane, the other three proteins are soluble proteins. Fis1p recruits Dnm1p from the cytosol to the mitochondria via the bridge proteins Mdv1p and Caf4p. Dnm1p forms clusters into large foci at the sites of fission. Furthermore, as MFN2 has the GTPase fission driving force, this mechanism induces fragmentation of mitochondria and several ultrastructural changes in cristae [10, 21, 22]. Only two out of the four proteins have a human counterpart. FIS1 is the orthologue protein of Fis1p whereas DRP1 (dynamin-related protein 1) is the orthologue of Dnm1p. FIS1 and DRP1 are also required for the division of peroxisomes [23, 24]. No mutations in the human *FIS1* gene has been reported in patients; by contrast, it has been recently reported a patient with microcephaly, abnormal brain development, optic atrophy and hypoplasia, persistent lactic acidosis, and heterozygous dominant lethal with a mutation in the DRP1 gene, *DNM1L* [25]. In mammals, a third protein, GDAP1 (ganglioside-induced differentiation associated protein 1) has been recently postulated to participate in the regulation of the mitochondrial fission network pathway. GDAP1 localizes to the mitochondrial outer membrane [26, 27], and its overexpression induces mitochondrial fragmentation [26, 27]. Interestingly, mutations in *GDAP1* gene are the cause of a severe form of Charcot-Marie-Tooth disease [28, 29].

## 9.3 Mitochondrial Dynamics and Charcot-Marie-Tooth Neuropathies

Mutations in *MFN2*, *OPA1* and *GDAP1* genes give us evidence of the important role mitochondrial dynamics have in the pathogenesis of neuropathies, involving either peripheral or cranial nerves, more specifically the optic nerve.

Mitofusin 2 is involved in the most common form of the autosomal dominat axonal Charcot-Marie-Tooth, which is CMT2A. In fact, mutations in *MFN2* are the cause of the CMT2A2 variant (OMIM 609260) [12], whereas the CMT2A1 (OMIM 118210) variant has only been described in one Japanese family segregating a mutation in the gene belonging to the kinesin family, *KIF1B* [30]. Mutations in *MFN2* have been associated to the classical axonal CMT (CMT2A), but it has also been reported to cause hereditary motor and sensory neuropathy VI (HMSN VI) which is an axonal neuropathy with optic atrophy [31]. Some patients with axonal CMT and pyramidal signs (HMSN VII) have also been reported in association with *MFN2* mutations. Typical findings in CMT2A patients include earlier involvement of lower limbs, which are also more severe than in the upper limbs, prominent motor deficits more frequently than sensory deficits, and normal or only slightly decreased nerve conduction velocities (NCVs) [12]. Some individuals are asymptomatic and they only show mild findings upon examination; these individuals might eventually develop the disease later on [11]. In fact, clinical and genetic studies performed in large series of patients and families showed that there are two main phenotypes with different natural history, that is, an early-onset severe CMT2A variant, and a late-onset mild CMT2A variant [32]. Patients with HMSN VI phenotype used to have an early onset. Some *MFN2* mutations have been reported several times in HMSN VI, but also in families segregating CMT2 neuropathy with no optic atrophy. This is the case for the R94W and R364W mutations [31, 32].

*MFN2* has 17 exons, and more than 50 mutations have been reported, being the missense mutations the most frequent ones (Inherited Peripheral Neuropathies Mutation Database, www.molgen.ua.ac.be/CMTMutations/Home/IPN.cfm, update December 17, 2007). These mutations segregate in families as an autosomal dominant trait, but putative de novo dominant mutations have also been described [32]. It could be postulated a gain-of-function effect either by a toxic mechanism or by a dominant negative mechanism. Such pathogenic mechanisms may involve the fusion/fission balance of mitochondria, but also the energy metabolism [33] or the axonal mitochondrial transport [34].

*GDAP1* neuropathies were originally reported in families with patients affected with either demyelinating CMT (CMT4A, OMIM 214400) [28] or axonal CMT with vocal cord paresis (ARCMT2 or CMT4C4, OMIM 607706) diseases [29]. In both types, the onset occurred in early infancy starting with distal limb weakness, progressing proximally and causing severe disability. Furthermore, motor NCVs are variable. Most affected individuals showed an axonal neuropathy with normal NCVs and reduced amplitudes [35], a fact that agrees with the observation of a prominent loss of medium-sized and large myelinated fibers [35–39]. Onion bulb formations, however, have also been observed in some cases, which correlates with the demyelinating phenotype associated with slowed NCVs [36, 37, 40], and intermediate phenotypes have also been reported [41]. Hoarse voice as consequence of vocal cord paresis is a frequent symptom, and a restrictive respiratory function after taking part the phrenic nerve has also been reported in a number of patients [35, 38, 39]. Involvement of laryngeal and respiratory muscles occurs late in the natural history of the disease when significant proximal upper limb weakness has developed. Left vocal cord innervated by the longer left recurrent laryngeal nerve is early involved as well, which suggests a length dependent pattern of neurodegeneration in *GDAP1* neuropathies [39].

Mutations have been described in every exon of *GDAP1* (www.molgen.ua.ac.be/CMTMutations/Home/IPN.cfm. update December 17, 2007, and data not published) and include missense, nonsense, splicing site, short deletions, and insertion mutations. Overexpression of wild type *GDAP1* induces fragmentation of mitochondria [26, 42]. Morphological studies of mitochondria were performed with the *GDAP1* missense mutations in a scale that considers different mitochondria patterns (defined as aggregated-tubular-vesicular-fragmented), and showed variable results. Whereas R120Q, R120W, R282C and R161H mutations showed a fragmentation pattern similar to the wild-type protein, 70 percent of mitochondria overexpressing T157P mutation showed the opposite perinuclear aggregation [42]. Niemann et al. [26] obtained different mitochondrial patterns as missense mutations induced less fragmentation than the wild-type protein. This is especially evident for the

R161H and the R310Q amino acid changes. In spite of *GDAP1* neuropathies used to segregate as an autosomal recessive trait in families, the disease is inherited as a dominant trait in other families [43, 44]. R120W mutation has been described in a complex heterozygous patient [37], but also in families with autosomal dominant mode of inheritance, which suggests a putative dominant negative mechanism. However, current data of cellular analyses does not explain how missense changes affect GDAP1 function, and whether mitochondrial fission plays a role in the cell and in the pathophysiology of the disease. By contrast, there is strong evidence that proteins expressing disease truncated alleles [26] or truncated constructs without the transmembrane domain located at its C-terminus [27] could not reach the mitochondrial outer membrane, suggesting a loss of function mechanism for such mutant alleles.

Some mutations have been described in specific populations for which a founder effect has been suggested. This is the case for the Q163X mutation in the Hispanic population [38]. This mutation is the most frequent pathogenic change in Spanish patients. Both Spanish and Hispanic patients share a common haplotype which suggest a unique origin in the Iberian Peninsula [43]. Haplotype analysis also showed a common origin for the S194X mutations in Spanish and Moroccan families [36, 43], and for the M116R mutation in three Italian families [45].

## 9.4 Mitochondrial Dynamics and Neurodegeneration

Neurodegeneration is associated with defects in mitochondrial-related proteins. Both mutant proteins with primary mitochondrial localization and mutant proteins potentially affecting mitochondrial functions are the cause of monogenic disorders of the nervous system. Disorders associated with a neurodegenerative pathophysiology and defects in mitochondrial function include Parkinson disease, Huntington disease, hereditary spastic paraplegia (HSP), and Charcot-Marie-Tooth neuropathy [3, 46]. Parkinson's genes associated with mitochondria and causing Mendelian disease include *PINK1*, a serine/threonine kinase, and *PARKIN*, an E3 ubiquitin ligase. Several genes causing either recessive or dominant spastic paraplegia are related to mitochondria or mitochondrial dynamics and mobility. Mutations in *SPG13* (HSP type 13), encoding the chaperoning heat-shock protein 60, may affect mitochondrial import, and mutations in the *SPG7* gene (HPS type 7) that encodes paraplegin, a m-AAA protease located in the inner mitochondrial membrane, may be relevant in mitochondrial quality control [46]. Genes related to microtubules are especially interesting as mitochondria must be attached to them while travelling from the cell body towards the axons and vice versa, and disruption of this connection should affect mitochondrial mobility. Examples of this category are the autosomal dominant HSP type 4 caused by mutations in spastin (*SPAST* gene), a microtubule-severing protein, and the autosomal recessive HSP type 20 (Troyer syndrome), which is the result of mutations in spartin (*SPG20* gene), a protein of the outer mitochondrial membrane that has interaction with microtubules and a trafficking domain.

Neurodegeneration in peripheral neuropathies affects long axons. An important mechanism defining the primary pathology in the axons involves axonal transport and also includes several cargos such as mitochondria [47]. Expression of mutant MFN2 proteins in cultured dorsal root ganglion neurons induces abnormal clustering of small fragmented mitochondria, and transport of mitochondria in axons is also impaired [34]. Interestingly, these studies also show that axonal transport reduction could not be attributable to energy defect in the neurons. No similar data obtained in neuron models are available with *GDAP1* mutations. As mentioned above, although most of *GDAP1* missense mutations induce fragmentation in such a way that they do not significantly differ from normal protein, there are mutations that have a different effect on the mitochondrial distribution in mammalian cells [42]. It is tentative to speculate that GDAP1 may be connected to the microtubule network and to transport in neurons, and that those GDAP1-associated neuropathies, CMT4A, CMT4C4 and CMT2K, might be also related with defects in the axonal transport as it has been proposed for CMT2A. Fusion and fission proteins involved in inherited peripheral neuropathies, such as MFN2 and GDAP1, beyond maintaining the morphology and dynamics of mitochondria could probably contribute to other biological pathways. Axonal transport has been mentioned above, but oxidative stress, cell death or synaptic maintenance are other processes that might be also relevant [4, 46]. The study of the biology of mitochondrial dynamics and the fusion/fission balance proteins interactions in a more complex physiological network in the axon in the peripheral nerve will contribute to better understand the pathophysiology of peripheral neuropathies, and to design proper treatment and give clues for prevention of these and other neurodegenerative disorders.

**Acknowledgements** This work is supported by the Spanish Ministry of Science and Innovation and the Fondo de Investigación Sanitaria. The CIBER de Enfermedades Raras is an initiative of the Instituto de Salud Carlos III.

## References

1. Combarros O, Calleja J, Polo JM, et al. Prevalence of hereditary motor and sensory neuropathy in Cantabria. *Acta Neurol Scand* 1987;**75**(1):9–12.
2. Skre H. Genetic and clinical aspects of Charcot-Marie-Tooth's disease. *Clin Genet* 1974;**6**(2):98–118.
3. DiMauro S, Schon EA. Mitochondrial disorders in the nervous system. *Annu Rev Neurosci* 2008;**31**:91–123.
4. Knott AB, Perkins G, Schwarzenbacher R, et al. Mitochondrial fragmentation in neurodegeneration. *Nat Rev Neurosci* 2008;**9**(7):505–518.
5. Lin MT, Beal MF. Mitochondrial dysfunction and oxidative stress in neurodegenerative diseases. *Nature* 2006;**443**(7113):787–795.
6. Deng H, Dodson MW, Huang H, et al. The Parkinson's disease genes pink1 and parkin promote mitochondrial fission and/or inhibit fusion in Drosophila. *Proc Natl Acad Sci U S A* 2008;**105**(38):14503–14508.
7. Yang Y, Ouyang Y, Yang L, et al. Pink1 regulates mitochondrial dynamics through interaction with the fission/fusion machinery. *Proc Natl Acad Sci U S A* 2008;**105**(19):7070–7075.
8. Cerveny KL, Tamura Y, Zhang Z, et al. Regulation of mitochondrial fusion and division. *Trends Cell Biol* 2007;**17**(11):563–569.

9. Chan DC. Mitochondrial fusion and fission in mammals. *Annu Rev Cell Dev Biol* 2006;**22**: 79–99.
10. Detmer SA, Chan DC. Functions and dysfunctions of mitochondrial dynamics. *Nat Rev Mol Cell Biol* 2007;**8**(11):870–879.
11. Lawson VH, Graham BV, Flanigan KM. Clinical and electrophysiologic features of CMT2A with mutations in the mitofusin 2 gene. *Neurology* 2005;**65**(2):197–204.
12. Zuchner S, Mersiyanova IV, Muglia M, et al. Mutations in the mitochondrial GTPase mitofusin 2 cause Charcot-Marie-Tooth neuropathy type 2A. *Nat Genet* 2004;**36**(5):449–451.
13. Alexander C, Votruba M, Pesch UE, et al. OPA1, encoding a dynamin-related GTPase, is mutated in autosomal dominant optic atrophy linked to chromosome 3q28. *Nat Genet* 2000;**26**(2):211–115.
14. Delettre C, Lenaers G, Griffoin JM, et al. Nuclear gene OPA1, encoding a mitochondrial dynamin-related protein, is mutated in dominant optic atrophy. *Nat Genet* 2000;**26**(2): 207–210.
15. Chen H, Detmer SA, Ewald AJ, et al. Mitofusins Mfn1 and Mfn2 coordinately regulate mitochondrial fusion and are essential for embryonic development. *J Cell Biol* 2003;**160**(2): 189–200.
16. Koshiba T, Detmer SA, Kaiser JT, et al. Structural basis of mitochondrial tethering by mitofusin complexes. *Science* 2004;**305**(5685):858–862.
17. de Brito OM, Scorrano L. Mitofusin 2: a mitochondria-shaping protein with signaling roles beyond fusion. *Antioxid Redox Signal* 2008;**10**(3):621–633.
18. de Brito OM, Scorrano L. Mitofusin 2 tethers endoplasmic reticulum to mitochondria. *Nature* 2008;**456**(7222):605–610.
19. Olichon A, Emorine LJ, Descoins E, et al. The human dynamin-related protein OPA1 is anchored to the mitochondrial inner membrane facing the inter-membrane space. *FEBS Lett* 2002;**523**(1–3):171–176.
20. Cipolat S, Martins de Brito O, Dal Zilio B, et al. OPA1 requires mitofusin 1 to promote mitochondrial fusion. *Proc Natl Acad Sci U S A* 2004;**101**(45):15927–15932.
21. Sesaki H, Jensen RE. Division versus fusion: Dnm1p and Fzo1p antagonistically regulate mitochondrial shape. *J Cell Biol* 1999;**147**(4):699–706.
22. Smirnova E, Griparic L, Shurland DL, et al. Dynamin-related protein Drp1 is required for mitochondrial division in mammalian cells. *Mol Biol Cell* 2001;**12**(8):2245–2256.
23. Koch A, Thiemann M, Grabenbauer M, et al. Dynamin-like protein 1 is involved in peroxisomal fission. *J Biol Chem* 2003;**278**(10):8597–8605.
24. Koch A, Yoon Y, Bonekamp NA, et al. A role for Fis1 in both mitochondrial and peroxisomal fission in mammalian cells. *Mol Biol Cell* 2005;**16**(11):5077–5086.
25. Waterham HR, Koster J, van Roermund CW, et al. A lethal defect of mitochondrial and peroxisomal fission. *N Engl J Med* 2007;**356**(17):1736–1741.
26. Niemann A, Ruegg M, La Padula V, et al. Ganglioside-induced differentiation associated protein 1 is a regulator of the mitochondrial network: new implications for Charcot-Marie-Tooth disease. *J Cell Biol* 2005;**170**(7):1067–1078.
27. Pedrola L, Espert A, Wu X, et al. GDAP1, the protein causing Charcot-Marie-Tooth disease type 4A, is expressed in neurons and is associated with mitochondria. *Hum Mol Genet* 2005;**14**(8):1087–1094.
28. Baxter RV, Ben Othmane K, Rochelle JM, et al. Ganglioside-induced differentiation-associated protein-1 is mutant in Charcot-Marie-Tooth disease type 4A/8q21. *Nat Genet* 2002;**30**(1):21–22.
29. Cuesta A, Pedrola L, Sevilla T, et al. The gene encoding ganglioside-induced differentiation-associated protein 1 is mutated in axonal Charcot-Marie-Tooth type 4A disease. *Nat Genet* 2002; **30**(1):22–25.
30. Zhao C, Takita J, Tanaka Y, et al. Charcot-Marie-Tooth disease type 2A caused by mutation in a microtubule motor KIF1Bbeta. *Cell* 2001;**105**(5):587–597.
31. Zuchner S, De Jonghe P, Jordanova A, et al. Axonal neuropathy with optic atrophy is caused by mutations in mitofusin 2. *Ann Neurol* 2006;**59**(2):276–281.

32. Chung KW, Kim SB, Park KD, et al. Early onset severe and late-onset mild Charcot-Marie-Tooth disease with mitofusin 2 (MFN2) mutations. *Brain* 2006;**129**(Pt 8):2103–2118.
33. Pich S, Bach D, Briones P, et al. The Charcot-Marie-Tooth type 2A gene product, Mfn2, up-regulates fuel oxidation through expression of OXPHOS system. *Hum Mol Genet* 2005;**14**(11):1405–1415.
34. Baloh RH, Schmidt RE, Pestronk A, et al. Altered axonal mitochondrial transport in the pathogenesis of Charcot-Marie-Tooth disease from mitofusin 2 mutations. *J Neurosci* 2007;**27**(2):422–430.
35. Sevilla T, Cuesta A, Chumillas MJ, et al. Clinical, electrophysiological and morphological findings of Charcot-Marie-Tooth neuropathy with vocal cord palsy and mutations in the GDAP1 gene. *Brain* 2003;**126**(Pt 9):2023–2033.
36. Nelis E, Erdem S, Van Den Bergh PY, et al. Mutations in GDAP1: autosomal recessive CMT with demyelination and axonopathy. *Neurology* 2002;**59**(12):1865–1872.
37. Ammar N, Nelis E, Merlini L, et al. Identification of novel GDAP1 mutations causing autosomal recessive Charcot-Marie-Tooth disease. *Neuromuscul Disord* 2003;**13**(9):720–728.
38. Boerkoel CF, Takashima H, Nakagawa M, et al. CMT4A: identification of a Hispanic GDAP1 founder mutation. *Ann Neurol* 2003;**53**(3):400–405.
39. Sevilla T, Jaijo T, Nauffal D, et al. Vocal cord paresis and diaphragmatic dysfunction are severe and frequent symptoms of GDAP1-associated neuropathy. *Brain* 2008;**131**(11):3051–3061.
40. De Sandre-Giovannoli A, Chaouch M, Boccaccio I, et al. Phenotypic and genetic exploration of severe demyelinating and secondary axonal neuropathies resulting from GDAP1 nonsense and splicing mutations. *J Med Genet* 2003;**40**(7):e87.
41. Senderek J, Bergmann C, Ramaekers VT, et al. Mutations in the ganglioside-induced differentiation-associated protein-1 (GDAP1) gene in intermediate type autosomal recessive Charcot-Marie-Tooth neuropathy. *Brain* 2003;**126**(Pt 3):642–649.
42. Pedrola L, Espert A, Valdes-Sanchez T, et al. Cell expression of GDAP1 in the nervous system and pathogenesis of Charcot-Marie-Tooth type 4A disease. *J Cell Mol Med* 2008;**12**(2):679–689.
43. Claramunt R, Pedrola L, Sevilla T, et al. Genetics of Charcot-Marie-Tooth disease type 4A: mutations, inheritance, phenotypic variability, and founder effect. *J Med Genet* 2005;**42**(4):358–365.
44. Chung KW, Kim SM, Sunwoo IN, Cho SY, Hwang SJ, Kim J, Kang SH, Park KD, Choi KG, Choi IS, Choi BO. A novel GDAP1 Q218E mutation in autosomal dominant Charcot-Marie-Tooth disease. *J Hum Genet* 2008;**53**(4):360–364.
45. Di Maria E, Gulli R, Balestra P, et al. A novel mutation of GDAP1 associated with Charcot-Marie-Tooth disease in three Italian families: evidence for a founder effect. *J Neurol Neurosurg Psychiatry* 2004;**75**(10):1495–1498.
46. Tatsuta T, Langer T. Quality control of mitochondria: protection against neurodegeneration and ageing. *EMBO J* 2008;**27**(2):306–314.
47. Baloh RH. Mitochondrial dynamics and peripheral neuropathy. *Neuroscientist* 2008;**14**(1):12–18.

# Chapter 10
# Pathogenesis and Treatment of Mitochondrial Disorders

**Salvatore DiMauro and Michio Hirano**

**Abstract** In the past 50 years, our understanding of the biochemical and molecular causes of mitochondrial diseases, defined restrictively as disorders due to defects of the mitochondrial respiratory chain (RC), has made great strides. Mitochondrial diseases can be due to mutations in mitochondrial DNA (mtDNA) or in nuclear DNA (nDNA) and each group can be subdivided into more specific classes. Thus, mtDNA-related disorders can result from mutations in genes affecting protein synthesis *in toto* or mutations in protein-coding genes. Mendelian mitochondrial disorders can be attributed to mutations in genes that (i) encode subunits of the RC ("direct hits"); (ii) encode assembly proteins or RC complexes ("indirect hits"); (iii) encode factors needed for mtDNA maintenance, replication, or translation (intergenomic signaling); (iv) encode components of the mitochondrial protein import machinery; (v) control the synthesis and composition of mitochondrial membrane phospholipids; and (vi) encode proteins involved in mitochondrial dynamics.

In contrast to this wealth of knowledge about etiology, our understanding of pathogenic mechanism is very limited. We discuss pathogenic factors that can influence clinical expression, especially ATP shortage and reactive oxygen radicals (ROS) excess.

Therapeutic options are limited and fall into three modalities: (i) symptomatic interventions, which are palliative but crucial for day-to-day management; (ii) radical approaches aimed at correcting the biochemical or molecular error, which are interesting but still largely experimental; and (iii) pharmacological means of interfering with the pathogenic cascade of events (e.g. boosting ATP production or scavenging ROS), which are inconsistently and incompletely effective, but can be safe and helpful.

**Keywords:** Mitochondrial diseases · Respiratory chain defects · Energy failure · Oxidative stress · Symptomatic therapy · Pharmacological therapy · Gene therapy

---

S. DiMauro (✉)
Department of Neurology, Columbia University Medical Center, 3-313 Russ Berrie Medical Science Pavilion, New York, NY, 10032 USA
e-mail: sd12@columbia.edu

C. Espinós et al. (eds.), *Inherited Neuromuscular Diseases*, Advances in Experimental Medicine and Biology 652, DOI 10.1007/978-90-481-2813-6_10,

In the 46 years since the description of the first mitochondrial disease, Luft syndrome [1] and in the 20 years since the discovery of the first pathogenic mitochondrial DNA (mtDNA) mutations [2, 3] the field of mitochondrial disorders has expanded (and continues expanding) at an unprecedented pace. In fact, the term "mitochondrial medicine" introduced in 2004 by Rolf Luft [4], is widely accepted and two books with this title have already appeared [5, 6]. Despite this extraordinary progress, our task is arduous because the two subjects of this review, pathogenesis and therapy, are the least developed, especially when it comes to mtDNA-related disorders. After a "taxonomic" introduction to the mitochondrial diseases, we will attempt to merge for each group of diseases, what little we understand about pathogenesis with what little we can do about treatment.

## 10.1 Classification of the Mitochondrial Diseases

Mitochondrial myopathies were described in the early 1960s, when systematic ultrastructural and histochemical studies revealed excessive proliferation of normal- or abnormal-looking mitochondria in muscle of patients with weakness or exercise intolerance [7, 8]. Because, with the modified Gomori trichrome stain, the areas of mitochondrial accumulation appeared crimson (Fig. 10.1), the abnormal fibers were dubbed "ragged-red fibers" (RRF) [9] and came to be considered the pathological hallmark of mitochondrial disease. However, it soon became apparent that in many patients with RRF, the myopathy is associated with symptoms and signs of brain involvement, and the term mitochondrial encephalomyopathies was introduced [10]. It also became clear that lack of RRF in the biopsy does not exclude a mitochondrial etiology, as exemplified by Leigh syndrome (LS), an encephalopathy of infancy or childhood invariably due to mitochondrial dysfunction but rarely accompanied by RRF.

According to the widely accepted "endosymbiotic hypothesis", mitochondria are the relics of protobacteria that populated anaerobic nucleated cells and endowed them with the precious gift of oxidative metabolism. Thus, mitochondria are the main source of energy for all human tissues and contain many metabolic pathways, only some of which (e.g. pyruvate dehydrogenase complex [PDHC], the carnitine cycle, the ß-oxidation "spirals", and the Krebs cycle [also known as tricarboxylic acid cycle, TCA] are shown in Fig. 10.2.

Although defects in all of these pathways are by definition mitochondrial diseases, the term "mitochondrial encephalomyopathy" has come to indicate disorders due to defects in the respiratory chain (RC). This is the "business end" of oxidative metabolism, where ATP is generated. Reducing equivalents produced in the Krebs cycle and in the ß-oxidation spirals are passed along a series of protein complexes embedded in the inner mitochondrial membrane (the electron transport chain). The electron transport chain consists of four multimeric complexes (I to IV) plus two small electron carriers, coenzyme Q (or ubiquinone) and cytochrome *c*. The energy generated by these reactions is used to pump protons from the mitochondrial matrix into the space between the inner and outer mitochondrial membranes. This

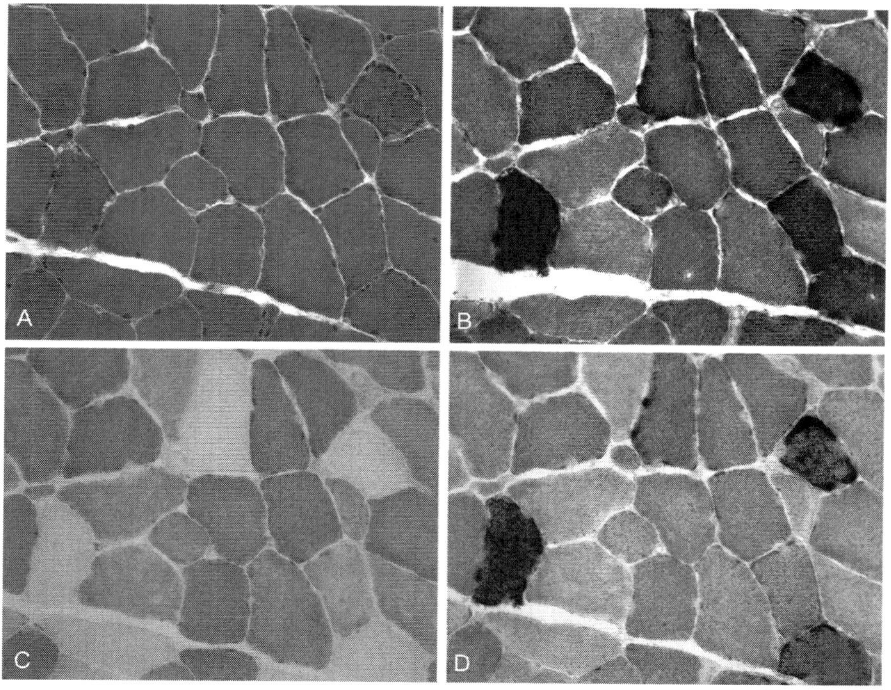


**Fig. 10.1** Serial cross-sections of the muscle biopsy from a patient with KSS and a large-scale mtDNA deletion. A. With the modified Gomori trichrome, ragged-*red* fibers (RRF) show irregular crimsom staining; B. With the succinate dehydrogenase (SDH) stain, the same fibers appear *dark blue* ("ragged-blue" fibers); C. With the cytochrome *c* oxidase (COX) stain, most RRF lack COX activity either completely or partially; D. By superimposing the SDH and COX stains, normal fibers appear brown, while COX-deficient fibers stand out as *blue* (Courtesy of Drs. Eduardo Bonilla and Kurenai Tanji, Columbia University Medical Center)

creates an electrochemical proton gradient, which is utilized by complex V (or ATP synthase) to produce ATP in a process known as oxidation/phosphorylation coupling.

A unique feature of the RC is its dual genetic control: mitochondrial DNA (mtDNA) encodes 13 of the approximately 80 proteins that compose the RC and nuclear DNA (nDNA) encodes all the others. As indicated by the different shadings in Fig. 10.2, complex II, coenzyme Q, and cytochrome *c* are exclusively encoded by nDNA. In contrast, complexes I, III, IV, and V contain some subunits encoded by mtDNA: seven for complex I (ND1, ND2, ND3, ND4, ND4L, ND5, and ND6), one for complex III (cytochrome *b*), three for complex IV (COX I, COX II, and COX III), and two for complex V (ATPase 6 and ATPase 8).

Human mtDNA (Fig. 10.3) is a 16,569-kb circular, double-stranded molecule, which contains 37 genes: 2 rRNA genes, 22 tRNA genes, and 13 structural genes encoding the respiratory chain subunits listed above. In the course of evolution, mtDNA has lost much of its original autonomy and now depends heavily on the

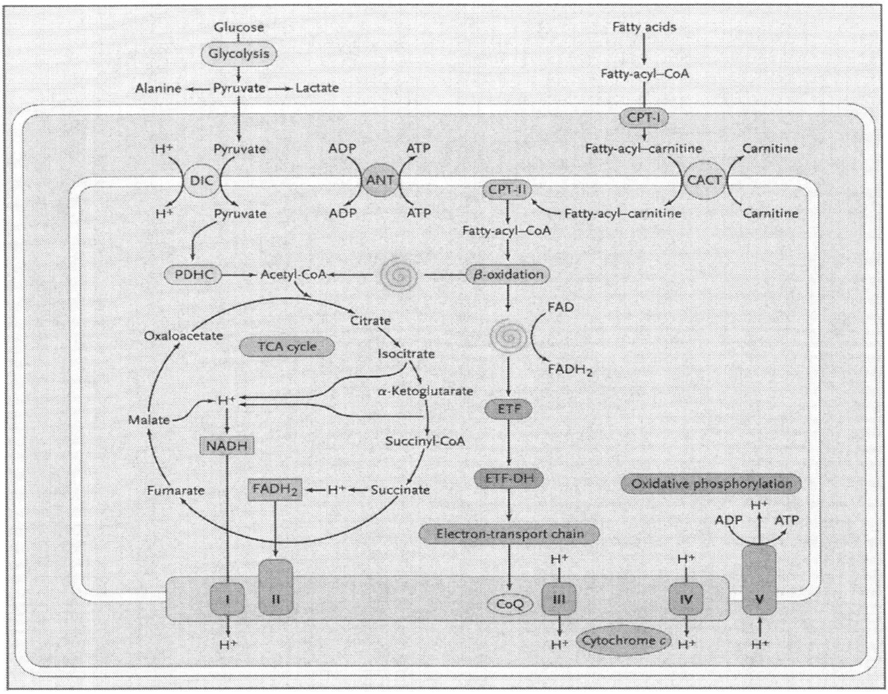


**Fig. 10.2** Scheme of the mitochondrion showing selected metabolic pathways. The spirals depict the reactions of the β-oxidation pathway, resulting in the liberation of acetyl-coenzyme A (CoA) and the reduction of flavoprotein. See the list of abbreviations at the end of the text

nuclear genome for the production of factors needed for mtDNA transcription, translation, and replication. Since 1988, the circle of mtDNA has become crowded with pathogenic mutations, and the principles of mitochondrial genetics should, therefore, be familiar to the practicing physician.

(i) *Heteroplasmy and threshold effect.* Each cell contains hundreds or thousands of mtDNA copies, which, at cell division, distribute randomly among daughter cells. In normal tissues, all mtDNA molecules are identical (homoplasmy). Deleterious mutations of mtDNA usually (but not always) affect some but not all mtDNAs and the clinical expression of a pathogenic mtDNA mutation is largely determined by the relative proportion of normal and mutant genomes in different tissues. A minimum critical number of mutant mtDNAs is required to cause mitochondrial dysfunction in a particular organ or tissue and mitochondrial disease in an individual (threshold effect).

(ii) *Mitotic segregation.* At cell division, the proportion of mutant mtDNAs in daughter cells may shift and the phenotype may change accordingly. This phenomenon, called mitotic segregation, explains how the clinical phenotype may change in certain patients with mtDNA-related disorders as they grow older.

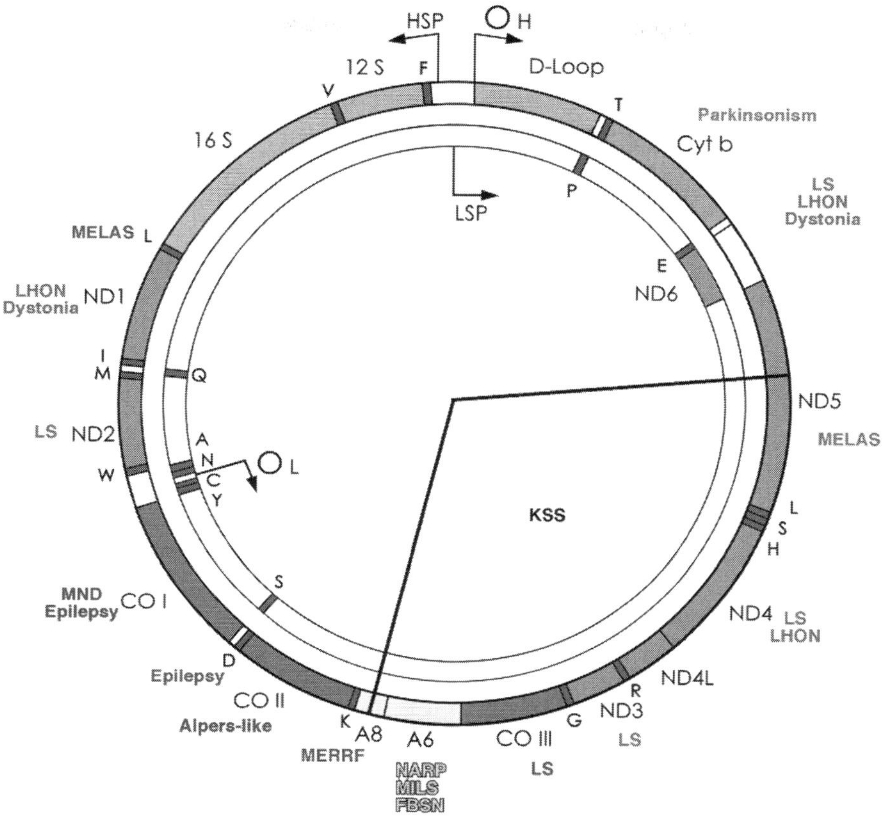


**Fig. 10.3** Schematic view of human mtDNA, showing the gene products for the 12S and 16S ribosomal RNAs, the subunits of NADH-coenzyme Q oxidoreductase (ND), cytochrome *c* oxidase (COX), cytochrome b (*cyt b*), and ATP synthase (A), and 22 tRNAs (1-letter amino acid nomenclature), the origins of heavy- and light-strand replication ($O_H$ and $O_L$) and the promoters of heavy- and light-strand transcription (HSP and LSP). Some pathogenic mutations are indicated. The arc removed by the "common deletion" is subtended by the two radii. For abbreviations and acronyms, see the list at the end of the text. (Reproduced from the *Annual Review of Neuroscience* [39] with permission)

(iii) *Maternal inheritance*. At fertilization, all mtDNA derives from the oöcyte. Therefore, the mode of transmission of mtDNA and of mtDNA point mutations (single deletions of mtDNA are usually sporadic events) differs from Mendelian inheritance. A mother carrying a mtDNA point mutation will pass it on to all her children (boys and girls), but only her daughters will transmit it to their progeny. The best way for a clinician to chart his course toward a diagnosis in the morass of mitochondrial encephalomyopathies is to use a classification that combines clinical features, muscle histochemistry and biochemistry, and genetics. From the genetic point of view, there are two major

**Table 10.1** Genetic classification of the mitochondrial diseases (RC defects)

| Mutations in mtDNA | Mutations in nDNA |
|---|---|
| Defects of mitochondrial protein synthesis | Mutations in RC subunits |
| mtDNA large-scale deletions | Mutations in RC assembly proteins |
| mutations in tRNA or rRNA genes | Defects of mtDNA maintenance, replication, translation |
| Mutations in protein-coding genes | Defects in protein importation |
| | Defects of membrane lipid composition |
| | Defects of mitochondrial dynamics |

categories, disorders due to defects of mtDNA and disorders due to defects of nDNA (Table 10.1).

1. *Disorders due to defects of mtDNA*. These include rearrangements (single deletions or duplications) and point mutations.

A. *mtDNA rearrangements*. Single deletions of mtDNA have been associated with three sporadic conditions: (i) Pearson syndrome, a usually fatal disorder of infancy characterized by sideroblastic anemia and exocrine pancreas dysfunction. (ii) Kearns-Sayre syndrome (KSS), a multisystem disorder with onset before age 20 of impaired eye movements (progressive external ophthalmoplegia, PEO), pigmentary retinopathy, and heart block. Frequent additional signs include ataxia, dementia, and endocrinopathies (diabetes mellitus, short stature, hypoparathyroidism). Lactic acidosis, elevated cerebrospinal fluid (CSF) protein (over 100 mg/dl), and scattered COX-negative RRF in the muscle biopsy are typical laboratory abnormalities. (iii) PEO with or without proximal limb weakness, often compatible with a normal lifespan. Deletions vary in size and location, but a "common" deletion of 5 kb is frequently seen in patients and in aged individuals (Fig. 10.3).

B. *Point mutations*. Over 200 pathogenic point mutations have been identified in mtDNA from patients with a variety of disorders [11], most of which are maternally inherited and multisystemic, but some are sporadic and tissue-specific (Fig. 10.3). Among the maternally inherited encephalomyopathies, four syndromes are more common. The first is MELAS (mitochondrial encephalomyopathy, lactic acidosis, and stroke-like episodes), which typically presents in children or young adults after normal early development. Symptoms include recurrent vomiting, migraine-like headache, and stroke-like episodes causing cortical blindness, hemiparesis, or hemianopia. MRI of the brain shows "infarcts" that do not correspond to the distribution of major vessels, raising the question of whether the strokes are vascular or metabolic in nature [12]. The most common mtDNA mutation is A3243G in the $tRNA^{Leu(UUR)}$ gene, but about a dozen other mutations have been associated with MELAS, most notably a mutation (G13513A) in the ND5 gene, which encodes subunit 5 of complex I [13]. It is important to note that most maternal relatives of a typical

MELAS patient carry the mutation in low abundance and are either mildly affected or unaffected. In agreement with this observation, two epidemiological studies have reported comparably high prevalence (about 1:750) of the A3243G mutation in the normal population [14, 15], and one of them has found that the prevalence of pathogenic mtDNA mutations is 1:200 individuals in Northern England [15].

The second syndrome is MERRF (myoclonus epilepsy with ragged red fibers), characterized by myoclonus, seizures, mitochondrial myopathy, and cerebellar ataxia. Less common signs include dementia, hearing loss, peripheral neuropathy, and multiple lipomas. Three mtDNA mutations have been associated with MERRF and all are in the $tRNA^{Lys}$ gene (A8344G; T8356C; G8363A).

The third syndrome comes in two flavors: (i) NARP (neuropathy, ataxia, retinitis pigmentosa) usually affects young adults and causes retinitis pigmentosa, dementia, seizures, ataxia, proximal weakness, and sensory neuropathy; (ii) maternally inherited Leigh syndrome (MILS) is a severe infantile encephalopathy with characteristic symmetrical lesions in the basal ganglia and the brainstem [16, 17].

The fourth syndrome, LHON (Leber hereditary optic neuropathy) is characterized by acute or subacute loss of vision in young adults, more frequently males, due to bilateral optic atrophy. Three mtDNA point mutations in ND genes account for more than 90% of LHON cases. These are G11778A in ND4, G3460A in ND1, and T14484C in ND6 [18].

Not surprisingly, syndromes associated with mtDNA mutations can affect every system in the body, including the eye (optic atrophy; retinitis pigmentosa; cataracts); hearing (sensorineural deafness); endocrine system (short stature; diabetes mellitus; hypoparathyroidism); heart (familial cardiomyopathies; conduction blocks); gastrointestinal tract (exocrine pancreas dysfunction; intestinal pseudo-obstruction; gastroesophageal reflux); and kidney (renal tubular acidosis) [6]. Any combination of the symptoms and signs listed above should raise the suspicion of a mitochondrial disorder, especially if there is evidence of maternal transmission.

On the other hand, point mutations in mtDNA protein-coding genes often escape the rules of mitochondrial genetics in that they affect single individuals and single tissues, most commonly skeletal muscle. Thus, patients with exercise intolerance, myalgia and, sometimes recurrent myoglobinuria, may have isolated defects of complex I, complex III, or complex IV, due to pathogenic mutations in genes encoding ND subunits, cytochrome *b*, or COX subunits [19]. The lack of maternal inheritance and the involvement of muscle alone suggest that mutations arose *de novo* in myogenic stem cells after germ-layer differentiation ("somatic mutations").

2. *Disorders due to defects of nDNA*. These are all transmitted by Mendelian inheritance and include three major subgroups.

A. *Mutations in genes encoding subunits or ancillary proteins of the respiratory chain*. As noted above, mtDNA encodes only 13 subunits of the respiratory chain, while nDNA encodes all subunits of complex II, most subunits

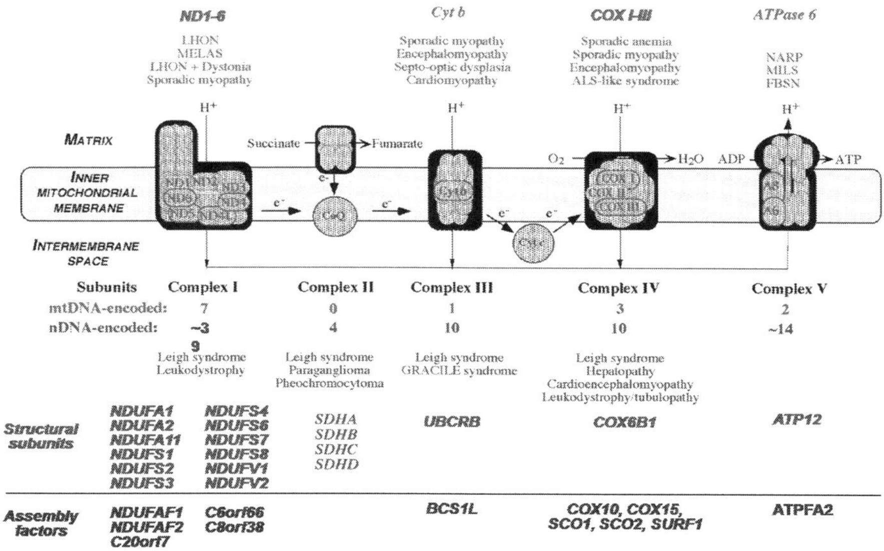


**Fig. 10.4** The mitochondrial respiratory chain (RC), showing nDNA-encoded subunits in blue and mtDNA-encoded subunits in *red*. Coenzyme Q and cytochrome *c* are electron ($e^-$) carriers. Diseases are listed according to the affected RC complex and divided in defects of RC subunits and defects of assembly proteins

of the other four complexes, as well as $CoQ_{10}$ and cytochrome *c*. Nuclear DNA mutations can affect respiratory chain complexes directly or indirectly (Fig. 10.4).

Direct "hits", that is, mutations in gene encoding respiratory chain subunits, were known until recently only for complex I and complex II, but the first mutation in a subunit of complex IV, COX VIB1, has now been reported [20]. These have been associated with autosomal recessive forms of LS. Primary or secondary coenzyme $Q_{10}$ ($CoQ_{10}$) deficiency can cause five major syndromes: (i) a predominantly myopathic disorder with recurrent myoglobinuria but also CNS involvement (seizures, ataxia, mental retardation) [21, 22, 23]; (ii) a predominantly encephalopathic disorder with ataxia and cerebellar atrophy [24, 25, 26]; (iii) an isolated myopathy, with RRF and lipid storage [27]; (iv) a generalized mitochondrial encephalomyopathy, usually with onset in infancy [28, 29, 30]; and (v) nephropathy alone or associated with encephalopathy [31]. Examples of secondary $CoQ_{10}$ deficiency include ataxia oculomotor apraxia (AOA1) associated with mutations in the apprataxin (*APTX*) gene [32], and the myopathic presentation of glutaric aciduria type II (GA II) due to mutations in the electron transfer flavoprotein dehydrogenase (*EFTDH*) gene [33]. Examples of primary CoQ10 deficiency include mutations in the biosynthetic genes, *COQ1* (*PDSS1* and *PDSS2*) [34, 35], *COQ2* [36], and *CABC1/ADCK3* [37, 38]. Irrespective of etiology, diagnosis is important because most patients with $CoQ_{10}$ deficiency respond to high-dose $CoQ_{10}$ supplementation.

Indirect "hits" are mutations in genes encoding proteins that are not components of the respiratory chain, but are needed for the proper assembly and function of respiratory chain complexes (Fig. 10.4). This "murder by proxy" mechanism has been identified in disorders due to defects in complexes I, III, IV, and V (for review, see [39], and is best illustrated by Mendelian defects of complex IV (COX). Mutations in genes encoding six ancillary proteins (*SURF1*, *SCO2*, *SCO1*, *COX10*, *COX15*, and *LRPPRC*) have been associated with COX-deficient LS or other multisystemic fatal infantile disorders in which encephalopathy is accompanied by cardiomyopathy (*SCO2; COX15*), nephropathy (*COX10*), or hepatopathy (*SCO1*).

This is a burgeoning field of research with important theoretical and practical implications. From an investigative point of view, these disorders are teaching us a lot about the structural and functional complexity of the respiratory chain. At a more practical level, identification of mutations in these genes allows prenatal diagnosis and suggests approaches to therapy.

B. *Defects of intergenomic signaling*. As noted above, the mtDNA is highly dependent for its proper function and replication on numerous factors encoded by nuclear genes. Mutations in these genes cause Mendelian disorders characterized by qualitative or quantitative alterations of mtDNA. Examples of qualitative alterations include autosomal dominant or recessive multiple deletions of mtDNA, usually accompanied clinically by progressive external ophthalmoplegia (PEO) plus a variety of other symptoms and signs. Four of these conditions have been characterized at the molecular level. Mutations in the gene (*TYMP*) for thymidine phosphorylase (TP) are responsible for an autosomal recessive multisystemic syndrome called MNGIE (mitochondrial neurogastrointestinal encephalomyopathy) [40, 41]. Mutations in the gene for one isoform of the adenine nucletide translocator (*ANT1*) have been identified in patients with autosomal dominant PEO [42]. Mutations in the *PEO1* gene, encoding Twinkle, a helicase, are also associated with autosomal dominant PEO [43], whereas mutations in the gene encoding polymerase $\gamma$ (*POLG*) may cause either autosomal dominant or autosomal recessive PEO [44] as well as a variety of symptoms and signs, including ataxia, peripheral neuropathy, parkinsonism, and MNGIE or MERRF-like syndromes [45]. Mutations in the gene (*POLG2*) encoding the accessory subunit of POLG, can also cause autosomal dominant PEO and multiple deletions [46]. Finally, mutations in *OPA1*, which encodes a protein involved in mitochondrial dynamics (see below), besides causing dominant optic atrophy (DOA) can also result in a syndrome that includes optic neuropathy, PEO, deafness, ataxia, and axonal neuropathy associated with multiple mtDNA deletions in the muscle biopsy [47, 48, 49, 50].

Examples of quantitative alterations of mtDNA include severe or partial mtDNA depletion, usually characterized clinically by congenital or childhood forms of autosomal recessively inherited myopathy or hepatopathy [51]. Mutations in eight genes, seven of them involved in mitochondrial nucleotide homeostasis, have been associated with mtDNA depletion syndromes, although they still do not explain all cases.

Mutations in the gene encoding thymidine kinase 2 (*TK2*) are typically seen in patients with myopathic mtDNA depletion syndromes [52], whereas mutations in the genes encoding the β subunit (*SUCLA2*) or the α subunit (*SUCLG1*) of the mitochondrial matrix enzyme succinyl-CoA synthetase, (SCS-A) cause both myopathy and encephalopathy [53, 54]. Mutations in *DGUOK*, encoding deoxyguanosine kinase, predominate in patients with hepatic or hepatocerebral mtDNA depletion syndromes [55, 56], but mutations in *POLG* are the major causes of Alpers-Huttenlocher syndrome, a severe hepatocerebral syndrome with vulnerability to valproic acid [45, 57]. Mutations in a gene (*MPV17*) not involved in nucleotide pool homeostasis, have been associated with hepatocerebral syndrome [58] and with the Navajo neurohepatopathy (NNH) syndrome, prevalent in the Navajo population of Southwestern Unites States [59]. An iatrogenic form of mtDNA depletion may follow treatment with nucleoside analogs, such as zidovudine (AZT).

A group of defects of intergenomic communication is due to mutations in genes encoding factors necessary for the faithful translation of mtDNA-encoded proteins [60], including *EFG1* (encoding elongation factor 1) [61], *MRPS16* (encoding small subunit protein) [62], *EFTu* (encoding elongation factor Tu), *TSFM* (controlling the expression of both EFTs and EFTu) [63], *PUS1* (encoding pseudouridine synthase 1) [64, 65]. The resulting disorders usually affect infants and cause severe encephalomyopathy, cardiomyopathy, or sideroblastic anemia. Typically, both quality and quantity of mtDNA are normal in these patients, but there are – not surprisingly – multiple respiratory chain defects involving all complexes containing mtDNA-encoded subunits. Mutations in *DARS2* (encoding mitochondrial aspartyl-tRNA synthetase) causes leukoencephalopathy with brain stem and spinal cord involvement and lactate elevation (LBSL) [66]: surprisingly, no defects of respiratory chain enzymes were found in patients with LBSL, at least in fibroblasts and lymphoblasts.

C. *Defects of mitochondrial motility, fusion, and fission*. Defects of mitochondrial dynamics are taking center stage as causes of neurodegenerative disorders [39, 67] and do belong to the mitochondrial diseases "sensu stricto" because impairment of oxidative phosphorylation has been documented in at least one form [47]. Mitochondria travel on microtubular rails, propelled by motor proteins, usually GTPases, called kinesins or dyneins. The first defect of mitochondrial motility was identified in a family with autosomal dominant hereditary spastic paraplegia and mutations in a gene (*KIF5A*) encoding one of the kinesins [68]. Interestingly, the mutation affects a region of the protein involved in microtubule binding.

Mutations in *OPA1* cause autosomal dominant optic atrophy, the Mendelian counterpart of LHON [69, 70]. Mutations in *MFN2*, encoding mitofusin 2, cause an autosomal dominant axonal variant of Charcot-Marie-Tooth (CMT) disease [71, 72]. Also, mutations in *GDAP1*, the gene encoding ganglioside-induced differentiation protein 1, which is located in the mitochondrial outer membrane and regulates the mitochondrial network, cause CMT type 4A, an autosomal recessive, severe, early-onset form of either demyelinating or axonal neuropathy [73, 74].

D. *Indirect causes of respiratory chain dysfunction*. We include here mutations in nuclear genes that affect neither respiratory chain subunits or ancillary proteins, nor mtDNA structure or copy number. For example, the function of the respiratory chain can be impaired by alterations in the lipid composition of the inner mitochondrial membrane or by defective importation of one or more subunits. The first situation is exemplified by Barth syndrome, an X-linked recessive disorder characterized by mitochondrial myopathy, cardiopathy, and leukopenia [75]. The gene responsible for this disorder (*TAZ*) [76] encodes a family of proteins ("tafazzins") involved in the synthesis of phospholipids, and biochemical analysis has shown altered amounts and composition of cardiolipin, the main phospholipid component of the inner mitochondrial membrane [77].

An example of defective protein importation is the X-linked deafness-dystonia syndrome (Mohr-Tranebjaerg syndrome), characterized by progressive sensorineural deafness, dystonia, cortical blindness, and psychiatric symptoms. This disorder is due to mutations in *TIMM8A*, which encodes the deafness-dystonia protein (DDP1) [78], a component of the mitochondrial protein import machinery located in the intermembrane space.

## 10.2 Pathogenic Mechanisms in Mitochondrial Diseases

From the standard textbook gloss that mitochondria are the "powerhouses" of the cell, most people would reasonably deduce that mitochondrial diseases are energy failures, blackouts or brownouts affecting various tissues. Although defective ATP production undoubtedly has an important pathogenic role, mitochondria perform multiple additional functions that are important for cell life and death, including the generation of reactive oxygen species (ROS), the control of calcium homeostasis, and the regulation of programmed cell death (apoptosis). It is likely that the pathogenesis of any mitochondrial disease will involve, at least to some extent, all these functions. An elegant verification of this concept was offered by studying the consequences of mutations in *OPA1*, which – as we have seen above – encodes a dynamin-related GTPase important for mitochondrial fusion. Besides exhibiting the predictable alteration of the mitochondrial network, mutant skin fibroblasts grown in galactose medium and forced to use oxidative metabolism also showed impaired ATP synthesis at the level of complex I and activation of apoptosis [47]. Co-immunoprecipitation experiments documented a direct interaction of OPA1 with subunits of the respiratory chain and with the apoptosis-inducing factor (AIF). Direct evidence that alteration of mitochondrial dynamics can cause loss of mtDNA, impaired RC function, and increased ROS production was provided by experiments in which HeLa cells were depleted of DrP1, another dynamin-related

GTPase like OPA1 [79]. Similar multiple interactions of pathogenic mechanisms have been shown in neurodegenerative disorders due to mutant mitochondrial proteins, such as Friedreich ataxia (FA), Parkinson disease (PD), Alzheimer disease (AD), amyotrophic lateral sclerosis (ALS), and hereditary spastic paraplegia (HSP) [39].

If we now focus on the "primary" mitochondrial diseases outlined above, i.e. defects of the RC, two pathogenic mechanisms are most commonly considered, impaired ATP synthesis and excessive ROS production. Due to the lack of animal models for mtDNA-related disorders, the biochemical and functional consequences of mtDNA mutations have been gleaned from studies of cybrid cell lines, that is, immortalized human cells that have been "emptied" of their mtDNA and repopulated with mutant mtDNA from patients [80]. In particular, cybrid cell lines harboring varying mutant loads have been used to identify the threshold of pathogenicity for various mtDNA mutations. For all major mutations, the thresholds assessed *in vitro* appeared to be both high and steep: for the A3243G MELAS mutation, as one example, the threshold was around 90% [81]. However, these data cannot be extrapolated to the *in vivo* situation, as shown by oligosymptomatic carriers of the A3243G mutation, who had abnormal $^{31}$P-MRS studies of muscle [82] and abnormal lactate peaks in both cerebrospinal fluid and brain parenchyma by $^{1}$H-MRS [83].

Let us consider separately each of the several factors that may contribute to the clinical expression of mutations (of mtDNA or nDNA) affecting the respiratory chain.

(i) *Heteroplasmy and mutation load*. Obviously, this applies only to mtDNA-related diseases. It is intuitive that the severity of clinical manifestations should show some correlation to the mutation load, a concept that has been adopted as one of the canonical criteria of pathogenicity for novel mtDNA mutations. This is best exemplified by the T8993G mutation, as noted by Tatuch et al. back in 1992 [84]: when the mutation load is high (about 90%), the disease manifests in infants or children as LS (maternally inherited LS, MILS); when the mutation load is lesser (about 70%), it causes a later-onset and milder encephalomyopathy (NARP). Accordingly, ATP synthesis was inversely correlated to the mutation load in platelet-derived submitochondrial particles [85] and in lymphocytes [86] harboring the T8993G mutation.

Different loads of the same mutation in different tissues may also explain, to a certain extent, differential tissue involvement. This may be especially true for a heterogeneous and highly specialized tissue like the brain. In an attempt to explain the distinctive neurological symptoms of patients with KSS, MERRF, and MELAS, the mutation typical of each disease has been "mapped" indirectly through immunohistochemical techniques. Consistent with clinical symptomatology and laboratory data, immunocytochemical evidence suggested that the single mtDNA deletions of KSS abound in the choroids plexus [87], the A3243G-MELAS mutation is abundant in the walls of cerebral arterioles [88], and the 8344G-MERRF mutation is abundant in the olivary nuclei of the cerebellum [89]. However, these data fail to explain what directs each mutation to a particular area of the brain.

Even if the high pathogenic threshold established by studies of cybrid cell lines does not correlate strictly with the *in vivo* threshold, it is nonetheless true that most mtDNA mutations have high and steep thresholds. The recent finding that the prevalence of pathogenic mtDNA mutations (at relatively low mutation loads) in the normal population is as high as 1:200 [15] bolsters this concept.

There is, thus far, one notable exception, a *de novo* mutation in $tRNA^{Trp}$ (C5545T), which caused a childhood mitochondrial encephalomyopathy at unusually low heteroplasmic levels (<25%) in affected tissues and whose pathogenic threshold in cybrids was as low as 4–8% [90]. This new concept of dominance in mitochondrial genetics, together with the unexpectedly high prevalence of mtDNA mutations in the general population, pose new diagnostic challenges, because low-level mutations may escape detection.

(ii) *Type of mutation.* Obviously, some mutations in the same gene can be more severely pathogenic than others and this applies to both mtDNA and nDNA changes. The best example of this concept is offered by two pairs of mutations in the ATPase 6 gene of mtDNA, which are not only in the same gene but also each at the very same site, T8993G/T8993C and T9176G/T9176C and each resulting in the same amino acid change, either Leu>Arg or Leu>Pro [91]. While T8993G and T7917G cause NARP/MILS at different mutant loads [16, 92, 93], T8993C and T7917C cause a generally (but not invariably) milder clinical syndrome [94]. Accordingly, ATP synthesis is much more severely impaired by the T8993G than by the T8993C mutation in primary fibroblasts [95], lymphocytes [96, 97], and cybrid cell lines [98]. Why should the Leu>Arg mutation interfere with the rotary engine's function of ATP synthase more severely than the Arg>Pro mutation is not completely clear, but assembly seems to be normal with both mutations [99, 100, 101] and it was postulated that the problem was either reduced proton travel through the A6 channel or reduced efficiency of the rotary coupling between subunit A6 (part of the stator) and subunit c (part of the rotor), or both [91]. Recent work suggests that the coupling of proton flux and ATP synthesis is slowed by the T-to-C mutations and virtually abolished by the T-to-G mutations [96].

The $tRNA^{Leu(UUR)}$ is a very hot spot, with over 20 mutations listed as pathogenic. It has been suggested that the site of the mutation in the cloverleaf structure of the gene may influence the phenotype, in that mutations causing the typical MELAS syndrome lack the normal taurine-containing modification (5-taurinomethyluridine) at the anticodon wobble position whereas mutations associated with other syndromes have normal 5-taurinomethyluridine modifications [102].

(iii) *Why tissue specificity?* It has been observed that some tRNA genes of mtDNA tend to affect certain tissues more than others: for example, mutations in $tRNA^{Ile}$ are often associated with cardiomyopathy and mutations in $tRNA^{Lys}$

are selectively, though not invariably, associated with multiple lipomas. More in general, it has been proposed that mutations in certain sites of the tRNA cloverleaf structure may play a role in determining the tissue specificity of the phenotype [103]. Interesting as they are, as of now these are mere associations in search of explanations.

In mtDNA-related disorders, real or apparent tissue-specificity can occur in two situations, best exemplified by mitochondrial myopathies. There are numerous examples of mutations in tRNA or protein-coding genes arising *de novo* and affecting selectively the progenitor cells of skeletal muscle [19]: these often create diagnostic conundrums because they contradict at least two "rules" of mitochondrial genetics, maternal inheritance and multisystem distribution. An example of what could be called "pseudo tissue specificity" is extreme skewed heteroplasmy of a generalized mutation: as the pathogenic threshold is surpassed only in skeletal muscle, the resulting phenotype will be a pure myopathy [104].

A more puzzling example of tissue specificity is offered by homoplasmic mtDNA mutations, such as most LHON-associated mutations in ND genes [18]: obviously, there is a special vulnerability of the retinal ganglion cells to the consequences of these mutations (be it ATP deprivation, ROS excess, or both – see below), probably exacerbated by the extraordinary dependence of these cells on oxidative metabolism, but – again – we are making associations rather than offering explanations. Similarly, a homoplasmic mutation in $tRNA^{Ile}$ [105] causes a selective cardiomyopathy whereas a homoplasmic mutation in $tRNA^{Glu}$ causes a selective myopathy [106], which can only be explained by the coexistence of modifier nuclear genes. Clearly, the cross-talk between the two genomes goes beyond the classical defects of intergenomic signaling and remains to be fully elucidated. It also goes beyond tissue specificity to affect, for example, gender vulnerability: the predominance of affected males in LHON has been attributed to two loci in the X-chromosome [107, 108].

Among the nDNA-related diseases, the most reasonable explanation for tissue specificity would be mutations in tissue-specific isoforms of RC subunits or assembly factors. While no clinical entity is referable to such mechanism as yet, selective tissue vulnerability is seen in many Mendelian mitochondrial diseases. This is best exemplified by the several mutations in COX-assembly proteins: while all affect the brain, usually resulting in LS-like symmetrical lesions, they affect other tissues differently. Thus, mutations in *SCO2* and *COX15* cause cardiomyopathy [109] whereas mutations in *SCO1* cause liver disease [110]. This is probably due to different levels of residual COX activity, as illustrated by COX histochemistry in skeletal muscle from children with *SCO2* or *SURF1* mutations (Fig. 10.5). Yet another example is offered by a Mendelian defect of intergenomic communication. Although the expression of mutations in the mitochondrial translation factor EFG1 is ubiquitous, the clinical presentation is a devastating hepatopathy. Accordingly, liver had the lowest residual assembly of various RC complexes whereas cardiac tissue had the highest, with other tissues falling in between [61]. The varying residual activities

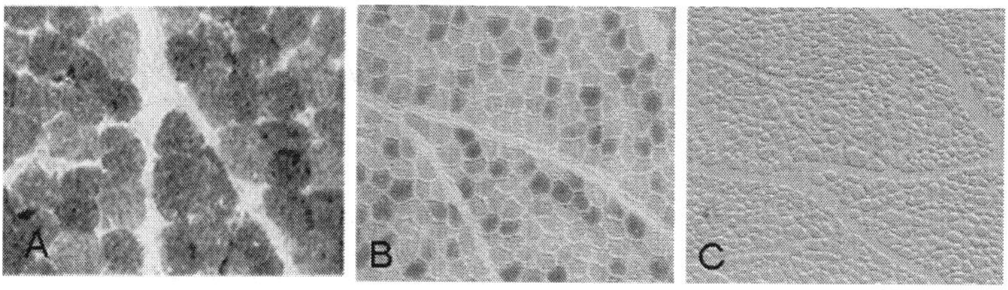


**Fig. 10.5** Cross-sections of human muscle stained for COX. A, normal; B. patient with LS due to mutations in *SURF1*; C. patient with encephalocardiomyopathy due to mutations in *SCO2*

of RC complexes in different tissues – an the severity of their clinical involvement – was apparently dictated by the ratio of EFTu:EFTs translation elongation factors [61].

IV. *Pathogenic importance of mtDNA haplotypes*. In their migration out of Africa, human beings have accumulated distinctive variations from the mtDNA of our ancestral "mitochondrial Eve", resulting in several haplotypes characteristic of different ethnic groups [111]. It has been suggested that different mtDNA haplogroups or subhaplogroups may modulate oxidative phosphorylation, thus influencing the overall physiology of individuals and predisposing them to – or protecting them from – certain diseases [111].

The best example of the pathogenic importance of the mitochondrial genetic background comes from studies of LHON, a monogenic mtDNA disorder notorious for its incomplete penetrance. Careful mtDNA haplogroup subtyping of 159 LHON families has established that the risk of visual loss is greater when patients harboring the A11778G mutation belong to haplogroup J2, patients harboring the T14484C belong to haplogroup J1, or patients with the G3460G mutation belong to haplogroup K [112]. Substitutions in the *cyt b* gene characteristic of these haplogroups may explain the increased penetrance through their influence on the supercomplex formation between complexes I and complex III.

V. *Defective ATP synthesis and excessive ROS production*. In mtDNA-related disorders, impaired ATP synthesis has been documented *in vivo* by $^{32}P$ NMR spectroscopy of skeletal muscle or brain in patients with LHON [113, 114], encephalomyopathy [115, 116] or myopathy [117].

Defective ATP synthesis has also been documented in easily accessible tissues, such as primary fibroblasts [95, 118], lymphocytes [86, 96, 97], platelets [85, 119, 120], or cybrid cell lines [98, 119]. The relative role of impaired ATP synthesis and excessive ROS production has been well documented in cybrid cell lines harboring the three mtDNA mutations associated with LHON. In a series of elegant studies, Carelli and collaborators have clearly shown that complex I is defective and ATP production is impaired, although these effects may require exposure to a medium containing galactose instead of glucose, thus "forcing" the cells to use oxidative metabolism [119, 121, 122]. They have shown that LHON mutations affect the ubiquinone (CoQ) binding site of complex I, leading to overproduction of unstable ubisequinones and attending ROS damage to DNA, proteins, and lipids [123]. The crucial role of the balance between ATP shortage and ROS excess was also documented by the same group in a study of lymphocytes from patients with NARP/MILS: interestingly, they showed that ATP shortage prevailed in the "severe" T8993G mutation whereas the "mild" T8993C mutation favored ROS overproduction [97]. Excessive ROS production was also the only biochemical alteration encountered in cybrid cell lines harboring a mutation (T9957C) in the mtDNA gene encoding COX III [119].

The relative contribution of ATP shortage and ROS excess to pathogenesis and clinical expression is also evident in Mendelian mitochondrial diseases. This is borne out by studies of complex I deficiency and primary $CoQ_{10}$ deficiency. In agreement with the concept that complex I and complex III are the major producers of ROS [124], complex I deficiency causes severe oxidative stress. This was discussed above for the ND mutations of LHON, but is even more evident in children with mutations in nDNA-coded subunits or assembly factors of complex I [125]. Not too surprisingly, there was an inverse relationship between amount and residual activity of complex I and ROS overproduction, although, contrary to a similar report published before the molecular era [126], there was no apparent relationship between severity of ROS production and clinical phenotype. In contrast, a correlation between biochemical data and severity of clinical presentation was observed in studies of fibroblasts from two patients with different primary CoQ10 deficiency, one due to mutations in *COQ2* and causing a severe but treatable infantile encephalomyopathy [28, 36], the other due to mutations in *PDSS2* and causing fatal infantile LS [34]. *PDSS2* mutant cells showed severely reduced ATP synthesis but no ROS overproduction or compensatory increase of antioxidant defense markers, whereas *COQ2* mutant cells showed only a partial defect of ATP synthesis but marked increased of ROS with attending oxidation of lipids and proteins.

Although it may be too soon for generalizations, which are risky anyway, data from both mtDNA and nDNA mutants suggest that severe blocks of the RC result in energy blackouts and mild oxidative stress, while partial blocks result in energy brownouts but severe oxidative stress.

VI. *Altered $Ca^{2+}$ handling*. Mitochondria have a central role in intracellular $Ca^{2+}$ homeostasis and are, in turn, very sensitive to $Ca^{2+}$ signals, which may trigger necrosis or apoptosis [127]. Different alterations of $Ca^{2+}$ signaling have been reported in different mtDNA-related disorders. In cybrid cells harboring the T8356C MERRF mutation, the transient mitochondrial $Ca^{2+}$ uptake and ATP production in response to physiological agonists were decreased, whereas both cytoplasmic and mitochondrial $Ca^{2+}$ handling were normal in cells harboring the T8993G NARP/MILS mutation [128]. In contrast, fibroblasts harboring the A3243G MELAS mutation were unable to effectively buffer intracellular $Ca^{2+}$, which resulted in a cytotoxic $Ca^{2+}$ effect [129]. Although the alterations of

$Ca^{2+}$ homeostasis appear to differ in different disorders and remain to be better defined, they certainly play crucial pathogenic roles, considering the importance of $Ca^{2+}$ in the control of ATP synthesis, mitochondrial dynamics, and apoptosis [130, 131].

## 10.3 Therapeutic Strategies

There are three strategies to be considered in treating mitochondrial diseases or, for that matter, any other group of diseases. The first strategy aims at modifying the consequences, i.e. the symptoms, of these disorders: this is palliative or "band-aid" (loosely translated from the Latin "pallium", cloak) therapy. The second strategy is both more ambitious and more difficult because it aims at attacking the causes of these disorders and is therefore based either on gene therapy or on enzyme replacement therapy. The third therapeutic approach falls between the other two and it aims at interrupting or modifying the pathogenic mechanisms, thus interrupting or, more likely, slowing the course of diseases.

I. *Symptomatic therapy*. The fact that mtDNA-related disorders can affect every tissue in the body requires the application of symptomatic therapy already used in different subspecialties of medicine[6, 132].

*Neurology*. Seizures are treated with conventional anticonvulsants, except that valproic acid should be avoided in children with Alpers-Huttenlocher syndrome and *POLG* mutations because it often triggers acute hepatic failure. Corticosteroids are helpful in the acute phase of strokes due to MELAS mutations and physical, occupational, respiratory, and speech therapy accelerates recovery from strokes and aids patients with other CNS problems [133]. Psychotropic drugs may be needed in patients with predominant or exclusive psychiatric symptoms [134]. Exercise intolerance is a most common complaint, which may lead to inactivity, deconditioning, and further deterioration of muscle function. Aerobic exercise under proper supervision has proven helpful in staving off this downhill course [135]. Patients with recurrent rhabdomyolysis and myoglobinuria should be hydrated and undergo renal dialysis in cases of renal failure. If the recurrent myoglobinuria is associated with muscle $CoQ_{10}$ deficiency, most patients will benefit from oral $CoQ_{10}$ supplementation.

*Ophthalmology*. Ptosis, often with PEO, is very common and debilitating both functionally and psychologically. Frontalis suspension is the preferred form of surgery because it protects from corneal exposure [136]. LHON is resistant to therapy, but a few patients appear to have responded to treatment with idebenone [18].

*ENT*. Neurosensory hearing loss is a common consequence of mtDNA-related disorders and, when severe, can be effectively treated in most patients with cochlear implants [137]. Chronic respiratory failure can initially be treated with noninvasive continuous positive air pressure (CPAP) or bilevel positive air pressure (Bi-PAP) but may eventually require tracheotomy [136].

*Endocrinology.* Diabetes mellitus should be treated by conventional means, including diet, sulfonylureas, and insulin, but metformin ought to be avoided because it has been associated with lactic acidosis. Patients with hypogonadism, hypothyroidism, or hypoparathyroidism should receive specific hormone replacement. Treatment of short stature with growth hormone is controversial, unless there is clear evidence of growth hormone deficiency.

*Cardiology.* In KSS, conduction blocks dominate the cardiological picture and timely placement of a pacemaker can be lifesaving. In mitochondrial disorders dominated by cardiomyopathy, cardiac transplantation is an option that should not be discarded *a priori* [138].

*Gastroenterology.* Feeding difficulties in infants and children can be alleviated by drugs or by surgical intervention, including percutaneous endoscopic gastrostomy (PEG) or fundoplication. Severe gastrointestinal dysmotility may necessitate parenteral nutrition in patients with MNGIE. Children with predominant or isolated liver failure, usually due to a mtDNA depletion syndrome, may benefit from liver transplantation, although this is still controversial [139].

*Hematology.* Infants with Pearson syndrome – dominated by sideroblastic anemia and due to single mtDNA deletions – may respond to repeated blood transfusion, although survivors often develop KSS later in life.

*Nephrology.* Renal tubular acidosis and Fanconi syndrome require readjustment of the electrolytic balance [140]. Primary $CoQ_{10}$ deficiency is often associated with – or dominated by – nephropathy [31] and early $CoQ_{10}$ supplementation may improve glomerular function and prevent neurological complications.

II. *Radical therapy.* The term implies tackling the root of the disease, i.e. its etiology. Ultimately, this implies replacing the mutant DNA with wild-type DNA. Gene therapy for mitochondrial diseases due to mutations in nDNA faces the same hurdles of gene therapy for other Mendelian disorders, including choice of optimal viral or non-viral vectors, effective delivery to the affected tissues, and avoidance of immunological rejection. For mtDNA-related diseases the formidable – and still unsolved – problem is that no investigator has been able to transfect DNA into mitochondria in a heritable manner.

A radical way of, if not correcting, at least avoiding the cause of a disease includes genetic counseling and prenatal diagnosis. As detailed in the first section of this review, the rapid progress in our knowledge of the molecular defects underlying Mendelian mitochondrial disorders has offered families, especially young families with fatal infantile conditions, the option of prenatal diagnosis. Unfortunately, prenatal diagnosis of most mtDNA-related diseases is impeded by two factors: (i) the mutation load in amniocytes of chorionic villi does not necessarily reflect that of other fetal tissues; and (ii) mutation loads measured in prenatal samples may shift due to mitotic segregation. Fortunately, there is good evidence that mutations in ATPase 6 associated with NARP/MILS do not undergo tissue- or age-related variations [141], a situation confirmed by the first pre-implantation diagnosis for a human mitochondrial disease [142].

Another promising way of preventing mtDNA-related disorders is currently ostracized by ethical concerns about the manipulation of germline cells. A woman carrying an mtDNA mutation, such as the common and potentially devastating A3243G MELAS change, could have her fertilized oöcytes cleansed in vitro of the cytoplasm and most mitochondria. The naked pronucleus would then be transferred to a normal enucleated host oöcyte and implanted in the woman's uterus: the result of this procedure would be a mitochondrially normal child carrying the nuclear traits of both parents. This technology has been successful in transmitochondrial "mitomice" [143] and has been approved for experimentation in the UK.

Another way of going to the root of the problem is stem cell therapy, which, for Mendelian disorders, offers real promise. For example, allogeneic stem cell transplantation (alloSCT) in a woman with MNGIE has improved her clinical condition and her nerve cell conduction velocities [144]. Biochemically, TP activity reached mutation carrier levels in her blood, where the concentrations of toxic compounds, thymidine and deoxyuridine, returned to normal.

Yet another way of circumventing the genetic defect in Mendelian RC disorders tries to imitate mother nature by promoting mitochondrial biogenesis (and the residual activity of a defective enzyme). Mitochondrial biogenesis is regulated by PPAR$\gamma$ coactivator $\alpha$ (PGC-1$\alpha$), which, in turn, is activated by bezafibrate, a drug already used in medicine. Interestingly, bezafibrate increased the activities of RC complexes both in normal cultured cells and in cells with RC enzyme defects [145]. Even more interestingly, in a mouse with COX deficiency myopathy due to an engineered mutation in the assembly gene *COX10*, treatment with bezafibrate increased residual COX activity and ATP production in muscle and delayed both the onset of myopathy and the time of death [146]. Because bezafibrate is already part of our pharmacological armamentarium, it could be tested relatively rapidly in patients with mitochondrial diseases.

Gene therapy for mtDNA-related diseases poses special problems because of polyplasmy and heteroplasmy and because we are still unable to transfect DNA into mitochondria. Of the many indirect strategies proposed, probably the most viable is heteroplasmic shifting, aimed at lowering the mutant mtDNA below the pathogenic threshold. Many different approaches have been tried, although their applicability to humans appears remote. These include: (i) inhibiting the replication of mutant mtDNA by selective hybridization with nucleic acid derivatives (such as peptide nucleic acids, PNAs) [147, 148]; (ii) importation of yeast tRNA to replace mutated human tRNA [149]; (iii) importation of wild-type polypeptides (either allotopically or xenotopically expressed) into mitochondria to replace mutated ones [150, 151, 152, 153] or to complement faulty function [154, 155]; (iv) importing specific restriction endonucleases or custom-designed zinc-finger nucleases that cut mutated but not wild-type mtDNA and act as "silver bullets" [156, 157, 158].

III. *Acting on pathogenesis*. The most logical intervention in any inborn error of metabolism appears to be *removing noxious compounds*. In most mitochondrial encephalomyopathies, the obvious culprit is lactic acid [83]. Bicarbonate therapy is common and almost "automatic", but should be used prudently [159].

> Dichloroacetate (DCA) inhibits pyruvate dehydrogenase (PDH) kinase, keeps PDH in the dephosphorylated, active form, thus favoring pyruvate metabolism and lactate oxidation [160]. Although oral DCA was well tolerated in randomized studies of children with congenital lactic acidosis and heterogeneous mitochondrial diseases (including PDH deficiency and mtDNA- or nDNA-related defects of the RC), it did not improve neurological outcome and was associated with evidence of peripheral neuropathy [161, 162]. On the basis of tissue culture studies, the peripheral neuropathy was attributed to a reversible inhibition of myelin-related proteins by DCA [163]. Despite these non-alarming data in children, a double-blind, placebo-controlled, randomized, cross-over trial of oral DCA in a large and homogeneous cohort of patients carrying the A3243G MELAS mutation had to be terminated because of peripheral nerve toxicity [164], which at the very least exacerbates the lurking peripheral neuropathy due to the mtDNA mutation [165].

In MNGIE, the toxic metabolites that accumulate in blood as a direct consequence of the TP defect are thymidine and deoxyuridine [166]: hemodialysis was only transiently effective in lowering blood levels, whereas the effect of allo-SCT was more substantial and permanent, as discussed above [144].

*Administration of electron acceptors* is most effective in disorders due to primary defects of such acceptors, best exemplified by primary $CoQ_{10}$ deficiencies. However, at least one child with LS and $CoQ_{10}$ deficiency due to mutations in the *PDSS2* gene did not respond to $CoQ_{10}$ administration [34], possibly because therapy was started too late, because the dose was inadequate, or because the energy defect was too drastic [167]. In patients with secondary $CoQ_{10}$ deficiency, the response to supplementation is generally good but unpredictable and often variable. For example, patients with the myopathic form of glutaric aciduria type II (GA II) due to electron transfer flavoprotein dehydrogenase (ETFDH) deficiency may need both $CoQ_{10}$ and riboflavin for optimal response [33].

*"Cocktails" of vitamins and cofactors* are the most widely used therapy in clinical practice. Their popularity is based on two principles: (i) their safety makes the doctor feel "Hippocratically correct" ("primum non nocere") even when he/she doubts their efficacy; and (ii) their use is predicated by the two most widely accepted pathogenic mechanisms, energy shortage and oxidative stress.

In the hope of facilitating ATP production by a sluggish RC, electron flux is "boosted" by addition of electron acceptors ($CoQ_{10}$, vitamin C, vitamin K, succinate). Alternatively, we try to boost the synthesis of a naturally occurring high-phosphate compound, phosphocreatine (PC) by administering creatine. L-carnitine is prescribed because plasma free carnitine tends to be lower and esterified carnitine higher than normal, probably reflecting a partial impairment of β-oxidation. Some encouragement for the use of vitamins and cofactors was provided by a study of ATP synthesis in lymphocytes from 12 patients with diverse RC disorders before and after 12 months of therapy with a "cocktail" that included $CoQ_{10}$, L-carnitine, vitamin B complex, vitamin C, and vitamin $K_1$ [168]. Although ATP synthesis increased significantly with treatment, none of the patients improved clinically, and *in vitro* exposure of control lymphocytes to the various components showed that only $CoQ_{10}$ increased ATP synthesis in a dose-dependent manner.

Folic acid deserves special mention because early observations had shown that it was decreased in the CSF of patients with KSS [169, 170] and a recent study documented both clinical and neuroradiological improvement in a child after 1 year of monotherapy with 2.2 mg folinic acid/kg daily [171], suggesting that early and aggressive administration of this compound should be tried in this devastating condition.

To *improve ATP synthesis*, creatine monohydrate supplementation has been used, but the only two randomized studies came to different conclusions: a smaller cohort of severely affected patients improved [172], whereas a larger cohort of less severely affected patients did not [173], possibly due to the difference in muscle phosphocreatine concentration between the two groups [174, 175].

*Scavenging excessive ROS* is probably the most common approach not just to mitochondrial diseases due to RC defects but also to late-onset neurodegenerative disorders (including ALS, PD, and AD), in which there is direct or indirect evidence of oxidative stress [39]. Antioxidants used in clinical practice include vitamin E, $CoQ_{10}$, idebenone, glutathione, and dihydrolipoate.

As mentioned above, $CoQ_{10}$ is useful in primary $CoQ_{10}$ deficiencies, but is also widely prescribed to patients with RC disorders. Although anecdotal reports (too numerous to be cited here) are generally positive, we still lack a rigorous double blind trial. However, clinical experience teaches that $CoQ_{10}$ needs to be administered at high doses (no less than 300 mg daily in adults), which fortunately have shown to be well tolerated in numerous studies of large cohorts.

Importantly, initial studies of idebenone had suggested a beneficial effect only on the cardiopathic component of FA, but a recent standardized study showed a dose-related beneficial effect also on the neurological component of the disease [176].

The *management of strokes* in MELAS is difficult because we still have an incomplete understanding of pathogenesis. Increasing evidence, however, suggests an underlying mitochondrial angiopathy with altered vascular contractility. As citrulline [177] and arginine [178], both precursors of nitric oxide (NO), were decreased in MELAS patients, and NO homeostasis can affect vascular function, L-arginine administration has been studied rather extensively by Koga and collaborators: they found that intravenous administration of L-arginine (0.5 g/kg) during the acute phase improved all stroke-like symptoms, whereas interictal oral administration (0.15–0.30 g/kg) diminished both frequency and severity of strokes [178, 179, 180, 181]. Although these results remain to be confirmed by rigorously controlled studies, they do offer some hope for counteracting at least the most devastating manifestation of this disease.

The past 50 years have seen mitochondrial medicine develop beyond anyone's wildest expectation and the plot is still thickening. However, our understanding of pathogenesis leaves much to be desired and progress in this area is indispensable for the development of rational therapeutic strategies. The fervor of research in laboratories around the world and the rise of national and international collaborative groups bodes well for the future.

**Acknowledgements** Supported in part by NIH grant HD32062 and by the Marriott Mitochondrial Disorder Clinical Research Fund (MMDCRF).

## List of abbreviations

| | |
|---|---|
| AD: | Alzheimer disease |
| Allo-SCT: | allogeneic stem cell transplantation |
| ALS: | amyotrophic lateral sclerosis |
| ANT1: | adenine nucleotide transporter 1 |
| AOA1: | Ataxia oculomotor apraxia type 1 |
| APTX: | aprataxin |
| AIF: | apoptosis inducing factor |
| AZT: | azidothymidine |
| BCS1L: | cytochrome *b-c* complex assembly protein (complex III) |
| Bi-PAP: | bilevel positive air pressure |
| CACT: | carnitine-acylcarnitine translocase |
| CMT: | Charcot-Marie-Tooth |
| CoQ: | coenzyme Q (ubiquinone) |
| COX: | cytochrome *c* oxidase |
| CPAP: | continuous positive air pressure |
| CPT: | carnitine palmitoyltransferase |
| CSF: | cerebrospinal fluid |
| *DARS2*: | gene encoding mitochondrial aspartyl-tRNA synthetase |
| DCA: | dichloroacetate |
| DIC: | dicarboxylate carrier |
| DDP1: | deafness dystonia protein 1 |
| DGUOK: | deoxyguanosine kinase |
| DOA: | dominant optic atrophy |
| *EFG1*: | gene encoding elongation factor 1 |
| ETF: | electron transfer flavoprotein |
| EFTDH: | electron transfer flavoprotein dehydrogenase |
| EFTu: | elongation factor Tu |
| ENT: | ear-nose-throat |
| FA: | Friedreich ataxia |
| FBSN: | familial bilateral striatal necrosis |
| GA: | glutaric aciduria |
| GDAP1: | ganglioside-induced differentiation protein 1 |
| HSP: | hereditary spastic paraplegia |
| KSS: | Kearns-Sayre syndrome |
| LBSL: | leukoencephalopathy. brain stem, spinal cord involvement and lactate elevation |
| LHON: | Leber hereditary optic neuropathy |
| LPRRC: | leucine-rich pentatricopeptide repeat-containing protein |
| LS: | Leigh syndrome |

| | |
|---|---|
| MELAS: | mitochondrial encephalopathy, lactic acidosis, and stroke-like episodes |
| MERRF: | myoclonus epilepsy and ragged-red fibers |
| MFN: | mitofusin |
| MILS: | maternally inherited Leigh syndrome |
| MND: | motor neuron disease |
| MNGIE: | mitochondrial neurogastrointestinal encephalomyopathy |
| MPV17: | MPV17 mitochondrial inner membrane protein (SYM1) |
| MRI: | magnetic resonance imaging |
| MRS: | magnetic resonance spectroscopy |
| mtDNA: | mitochondrial DNA |
| NARP: | neuropathy, ataxia, retinitis pigmentosa |
| ND: | NADH-coenzyme Q oxidoreductase |
| nDNA: | nuclear DNA |
| NNH: | Navajo neurohepatopathy |
| NO: | nitric oxide |
| OPA1: | dynamin-related GTPase mutated in autosomal dominant optic atrophy |
| PD: | Parkinson disease |
| PDHC: | pyruvate dehydrogenase complex |
| PDSS2: | decaprenyl diphosphate synthase subunit 2 |
| PEO: | progressive external ophthalmoplegia |
| PNAS: | peptide nucleic acids |
| POLG: | polymerase $\gamma$ |
| PUS1: | pseudouridine synthase 1 |
| RC: | respiratory chain |
| RRF: | ragged-red fibers |
| ROS: | reactive oxygen species |
| SCO: | synthesis of cytochrome *c* oxidase |
| SCS-A: | succinyl-CoA synthetase |
| SDH: | succinate dehydrogenase |
| *SUCLA2*: | gene encoding the $\beta$ subunit of succinyl-CoA synthetase |
| SUCLG1: | gene encoding the $\alpha$ subunit of succinyl-CoA synthetase |
| SURF1: | surfeit gene 1 |
| TAZ: | tafazzin |
| TCA: | tricarboxylic acid cycle (Krebs cycle) |
| TIM: | translocase of the inner membrane |
| TK2: | thymidine kinase 2 |
| TP: | thymidine phosphorylase |
| *TYMP*: | the gene encoding thymidine phosphoylase |

# References


1. Luft R, Ikkos D, Palmieri G, et al. A case of severe hypermetabolism of nonthyroid origin with a defect in the maintenance of mitochondrial respiratory control: A correlated clinical, biochemical, and morphological study. J Clin Invest. 1962;41:1776–1804.
2. Wallace DC, Singh G, Lott MT, et al. Mitochondrial DNA mutation associated with Leber's hereditary optic neuropathy. Science 1988;242:1427–1430.
3. Holt IJ, Harding AE, Morgan Hughes JA. Deletions of muscle mitochondrial DNA in patients with mitochondrial myopathies. Nature 1988;331:717–719.
4. Luft R. The development of mitochondrial medicine. Proc. Natl. Acad. Sci. USA 1994;91:8731–8738.
5. Gvozdjáková A, ed. Mitochondrial Medicine: Springer, 2008.
6. DiMauro S, Hirano M, Schon EA, eds. Mitochondrial Medicine. Abingdon, UK: Informa Healthcare, 2006.
7. Shy GM, Gonatas NK. Two childhood myopathies with abnormal mitochondria: I. Megaconial myopathy; II. Pleoconial myopathy. Brain 1966;89:133–158.
8. Shy GM, Gonatas NK. Human myopathy with giant abnormal mitochondria. Science 1964;145:493–496.
9. Engel WK, Cunningham CG. Rapid examination of muscle tissue: An improved trichrome stain method for fresh-frozen biopsy sections. Neurology 1963;13:919–923.
10. Shapira Y, Harel S, Russell A. Mitochondrial encephalomyopathies: a group of neuromuscular disorders with defects in oxidative metabolism. Isr. J. Med. Sci. 1977;13:161–164.
11. Mitomap. MITOMAP: A human mitochondrial genome database. http://www.mitomap.org. 2003.
12. Sproule D, KaufmannP. Mitochondrial encephalomyopathy, lactic acidosis and strokelike episodes (MELAS): A review of basic concepts, the clinical phenotype, and therapeutic management. Ann. NY Acad. Sci. 2008;1142:133–158.
13. Shanske S, Coku J, Lu J, et al. The G13513A mutation in the ND5 gene of mitochondrial DNA as a common cause of MELAS or Leigh syndrome. Arch. Neurol. 2008;65: 368–372.
14. Manwaring N, Jones MM, Wang JJ, et al. Population prevalence of the MELAS A3243G mutation. Mitochondrion 2007;7:230–233.
15. Elliott HR, Samuels DC, Eden JA, et al. Pathogenic mitochondrial DNA mutations are common in the general population. Am. J. Hum. Genet. 2008;83:254–260.
16. Santorelli FM, Shanske S, Macaya A, et al. The mutation at nt 8993 of mitochondrial DNA is a common cause of Leigh syndrome. Ann Neurol. 1993;34:827–834.
17. Holt IJ, Harding AE, Petty RK, Morgan Hughes JA. A new mitochondrial disease associated with mitochondrial DNA heteroplasmy. Am J Hum Genet. 1990;46:428–433.
18. Carelli V, Barboni P, Sadun AA. Mitochondrial ophthalmology. In: DiMauro S, Hirano M, Schon EA, eds. Mitochondrial Medicine. London: Informa Healthcare, 2006:105–142.
19. Andreu AL, Hanna MG, Reichmann H, et al. Exercise intolerance due to mutations in the cytochrome b gene of mitochondrial DNA. New Engl. J. Med. 1999;341:1037–1044.
20. Massa V, Fernandez-Vizarra E, Alshahwan S, et al. Severe infantile encephalomyopathy caused by a mutation in COX6B1, a nucleus-encoded subunit of cytochrome c oxidase. Am. J. Hum. Genet. 2008;doi:10.1016/j.ajhg.2008.05.002.
21. Di Giovanni S, Mirabella M, Spinazzola A, et al. Coenzyme Q10 reverses pathological phenotype and reduces apoptosis in familial CoQ10 deficiency. Neurology 2001;57:515–518.
22. Sobreira C, Hirano M, Shanske S, et al. Mitochondrial encephalomyopathy with coenzyme Q10 deficiency. Neurology 1997;48:1238–1243.
23. Ogasahara S, Engel AG, Frens D, Mack D. Muscle coenzyme Q deficiency in familial mitochondrial encephalomyopathy. Proc Nat Acad Sci USA 1989;86:2379–2382.
24. Gironi M, Lamperti C, Nemni R, et al. Late-onset cerebellar ataxia with hypogonadism and muscle coenzyme Q10 deficiency. Neurology 2004;62:818–820.

25. Lamperti C, Naini A, Hirano M, et al. Cerebellar ataxia and coenzyme Q10 deficiency. Neurology 2003;60:1206–1208.
26. Musumeci O, Naini A, Slonim AE, et al. Familial cerebellar ataxia with muscle coenzyme Q10 deficiency. Neurology 2001;56:849–855.
27. Lalani S, Vladutiu GD, Plunkett K, et al. Isolated mitochondrial myopathy associated with muscle coenzyme Q10 deficiency. Arch Neurol 2005;62:317–320.
28. Salviati L, Sacconi S, Murer L, et al. Infantile encephalomyopathy and nephropathy with CoQ10 deficiency: a CoQ10-responsive condition. Neurology 2005;65:606–608.
29. Rotig A, Appelkvist E-L, Geromel V, et al. Quinone-responsive multiple respiratory-chain dysfunction due to widespread coenzyme Q10 deficiency. Lancet 2000;356:391–395.
30. Van Maldergem L, Trijbels F, DiMauro S, et al. Coenzyme Q-responsive Leigh's encephalopathy in two sisters. Ann Neurol 2002;52:750–754.
31. Diomedi-Camassei F, Di Giandomenico S, Santorelli F, et al. COQ2 nephropathy: A newly described inherited mitochondriopathy with primary renal involvement. J Am Soc Nephrol 2007;18:2773–2780.
32. Quinzii C, Kattah AG, Naini A, et al. Coenzyme Q deficiency and cerebellar ataxia associated with an aprataxin mutation. Neurology 2005;64:539–541.
33. Gempel K, Topaloglu H, Talim B, et al. The myopathic form of coenzyme Q10 deficiency is caused by mutations in the electron-transferring-flavoprotein dehydrogenase (ETFDH) gene. Brain 2007;130:2037–2044.
34. Lopez LC, Quinzii C, Schuelke M, et al. Leigh syndrome with nephropathy and CoQ10 deficiency due to decaproneyl diphosphate synthase subunit 2 (PDSS2) mutations. Am. J. Hum. Genet. 2006;79:1125–1129.
35. Mollet J, Giurgea I, Schlemmer D, et al. Prenyldiphosphate synthase (PDSS1) and OH-benzoate prenyltransferase (COQ2) mutations in ubiquinone deficiency and oxidative phosphorylation disorders. J. Clin. Invest. 2007;117:765–772.
36. Quinzii C, Naini A, Salviati L, et al. A mutation in para-Hydoxybenzoate-polyprenyl transferase (COQ2) causes primary coenzyme Q10 deficiency. Am. J. Hum. Genet. 2006;78:345–349.
37. Lagier-Tourenne C, Tazir M, Lopez LC, et al. ADSK3, an ancestral kinase, is mutated in a form of recessive ataxia associated with coenzyme Q10 deficiency. Am. J. Hum. Genet. 2008;82:661–672.
38. Mollet J, Delahodde A, Serre V, et al. CABC1 gene mutations cause ubiquinone deficiency with cerebellar ataxia and seizures. Am. J. Hum. Genet. 2008;82:623–630.
39. DiMauro S, Schon EA. Mitochondrial disorders in the nervous system. Annu. Rev. Neurosci. 2008;31:91–123.
40. Nishino I, Spinazzola A, Hirano M. Thymidine phosphorylase gene mutations in MNGIE, a human mitochondrial disorder. Science 1999;283:689–692.
41. Hirano M, Silvestri G, Blake D, et al. Mitochondrial neurogastrointestinal encephalomyopathy (MNGIE): Clinical, biochemical and genetic features of an autosomal recessive mitochondrial disorder. Neurology 1994;44:721–727.
42. Kaukonen J, Juselius JK, Tiranti V, et al. Role of adenine nucleotide translocator 1 in mtDNA maintenance. Science 2000;289:782–785.
43. Spelbrink JN, Li FY, Tiranti V, et al. Human mitochondrial DNA deletions associated with mutations in the gene encoding Twinkle, a phage T7 gene 4-like protein localized in mitochondria. Nature Genet. 2001;28:200–201.
44. Van Goethem G, Dermaut B, Lofgren A, et al. Mutation of POLG is associated with progressive external ophthalmoplegia characterized by mtDNA deletions. Nature Genet. 2001;28:211–212.
45. Graziewicz MA, Longley MJ, Copeland WC. DNA polymerade gamma in mitochondrial DNA replication and repair. Chem. Rev. 2006;106:385–405.
46. Longley MJ, Clark S, Man CYW, et al. Mutant POLG2 disrupts DNA polymerase gamma subunits and causes progressive external ophthalmoplegia. Am. J. Hum. Genet. 2006;78:1026–1034.

47. Zanna C, Ghelli A, Porcelli AM, et al. OPA1 mutations associated with dominant optic atrophy impair oxidative phosphorylation and mitochondrial fusion. Brain 2008;131: 352–367.
48. Ferraris S, Clark S, Garelli E, et al. Perogressive external ophthalmoplegia and vision and hearing loss in a patient with mutations in *POLG2* and *OPA1*. Arch. Neurol. 2008;65: 125–131.
49. Hudson G, Amati-Bonneau P, Blakeley E, et al. Mutation of OPA1 causes dominant optic atrophy with external ophthalmoplegia, ataxia, deafness and multiple mitochondrial DNA deletions: a novel disorder of mtDNA maintenance. Brain 2007;131:329–337.
50. Amati-Bonneau P, Guichet A, Olichon A, et al. OPA1 R445H mutation in optic atrophy associated with sensorinaural deafness. Ann. Neurol. 2005;58:958–963.
51. Spinazzola A, Zeviani M. Disorders of nuclear-mitochondrial intergenomic signaling. Gene 2005;354:162–168.
52. Oskoui M, Davidzon G, Pascual J, et al. Clinical spectrum of mitochondrial DNA depletion due to mutations in the thymidine kinase 2 gene. Arch. Neurol. 2006;63: 1122–1126.
53. Ostergaard E, Christensen E, Kristensen E, et al. Deficiency of the alpha subunit of succinate-coenzyme A ligase causes fatal infantile lactic acidosis with mitochondrial DNA depletion. Am. J. Hum. Genet. 2007;81:383–387.
54. Elpeleg O, Miller C, Hershkovitz E, et al. Deficiency of the ADP-forming succinyl-CoA synthase activity is associated with encephalomyopathy and mitochondrial DNA depletion. Am. J. Hum. Genet. 2005;76:1081–1086.
55. Freisinger P, Futterer N, Lankes E, et al. Hepatocerebral mitochondrial DNA depletion syndrome caused by deoxyguanosine kinase (DGUOK) mutations. Arch. Neurol. 2006;63:1129–1134.
56. Mandel H, Szargel R, Labay V, et al. The deoxyguanosine kinase gene is mutated in individuals with depleted hepatocerebral mitochondrial DNA. Nature Genet. 2001;29: 337–341.
57. Naviaux RK, Nguyen KV. *POLG* mutations associated with Alpers' syndrome and mitochondrial DNA depletion. Ann Neurol. 2004;55:706–712.
58. Spinazzola A, Viscomi C, Fernandez-Vizarra E, et al. MPV17 encodes an inner mitochondrial membrane protein and is mutated in infantile hepatic mitochondrial DNA depletion. . Nature Genet. 2006;38:570–575.
59. Karadimas CL, Vu TH, Holve SA, et al. Navajo neurohepatopathy is caused by a mutation in the MPV17 gene. Am. J. Hum. Genet. 2006;79:544–548.
60. Jacobs HT, Turnbull DM. Nuclear genes and mitochondrial translation: a new class of genetic disease. Trends Genet. 2005;21:312–314.
61. Antonicka H, Sasarman F, Kennaway NG, Shoubridge EA. The molecular basis for tissue specificity of the oxidative phosphorylation deficiencies in patients with mutations in the mitochondrial translation factor EFG1. Hum. Mol. Genet. 2006;15:1835–1846.
62. Miller C, Saada A, Shaul N, et al. Defective mitochondrial translation caused by a ribosomal protein (MRPS16) mutation. Ann. Neurol. 2004;56:734–738.
63. Smeitink JAM, Elpeleg O, Antonicka H, et al. Distinct clinical phenotypes associated with a mutation in the mitochondrial translation elongation factor EFTs. Am. J. Hum. Genet. 2006;79:869–877.
64. Fernandez-Vizarra E, Berardinelli A, Valente L, et al. Nonsense mutation in pseudouridylate synthase 1 (PUS1) in two brothers affected by myopathy, lactic acidosis and sideroblastic anemia (MLASA). J. Med. Genet. 2007;44:173–180.
65. Bykhovskaya Y, Casas KA, Mengesha E, et al. Missense mutation in pseudouridine synthase 1 (*PUS1*) causes mitochondrial myopathy and sideroblastic anemia (MLASA). Am. J. Hum. Genet. 2004;74:1303–1308.
66. Scheper GC, van der Klok T, van Andel RJ. et al. Mitochondrial aspartyl-tRNA synthetase deficiency causes leukoencephalopathy with brain stem and spinal cord involvement and lactate elevation. Nature Genet. 2007;39:534–538.

67. Chan DC. Mitochondrial dynamics in disease. New Engl. J. Med. 2007;356:1707–1709.
68. Fichera M, Lo Giudice M, Falco M, et al. Evidence of kinesin heavy chain (*KIF5A*) involvement in pure hereditary spastic paraplegia. Neurology 2004;63: 1108–1110.
69. Delettre C, Lenaers G, Griffoin J-M, et al. Nuclear gene OPA1, encoding a mitochondrial dynamin-related protein, is mutated in dominant optic atrophy. Nature Genet. 2000;26: 207–210.
70. Alexander C, Votruba M, Pesch UEA, et al. OPA1, encoding a dynamin-related GTPase, is mutated in autosomal dominant optic atrophy linked to chromosome 3q28. Nature Genet. 2000;26:211–215.
71. Lawson VH, Graham BV, Flanigan KM. Clinical and electrophysiologic features of CMT2A with mutations in the mitofusin 2 gene. Neurology 2005;65:197–204.
72. Zuchner S, Mersiyanova IV, Muglia M, et al. Mutations in the mitochondrial GTPase mitofusin 2 cause Charcot-Marie-Tooth neuropathy type 2A. Nature Genet. 2004;36: 449–451.
73. Pedrola L, Espert A, Wu X, et al. GDAP1, the protein causing Charcot-Marie-Tooth disease type 4A, is expressed in neurons and is associated with mitochondria. Hum. Mol. Genet. 2005;14:1087–1094.
74. Niemann A, Ruegg M, La Padula V, et al. Ganglioside-induced differentiation associated protein 1 is a regulator of the mitochondrial network: new implications for Charcot-Marie-Tooth disease. J. Cell Biol. 2005;170:1067–1078.
75. Barth PG, Wanders RJA, Vreken P, et al. X-linked cardioskeletal myopathy and neutropenia (Barth syndrome) (MIM 30260). J. Inher. Metab. Dis. 1999;22:555–567.
76. Bione S, D'Adamo P, Maestrini E, et al. A novel X-linked gene, G4.5, is responsible for Barth syndrome. Nature Genet. 1996;12:385–389.
77. Schlame M, Ren M. Barth syndrome, a human disorder of cardiolipin metabolism. FEBS Lett. 2006;580:5450–5455.
78. Roesch K, Curran SP, Tranebjaerg L, Koehler CM. Human deafness dystonia syndrome is caused by a defect in assembly of the DDP1/TIMM8a-TIMM13 complex. Hum. Mol. Genet. 2002;11:477–486.
79. Parone PA, Da Cruz S, Tondera D, et al. Preventing mitochondrial fission impairs mitochondrial function and leads to loss of mitochondrial DNA. PLoS ONE 2008;3:e 3257.
80. King MP, Attardi G. Human cells lacking mtDNA: repopulation with exogenous mitochondria by complementation. Science 1989;246:500–503.
81. Chomyn A, Martinuzzi A, Yoneda M, et al. MELAS mutation in mtDNA binding site for transcription termination factor causes defects in protein synthesis and in respiration but no change in levels of upstream and downstream mature transcripts. Proc. Natl. Acad. Sci. USA 1992;89:4221–4225.
82. Chinnery PF, Taylor DJ, Brown DT, et al. Very low levels of the mtDNA A3243G mutation associated with mitochondrial dysfunction in vivo. Ann. Neurol. 2000;47:381–384.
83. Kaufmann P, Shungu D, Sano MC, et al. Cerebral lactic acidosis correlates with neurological impairment in MELAS. Neurology 2004;62:1297–1302.
84. Tatuch Y, Christodoulou J, Feigenbaum A, et al. Heteroplasmic mtDNA mutation (T>G) at 8993 can cause Leigh disease when the percentage of abnormal mtDNA is high. Am. J. Hum. Genet. 1992;50:852–858.
85. Carelli V, Baracca A, Barogi S, et al. Biochemical-clinical correlation in patients with different loads of the mitochondrial DNA T8993G mutation. Arch. Neurol. 2002;59: 264–270.
86. Sgarbi G, Baracca A, Lenaz G, et al. Inefficient coupling between proton transport and ATP synthesis may be the pathogenic mechanism for NARP and Leigh syndrome resulting from the T8993G mutation in mtDNA. Biochem. J. 2006;395:493–500.
87. Tanji K, Schon EA, DiMauro S, Bonilla E. Kearns-Sayre syndrome: oncocytic transformation of choroid plexus epithelium. J. Neurol. Sci. 2000;178:29–36.

88. Betts J, Jaros E, Perry RH, et al. Molecular neuropathology of MELAS: level of heteroplasmy in individual meurones and evidence of extensive vascular involvement. Neuropath Appl. Neurobiol. 2006;32:359–373.
89. Tanji K, Kunimatsu T, Vu TH, Bonilla E. Neuropathological features of mitochondrial disorders. Cell Develop. Biol. 2001;12:429–439.
90. Sacconi S, Salviati L, Nishigaki Y, et al. A functionally dominant mitochondrial DNA mutation. Hum. Mol. Genet. 2008;17:1814–1820.
91. Schon EA, Santra S, Pallotti F, Girvin ME. Pathogenesis of primary defects in mitochondrial ATP synthesis. Cell Develop. Biol. 2001;12:441–448.
92. Tatuch Y, Pagon RA, Vlcek B, et al. The 8993 mtDNA mutation: heteroplasmy and clinical presentation in three families. Eur. J. Hum. Genet. 1994.
93. Carrozzo R, Tessa A, Vazquez-Memije ME, et al. The T9176G mtDNA mutation affects ATP production and results in Leigh syndrome. Neurology 2001;56:687–690.
94. Thyagarajan D, Shanske S, Vazquez-Memije M, et al. A novel mitochondrial ATPase 6 point mutation in familial bilateral striatal necrosis. Ann. Neurol. 1995;38: 468–472.
95. Vazquez-Memije ME, Shanske S, Santorelli FM, et al. Comparative biochemical studies of ATPases in cells from patients with the T8993G or T8993C mitochondrial DNA mutations. J. Inher. Metab. Dis. 1998;21:829–836.
96. Solaini G, Harris DA, Lenaz G, et al. The study of the pathogenic mechanism of mitochondrial diseases provides information on basic bioenergetics. Biochim. Biophys. Acta. 2008;1777:941–945.
97. Baracca A, Sgarbi G, Mattiazzi M, et al. Biochemical phenotypes associated with the mitochondrial ATP6 gene mutations at nt8993. Biochim. Biophys. Acta. 2007;1767: 913–919.
98. Pallotti F, Baracca A, Hernandez-Rosa E, et al. Biochemical analysis of respiratory function in cybrid cell lines harbouring mitochondrial DNA mutations. Biochem. J. 2004;384: 287–293.
99. Carrozzo R, Murray J, Santorelli F, Capaldi RA. The T9176G mutation of human mtDNA gives fully assembled but inactive ATP synthase when modeled in *Escherichia Coli*. FEBS Lett. 2000;486:297–299.
100. Cortes-Hernandez P, Vazquez-Memije ME, Garcia JJ. ATP6 homoplasmic mutation inhibit and destabilize the human F1F0-ATP synthease without preventing enzyme assembly and oligomerization. J. Biol. Chem. 2007;282:1051–1058.
101. Garcia JJ, Ogilvie I, Robinson BH, Capaldi RA. Structure, function, and assembly of the ATP synthase in cells from patients with the T8993G mitochondrial DNA mutation. J. Biol. Chem. 2000;275:11075–11081.
102. Kirino Y, Yasukawa T, Ohta S, et al. Codon-specific translational defect caused by a wobble modification deficiency in mutant tRNA from a human mitochondrial disease. Proc. Natl. Acad. Sci. USA 2004;101:15070–15075.
103. Schon EA, Bonilla E, DiMauro S. Mitochondrial DNA mutations and pathogenesis. J. Bioenerg. Biomembr. 1997;29:131–149.
104. DiMauro S, Bonilla E, Mancuso M, et al. Mitochondrial myopathies. Basic Appl. Mol. 2003;13:145–155.
105. Taylor RW, Giordano C, Davidson MM, et al. A homoplasmic mitochondrial transfer ribonucleic acid mutation as a cause of maternally inherited cardiomyopathy. J. Am. Coll. Cardiol. 2003;41:1786–1796.
106. McFarland R, Schaefer AM, Gardner A, et al. Familial myopathy: New insights into the T14709C mitochondrial tRNA mutation. Ann. Neurol. 2004;55:478–484.
107. Shankar SP, Fingert JH, Carelli V, et al. Evidence for a novel X-linked modifier locus for Leber hereditary optic neuropathy. Opthal. Genet. 2008;29:17–24.
108. Hudson G, Keers S, Man PYW, et al. Identification of an X-chromosomal locus and haplotype modulating the phenotype of a mitochondrial DNA disorder. Am. J. Hum. Genet. 2005;77:1086–1091.

109. Papadopoulou LC, Sue CM, Davidson MM, et al. Fatal infantile cardioencephalomyopathy with COX deficiency and mutations in *SCO2*, a COX assembly gene. Nature Genet. 1999;23:333–337.
110. Valnot I, Osmond S, Gigarel N, et al. Mutations of the SCO1 gene in mitochondrial cytochrome c oxidase deficiency with neonatal-onset hepatic failure and encephalopathy. Am. J. Hum. Genet. 2000;67:1104–1109.
111. Wallace DC. The mitochondrial genome in human adaptive radiation and disease: On the road to therapeutics and performance enhancement. Gene 2005;354:169–180.
112. Hudson G, Carelli V, Spruijt L, et al. Clinical expression of Leber Hereditary optic neuropahy is affected by the mitochondrial DNA-haplogroup background. Am. J. Hum. Genet. 2007;81:228–233.
113. Lodi R, Carelli V, Cortelli P, et al. Phosphorus MR spectroscopy shows a tissue specific in vivo distribution of biochemical expression of the G3460A mutation in Leber's hereditary optic neuropathy. J. Neurol. Neurosurg. Psychiat. 2002;72:805–807.
114. Cortelli P, Montagna P, Pierangeli G, et al. Clinical and brain bioenergetics improvement with idebenone in a patient with Leber's hereditary optic neuropathy: a clinical amd 31P-MRS study. J. Neurol. Sci. 1997;148:25–31.
115. Valentino ML, Barboni P, Rengo C, et al. The 13042G>A/ND5 mutation in mtDNA is pathogenic and can be associated also with a prevalent ocular phenotype. J. Med. Genet. 2006;43:e38.
116. Barbiroli B, Iotti S, Lodi R. Improved brain and muscle mitochondrial respiration with CoQ. An in vivo study by 31P-MR spectroscopy in patients with mitochondrial cytopathies. BioFactors 1999;9:253–260.
117. Haller RG, Vissing J. Functional evaluation of metabolic myopathies. In: Engel AG, Franzini-Armstrong C, eds. Myology. Vol. 1. New York: McGraw-Hill, 2004:665–679.
118. Vazquez-Memije ME, Shanske S, Santorelli FM, et al. Comparative biochemical studies in fibroblasts from patients with different forms of Leigh syndrome. J. Inher. Metab. Dis. 1996;19:43–50.
119. Lenaz G, Baracca A, Carelli V, et al. Bioenergetics of mitochondrial diseases associated with mtDNA mutations. Biochim. Biophys. Acta. 2004;1658:89–94.
120. Baracca A, Barogi S, Carelli V, et al. Catalytic activities of mitochondrial ATP synthase in patients with mitochondrial DNA T8993G mutation in the ATPase 6 gene. J. Biol. Chem. 2000;275:4177–4182.
121. Baracca A, Solaini G, Sgarbi G, et al. Severe impairment of complex I-driven ATP synthesis in Leber's hereditary optic neuropathy cybrids. Arch. Neurol. 2005;62:730–736.
122. Carelli V, Rugolo M, Sgarbi G, et al. Bioenergetics shapes cellular death pathways in Leber's hereditary optic neuropathy: a model of mitochondrial neurodegeneration. Biochim. Biophys. Acta. 2004;1658:172–179.
123. Hirano M, DiMauro S. Leber's hereditary optic neuropathy: biochemical lights in a blurry scenario. Arch. Neurol. 2005;62:711–712.
124. Droge W. Free radicals in the physiologycal control of cell function. Physiol. Rev. 2002;82:47–95.
125. Verkaart S, Koopman WJH, van Emst-De Vries SE, et al. Superoxide production is inversely related to complex I activity in inherited complex I deficiency. Biochim. Biophys. Acta. 2007;1772:373–381.
126. Pitkanen S, Robinson BH. Mitochondrial complex I deficiency leads to increased production of superoxide radicals and induction of superoxide dismutase. J. Clin. Invest. 1996;98: 345–351.
127. Rimessi A, Giorgi C, Pinton P, Rizzuto R. The versatility of mitochondrial calcium signals: From stimulation of cell metabolism to induction of cell death. Biochim. Biophys. Acta. 2008;1777:808–816.
128. Brini M, Pinton P, King MP, et al. A calcium signaling defect in the pathogenesis of a mitochondrial DNA inherited oxidative phosphorylation deficiency. Nature Med. 1999;5:951–954.

129. Moudy AM, Handran SD, Goldberg MP, et al. Abnormal calcium homeostasis and mitochondrial polarization in a human encephalomyopathy. Proc. Natl. Acad. Sci. USA 1995;92:729–733.
130. Szadbadkai G, Simoni AM, Bianchi K, et al. Mitochondrial dynamics and Calcium signaling. Biochim. Biophys. Acta. 2006;1763:442–449.
131. Bianchi K, Rimessi A, Prandini A, et al. Calcium and mitochondria: mechanisms and functions of a troubled relationship. Biochim. Biophys. Acta. 2004;1742:119–131.
132. DiMauro S, Hirano M, Schon EA. Approaches to the treatment of mitochondrial diseases. Muscle Nerve 2006;34:265–283.
133. Millhouse-Flourie TJ. Physical, occupational, respiratory, speech, equine, and pet therapies for mitochondrial disease. Mitochondrion 2004;4:549–558.
134. DiMauro S, Hirano M, Kaufmann P, Mann JJ. Mitochondrial psychiatry. In: DiMauro S, Hirano M, Schon EA, eds. Mitochondrial Medicine. London: Informa Healthcare, 2006:261–277.
135. Taivassalo T, Haller RG. Exercise and training in mitochondrial myopathies. Med. Sci. Sports Exerc. 2005;37:2094–2101.
136. Edmonds JL. Surgical and anesthetic management of patients with mitochondrial dysfunction. Mitochondrion 2004;4:543–548.
137. Sinnathurai AR, Raut V, Awa A, et al. A review of cochlear implantation in mitochondrial sensorineural hearing loss. Otolol. Neurol. 2003;24:418–426.
138. Bonnet D, Rustin P, Rotig A, et al. Heart transplantation in children with mitochondrial cardiomyopathy. Heart 2001;86:570–573.
139. Lee WS, Sokol RJ. Mitochondrial hepatopathies: Advances in genetics and pathogenesis. Hepatology 2007;45:1555–1565.
140. Rotig A, Niaudet P. Mitochondrial nephrology. In: DiMauro S, Hirano M, Schon EA, eds. Mitochondrial Medicine. London: Informa Healthcare, 2006:197–207.
141. White SL, Shanske S, McGill JJ, et al. Mitochondrial DNA mutations at nucleotide 8993 show a lack of tissue- or age-related variation. J. Inher. Metab. Dis. 1999;22: 899–914.
142. Blok RB, Gook DA, Thorburn DR, Dahl H-HM. Skewed segregation of the mtDNA nt 8993 (T>G) mutation in human oocytes. Am. J. Hum. Genet. 1997;60:1495–1501.
143. Sato A, Kono T, Nakada K, et al. Gene therapy for progeny of mito-mice carrying pathogenic mtDNA by nuclear transplantation. Proc. Natl. Acad. Sci. USA 2005;102: 16765–16770.
144. Hirano M, Marti R, Casali C, et al. Allogeneic stem cell transplantation corrects biochemical derangements in MNGIE. Neurology 2006;67:1458–1460.
145. Bastin J, Aubey F, Rotig A, et al. Activation of peroxisome proliferator-activated receptor pathway stimulates the mitochondrial respiratory chain and can correct deficiencies in patients' cells lacking its components. J. Clin. Endocrinol. Metab. 2008;93:1433–1441.
146. Wenz T, Diaz F, Spiegelman BM, Moraes CT. Activation of the PPAR/PGC-1alpha pathway prevents a bioenergetic deficit and effectively improves a mitochondrial myopathy phenotype. Cell Metab. 2008;8:249–255.
147. Chinnery PF, Taylor RW, Diekert K, et al. Peptide nucleic acid delivery to human mitochondria. Gene Therapy 1999;6:1919–1928.
148. Taylor RW, Chinnery PF, Turnbull DM, Lightowlers RN. Selective inhibition of mutant human mitochondrial DNA replication in vitro by peptide nucleic acids. Nature Genet. 1997;15:212–215.
149. Kolesnikova OA, Entelis NS, Jacquin-Becker C, et al. Nuclear DNA-encoded tRNAs targeted into mitochondria can rescue a mitochondrial DNA mutation associated with the MERF syndrome in cultured human cells. Hum. Mol. Genet. 2004;13: 2519–2534.
150. Bai Y, Hajek P, Chomyn A, et al. Lack of complex I activity in human cells carrying a mutation in MtDNA-encoded ND4 subunit is corrected by the *Saccharomyces*

*cerevisiae* NADH-quinone oxidoreductase (NDI1) gene. J. Biol. Chem. 2001;276: 38808–38813.
151. Manfredi G, Fu J, Ojaimi J, et al. Rescue of a deficiency in ATP synthesis by transfer of *MTATP6*, a mitochondrial DNA-encoded gene, to the nucleus. Nature Genet. 2002;30: 394–399.
152. Bonnet C, Augustin S, Ellouze S, et al. The optimized allotopic expression of ND1 or ND4 genes restores respiratory chain complex I activity in fibroblasts harboring mutations in these genes. Biochim. Biophys. Acta. 2008;1783:1707–1717.
153. Ellouze S, Augustin S, Bouaita A, et al. Optimized allotopic expression of the human mitochondrial ND4 prevents blindness in a rat model of mitochondrial dysfunction. Am. J. Hum. Genet. 2008;83:373–387.
154. Hakkaart GA, Dassa EP, Jacobs HT, Rustin P. Allotopic expression of a mitochondrial alternative oxidase confers cyanide resistance to human cell respiration. EMBO Rep. 2006;7:341–345.
155. Seo BB, Nakamura-Ogiso E, Flotte TR, et al. A single-subunit NADH-quinone oxidoreductase renders resistance to mammalian nerve cells against complex I inhibition. Mol. Ther. 2002;6:336–341.
156. Bacman SR, Williams SL, Hernandez D, Moraes CT. Modulating mtDNA heteroplasmy by mitochondria-targeted restriction endonucleases in a “differential multiple cleavage-site” model. Gene Ther. 2007;14:1309–1318.
157. Tanaka M, Borgeld HJ, Zhang J, et al. Gene therapy for mitochondrial disease by delivering restriction endonuclease SmaI into mitochondria. J. Biomed. Sci. 2002;9:534–541.
158. Minczuk M, Papworth MA, Miller JG, et al. DEvelopment of a single-chain, quasi-dimeric zinc-finger nuclease for the selective degradation of mutated human mitochondrial DNA. Nucleic Acids Res. 2008;36:3926–3938.
159. Sabatini S, Kurtzman NA. Bicarbonate therapy in severe metabolic acidosis. J. Am. Soc. Nephrol. 2008;doi:10.1681/ASN.2007121329.
160. Stacpoole PW. The pharmacology of dichloroacetate. Metabolism 1989;38:1124–1144.
161. Stacpoole PW, Gilbert LR, Neiberger RE, et al. Evaluation of long-term treatment of children with congenital lactic cidosis with dichloroacetate. Pediatrics 2008;121:e1223-e1228.
162. Stacpoole PW, Kerr DS, Barnes C, et al. Controlled clinical trial of dichloroacetate for treatment of congenital lactic acidosis in children. Pediatrics 2006;117:1519–1531.
163. Felitsyn N, Stacpoole PW, Notterpek L. Dichloroacetate causes reversible demyelination in vitro: potential mechanism for its neuropathic effect. J. Neurochem. 2007;100:429–436.
164. Kaufmann P, Engelstad K, Wei Y-H, et al. Dichloroacetate causes toxic neuropathy in MELAS: A randomized, controlled clinical trial. Neurology 2006;66:324–330.
165. Kaufmann P, Anziska Y, Gooch CL, et al. Nerve conduction annormalities in MELAS/3243 patients. Arch. Neurol. 2006;63:746–748.
166. Spinazzola A, Marti R, Nishino I, et al. Altered thymidine metabolism due to defects of thymidine phosphorylase. J. Biol. Chem. 2002;277:4128–4132.
167. Quinzii C, Lopez de Munain A, Von-Moltke J, et al. Respiratory chain dysfunction and oxidative stress correlate with severity of primary CoQ10 deficiency. FASEB J. 2008; in press.
168. Marriage BJ, Clandinin MT, Macdonald IM, Glerum DM. Cofactor treatment improves ATP synthetic capacity in patients with oxidative phosphorylation disorders. Mol. Genet. Metab. 2004;81:263–272.
169. Dougados M, Zitoun J, Laplane D, Castaigne P. Folate metabolism disorder in Kearns-Sayre syndrome. Ann. Neurol. 1983;13:687.
170. Allen RJ, DiMauro S, Coulter DL, et al. Kearns-Sayre syndrome with reduced plasma and cerebrospinal fluid folate. Ann Neurol. 1983;13:679–682.
171. Pineda M, Ormazabal A, Lopez-Gallardo E, et al. Cerebral folate deficiency and leukoencephalopathy caused by a mitochondrial DNA deletion. Ann. Neurol. 2006;59: 394–398.

172. Tarnopolsky MA, Roy BD, MacDonald JR. A randomized, controlled trial of creatine monohydrate in patients with mitochondrial cytopathies. Muscle Nerve 1997;20:1502–1509.
173. Klopstock T, Schlamp V, Schmidt F, et al. Creatine monohydrate in mitochondrial diseases: a double-blind, placebo-controlled, cross-over study in 16 patienrs with progressive external ophthalmoplegia or mitochondrial myopathy. Neurology 1999;52:A543-A544.
174. Tarnopolsky MA, Parise G. Direct measurement of high-energy phosphate compounds in patients with neuromuscular disease. Muscle Nerve 1999;22:1228–1233.
175. Baker SK, Tarnopolsky MA. Targeting cellular energy production in neurological disorders. Expert Opin. Investig. Drugs. 2003;12:1–24.
176. Di Prospero N, Baker A, Jeffries N, Fischbeck KH. Neurological effects of high-dose idebenone in patients with Friedreich's ataxia: a randomised, placebo-controlled trial. Lancet Neurol. 2007;6:878–886.
177. Naini A, Kaufmann P, Shanske S, et al. Hypocitrullinemia in patients with MELAS: an insight into the "MELAS paradox". J. Neurol. Sci. 2005;229–230:187–193.
178. Koga Y, Akita Y, Nishioka J, et al. L-Arginine improves the symptoms of strokelike episodes in MELAS. Neurology 2005;64:710–712.
179. Koga Y, Akita Y, Junko N, et al. Endothelial dysfunction in MELAS improved by L-arginine supplementation. Neurology 2006;66:1766–1769.
180. Koga Y, Ishibashi M, Ueki I, et al. Effects of L-arginine on the acute phase of strokes in three patients with MELAS. Neurology 2002;58:827–828.
181. Kubota M, Sakakihara Y, Mori M, et al. Beneficial effect of L-arginine for stroke-like episode in MELAS. Brain Develop. 2004;26:481–483.

# Chapter 11
# Biology of Peripheral Inherited Neuropathies: Schwann Cell Axonal Interactions

**Michael E. Shy**

**Abstract** Development and maintenance of PNS myelin depends on continual signaling from axons ensheathed by myelin. Recent advances have demonstrated the roles of neuregulin 1 type III, Erb2/3 and intracellular signal transduction pathways in inducing Schwann cell myelination. Alternatively, maintenance of myelinated axons depends on healthy myelinating Schwann cells. Axonal degeneration is a feature of virtually all inherited demyelinating neuropathies and in many cases is more responsible for clinical impairment than the primary demyelination. Signaling mechanisms through which demyelinating Schwann cells damage axons are not well understood. In this review several examples of potential mechanisms by which demyelinating neuropathies damage axons will be presented. Understanding the molecular basis of Schwann cell-axonal interactions will not only increase the understanding of PNS biology but also identify therapeutic targets for inherited neuropathies.

**Keywords:** Myelin · PNS · Schwann cell · Axon · Genetic · Neuropathy

## 11.1 Introduction

Charcot-Marie-Tooth disease (CMT) refers to inherited peripheral neuropathies named for three investigators who described them in the late 1800s [1, 2]. CMT neuropathies affect approximately one in 2500 people [3], and are among the most common inherited neurological disorders. The majority of CMT patients have autosomal dominant inheritance, although X-linked dominant and autosomal recessive forms also exist. Apparent sporadic cases occur, since dominantly inherited disorders may begin as a new mutation in a given patient. The majority of CMT neuropathies are demyelinating, although up to one third appear to be primary axonal disorders. Most patients have a "typical" CMT phenotype characterized by onset in childhood or early adulthood, distal weakness, sensory loss, foot deformities (pes

---

M.E. Shy (✉)
Department of Neurology. Wayne State University School of Medicine, Detroit, MI, USA
e-mail: m.shy@wayne.edu

C. Espinós et al. (eds.), *Inherited Neuromuscular Diseases*, Advances in Experimental Medicine and Biology 652, DOI 10.1007/978-90-481-2813-6_11,

cavus and hammer toes), and absent reflexes. However, some patients develop severe disability in infancy (Dejerine Sottas Disease or congenital hypomyelination), while others develop few if any symptoms of disease. Thus far at least 30 genes are known to cause inherited neuropathies, and more than 45 distinct loci have been identified. A summary of the genes associated with CMT neuropathies can be found online at http://molgen-www.uia.ac.be/CMTMutations/. Genetic testing for several forms of CMT is now available, which, in addition to providing accurate diagnosis, also provides for genotypic-phenotypic correlations. Progress has been made toward understanding how particular mutations cause disease, but pathogenic mechanisms remain largely unknown. One group of mechanisms involves interactions between Schwann cells and axons. Signals from axons determine whether or not a Schwann cell will alter its phenotype and make myelin. Alternatively, Schwann cell abnormalities may induce axonal degeneration with or without demyelination. We will begin this presentation with a brief review of how axonal signals induce Schwann cells to myelinate.

## 11.2 Axonal Signals Regulate Myelination

Most peripheral nerves are mixed, containing both motor and sensory axons that are ensheathed along their length by Schwann cells. During development, Schwann cell precursors from the neural crest migrate out and contact the developing peripheral axons [4, 5]. These "immature" Schwann cells then ensheath bundles of developing axons, a process called "radial sorting", and further differentiate into myelinating or non-myelinating Schwann cells [6]. Schwann cells that establish a one to one association with an axon, called the "promyelinating stage" of Schwann cell development, initiate the program of myelination and become myelinating Schwann cells [6, 7]. In contrast, Schwann cells that do not establish this relationship with an axon do not activate the program of myelin gene expression, and become non-myelinating Schwann cells [6, 8]. Interestingly, this decision process is directed by axons, so that all immature Schwann cells have the potential to become either myelinating or nonmyelinating cells. Maintenance of myelin also depends on the axon and axonal integrity. If an axon is transected and undergoes Wallerian degeneration, Schwann cells that previously myelinated the axon will alter their molecular phenotype and resemble pre-myelinating Schwann cells. If the axon is crushed, rather than transected, this will allow the axon to regenerate following Wallerian degeneration. As the growing axon re-contacts denervated Schwann cells they will again myelinate demonstrating the necessity of axonal signals to induce and maintain myelination [7].

Axonal signaling to induce mayelination is directed, at least in part, by the concentration of neuregulin 1 type III (Nrg1*) on the axonal surface. Nrg1* is one of more than 15 epidermal growth factor like ligands that form the neuregulin family. Nrg1 isoforms are subdivided based on distinct amino-termini [9, 10]. Transmembrane domains of Nrg1 undergo proteolytic cleavage such that Nrg1 types I and II are shed from the neuronal cell surface and function as paracrine signaling molecules. However, Nrg1 type III (Nrg1*) has a cysteine rich domain (CRD) which functions as a second transmembrane domain. Therefore, Nrg1∗ remains tethered to neurons (axons) following proteolytic cleavage and functions as a juxtacrine signal [11]. Nrg1 isoforms mediate their effects through binding to ErbB receptors, themselves members of an EGF receptor superfamily [12]. Nrg1 binds to either ErbB3, which lacks an active kinase domain or ErbB4 which has a kinase region; each can heterodimerize with ErbB2 which can't bind Nrg1 directly but also has an active kinase domain (reviewed in [13]). Myelinating Schwann cells principally express ErbB2 and ErbB3.

Axonal bound Nrg1∗ has recently been shown to dictate the myelinating Schwann cell phenotype by ErbB 2/3 activation of PI3-kinase [14]. Nrg1∗ has also been shown to regulate the thickness of the PNS myelin sheath. Nrg1/ErbB2 compound heterozygous deficient mice developed abnormally thin myelin with elevated g-ratios [15]. Alternatively, mice that over-expressed Nrg1 type III (Nrg1∗ +/-) develop hypermyelinated sciatic nerves. Signal transduction pathways activated by ErbB2/3 result in changes in transcription that cause Schwann cells to myelinate axons. Myelinating Schwann cells upregulate, in a tightly coordinated manner, a group of genes including those that encode myelin specific proteins PMP22, MPZ, and Cx32, all of which cause CMT1 when mutated (reviewed in [16]). Interestingly, mutations in two transcription factors, EGR2/Krox20 [17] and Sox10 [18], that help regulate myelination, also cause demyelinating forms of CMT.

Axons also induce Schwann cells to localize specific molecules appropriately within the myelin sheath. The myelin sheath has two regions, compact and noncompact, each of which contains a unique non-overlapping set of protein constituents. The compact region contains the myelin structural proteins, MPZ, PMP22, and MBP, which participate in forming the highly organized myelin sheath and in electrically insulating axons. The noncompact region is composed of two subdomains, the paranode and the juxtaparanode. The paranodal region, the loops of Schwann cell membrane and interacting axonal membrane adjacent to the node of Ranvier, contains the Schwann cell proteins MAG, Connexin-32, Neurofascin-155, and the axolemmal proteins Caspr and Contactin. These proteins participate in Schwann cell-axonal or Schwann cell-Schwann cell interactions, and act to electrically isolate the nodal region. The juxtaparanodal region, the portion of Schwann cell and interacting axonal membrane adjacent to the paranode, contains potassium channels and Caspr-2, both expressed by axons (see [19, 20]). The complex cellular structures formed by myelinating Schwann cells and their axons are thus analogous in many respects to the neuromuscular junction formed between motor axons and muscle cells: both are highly ordered multicomponent systems formed by the interaction of two distinct cell types in order to carry out a specific biological function related to nerve transmission.

## 11.3 Demyelinating Schwann Cells Disrupt Axons

The processes of nerve development and associated Schwann cell differentiation, myelination and establishment of an electrically insulated node of Ranvier capable of salutatory conduction are all examples of the influence of axons on Schwann cells. However, the inherited demyelinating neuropathies provide examples of how the axons are also dependent on Schwann cells. These disorders end up disrupting the critical interactions between Schwann cells and axons in ways that damage the axon as well as, and sometimes independently of, the myelin sheath. Examples illustrating this point are provided below.

## 11.4 CMT1A and Axonal Degeneration

The most common form of CMT, CMT1A, is caused by a 1.4 Mb duplication on chromosome 17 in the region carrying the gene encoding *PMP22* [21–23]. Duplication of the *PMP22* gene within the region is the likely cause of the disease, since mice and rats with extra transgenic copies of the *Pmp22* gene develop a similar demyelinating neuropathy [24–26], as do some patients with *PMP22* point mutations [27–29]. Although demyelination is the pathological and physiological hallmark of CMT1A, the clinical signs and symptoms of this disease, progressive weakness and sensory loss, are produced by axonal degeneration [30]. Children with CMT1A, for example, have slow nerve conduction velocities before the onset of symptoms, and the velocities do not change appreciably as the disease progresses; thus, demyelination per se does not seem to be sufficient to cause the neurological signs and symptoms [31]. In addition, Krajewski and coworkers have shown that the amplitudes of compound motor action potentials, not nerve conduction velocities, are most closely related to weakness in patients with CMT1A [30], again suggesting that axonal loss is the cause of weakness. Finally, there is anatomical evidence of progressive length-dependent axonal loss in mouse and rat models of CMT1A [32]. The molecular basis for the progressive axonal degeneration is under investigation and a number of studies have demonstrated that they are likely to be the result of abnormalities in Schwann cell-axonal interactions. Brady and co-workers, for example, have demonstrated that Trembler mice, which have a dysmyelinating peripheral neuropathy due to a point mutation in *pmp22* similar to CMT1A, have significant changes in both axonal structure and function, including alterations in neurofilament phosphorylation, increased neurofilament density, and decreased axonal transport [33] and similar changes have been found in patients with CMT1A [34]. Transplantation of a segment of Trembler and CMT1A nerve into normal nerve also produces similar changes in axons that have regenerated through the nerve graft, but not in the surrounding nerve [33], demonstrating that this effect is induced by contact with the abnormal Schwann cells. Finally, neurofilament packing density is increased and axonal caliber is decreased at the node of Ranvier of normal nerve, where there is no axonal contact with Schwann cells, compared to the adjacent myelinated internode [33]. These data thus convincingly demonstrate that axonal contact with myelinating Schwann cells has a significant effect on underlying axonal physiology in both normal and abnormal nerves. Although the nature of the Schwann-cell-derived signaling that modulates neurofilament packing density, neurofilament phosphorylation, and axonal transport is not known, it must be altered or modified in dysmyelinating Schwann cells. In addition, it may be expressed as a part of the coordinated program of myelin-specific gene expression, since this signal is also produced by normal Schwann cells. An attractive location for the origin of this signaling pathway is the paranodal region of the myelinating Schwann cell, where specializations occur in both the Schwann cell and its underlying axolemma (see review by Salzer [20]). Further definition of the molecular architecture of the paranode and surrounding region of axolemma may further delineate the molecular pathways by which Schwann cells and axons communicate and contribute to understanding of axonal degeneration in demyelinating neuropathies.

It is also important to note that disruption of particular Schwann cell-axonal signaling pathways may not be the only mechanism by which demyelinating Schwann cells produce axonal degeneration. Increased energy demands on the neuron to propagate action potentials, and decreased trophic factor support from denervated schwann cells or muscle are other potential mechanisms that may also contribute to axonal degeneration in CMT1 (reviewed in [35]).

## 11.5 MPZ, CMT1B and Length Dependent Gliopathies

CMT1B is caused by mutations in the major myelin protein zero (*MPZ*), which comprises approximately 50% of myelin protein, and is necessary for both normal myelin structure and function [36, 37]. MPZ is a transmembrane protein of 219 amino acids and is a member of the immunoglobulin supergene family. It has a single immunoglobulin-like extracellular domain of 124 amino acids, a single transmembrane domain of 25 amino acids, and a single cytoplasmic domain of 69 amino acids [38, 39]. The first 29 amino acids of MPZ serve as a leader peptide that is cleaved prior to insertion of MPZ into the myelin sheath. MPZ is also posttranslationally modified by the addition of an N-linked oligosaccharide at a single asparagine residue in the extracellular domain, as well by the addition of sulfate, acyl and phosphate groups [40, 41]. MPZ, like other members of the immunoglobulin superfamily, is a homophilic adhesion molecule [42]. Heterologous cells expressing MPZ adhere to each other in an in vitro cell interaction assay [43], while absence of MPZ expression in vivo in *Mpz* knockout mice produces poorly compacted myelin sheaths [44]. Interestingly, overexpression of MPZ disrupts myelination as well, by inhibiting Schwann cell wrapping of axons, also consistent with an adhesive function for the protein [45]. Taken together, these data demonstrate that MPZ plays an essential role in myelination, probably by holding together adjacent wraps of myelin membrane through MPZ-mediated homotypic interactions.

To date there are over 100 different mutations in MPZ known to cause CMT1B in patients. In a recent study [46], we personally evaluated 13 individuals with CMT1B from 12 families with 8 unique MPZ mutations, and reviewed the data from 64 cases from the literature in which there was sufficient information to determine the patient's clinical phenotype. These data show that 90% of these patients fall into two distinct phenotypic groups: one with extremely slow nerve conduction velocities and onset of symptoms during the period of motor development; and a second with essentially normal nerve conduction velocities and the onset of symptoms as adults. Interestingly, there is little overlap between the early onset and late onset groups. In addition, both groups of patients are also clinically and physiologically distinct from patients with CMT1A caused by a duplication of the PMP22 gene region. Although the precise molecular mechanisms are not known, these data suggest that MPZ mutations causing CMT1B also fall into two functional groups: one which effects both the processes of myelin development, causing delayed motor development and markedly slowed nerve conductions; and a second which allows developmental myelination, but eventually leads to axonal degeneration and weakness in later life, with minimal evidence of demyelination.

## 11.6 H10P Causes a "Dying Back Gliopathy" and Accumulation of Proteinaceous Material Within the Intralaminar Space But Not Demyelination

How might *MPZ* mutations causing late onset neuropathy interfere with axonal function? One attractive possibility is that these mutations produce a subtly abnormal myelin sheath that causes alteration in Schwann cell-axonal interactions, and thereby leads to axonal degeneration. Most patients with late onset neuropathy, for example, have relatively normal or only slightly slowed nerve conductions, and show predominant axonal loss with mild or no demyelination on nerve biopsy. Clues to how this might occur were provided in an autopsy on a 73 year old woman with a late onset form of CMT1B caused by a H10P mutation. We found that unlike R69C, there was little evidence of segmental demyelination. Rather, length dependent axonal loss dominated the pathology. Additionally, we identified large amounts of amorphous material that had accumulated in the inner intralaminar spaces of proximal, large diameter myelinated nerves, particularly in dorsal roots. At high EM magnification, this material could be seen in the space between two split leaflets of major dense-line which suggest an intracellular accumulation of Schwann cell. This preferential involvement of adaxonal myelin wraps was frequent, consistent with a dying back pathology of the distal Schwann cell. Finally, we identified abnormalities in the molecular organization of the paranodal and juxtaparanodal regions of axolemma, including the spreading of voltage gated potassium channels and Caspr. Taken together, we hypothesized that a dying back of adaxonal myelin resulted in a breakdown of myelin axonal interactions that ultimately lead the axonal degeneration in our patient [47]. Whether these abnormalities are widespread in patients with late onset CMT1B is not yet known. Recently we have had the opportunity to re-examine an autopsy of a patient with a T95M mutation who also had a late onset neuropathy. Unfortunately, this autopsy did not include either ventral or dorsal roots. In general the findings were similar to those observed in the H10P autopsy. We did identify two myelinated fibers from the T95M sciatic nerve that contained a large amount of vesicular material in the periaxonal space [48].

## 11.7 Myelin in R69C Heterozygous Mice Refractory to Axonal Signaling

R69C invariably causes severe early onset CMT1B [49, 50], including one patient that died at 22 months of age due to respiratory failure from the CMT [51]. We have performed a sural nerve biopsy on a 50 year-old patient with an R69C mutation who was severely affected in infancy and remains unable to ambulate independently. We compared the results of this biopsy with results from a previous sural nerve biopsy she had obtained on the other ankle 20 years previously. Morphometric analysis of the biopsies revealed a loss of large diameter myelinated fibers that was similar in both cases suggesting that there had been little increase in axonal loss over the past 20 years. G-ratios were also virtually identical for the two biopsies. In the biopsy we performed, segmental demyelination and numerous "onion bulbs" were conspicuous although axonal loss of large diameter nerve fibers was also severe (no fibers > 7 μm). Macrophages were found in areas of segmental demyelination but only in a few nerve fibers (<1%). Teased fiber immunohistochemistry shows voltage-gated sodium channel subtype 1.8 (Nav1.8) expressed at the nodes of Ranvier around the areas of segmental demyelination. Occasionally, Nav1.8 formed hemi-nodes reminiscent of the expression pattern of Nav1.2 during myelin development. Interestingly, smaller diameter myelinated fibers did not show segmental demyelination, had uniform but extremely short internodes, and no nodal Nav 1.8 expression. We believe that these shorter internodes never achieved their normal developmental length, consistent with the developmental abnormalities caused by this mutation [12].

To investigate this further, in collaboration with Larry Wrabetz, we utilized cre-lox techniques to "knockin" the R69C mutation into PNS myelin of mice. Heterzygous (R69C+/−) mice did not develop an overt clinical phenotype but had NCV reduced by 50% and marked morphological abnormalities of their myelin sheath. Interestingly, in heterozygous mice, thin myelin did not appear to be uniformly present for all axon calibers; specifically, smaller diameter fibers appeared much less affected that large diameter axons. Myelin thickness increased, as expected, between 0.3 and 1.2 mm microns in proportion to axonal diameters in wild-type animals. However, the proportional increase in myelin thickness/axonal caliber only held true for smaller myelinated axons in R98C/+ nerves. For small diameter axons (<2.0 μm), the distribution of myelin thickness in heterozygous nerves is similar to that of wild type nerves. However, as R98C/+ axonal calibers increase there was a peak in myelin thickness at 0.3 mm, not present in wild type nerves, and myelin never became thicker than 0.5 mm regardless of axonal size in heterozygous nerves. Taken together, these data suggest that large diameter R69C/+ nerves become refractory to axonal signaling that regulates axonal thickness. To identify mechanisms for this refractoriness we are crossing the R69C mutant mice with mice engineered by Klaus Nave and his colleagues that over-express Nrg1∗. We hypothesize that these experiments will either provide a therapeutic strategy to increase myelination in this and similar mutations or permit an identification of molecular pathways that prevent normal myelin thickness in the mutant mice.

## 11.8 GDAP1 Mutations and the Role of Ubiquitously Expressed Genes in Axo-Glial Interactions

Mutations in the gene ganglioside-induced-differentiation-associated-protein-1 (*GDAP1* ) cause the recessively inherited CMT4A. When GDAP1 mutations were first identified there was some confusion about the exact phenotype of CMT4A patients. Some patients were described with a demyelinating phenotype [52] whereas other patients were described as having axonal neuropathies [53]. This apparent discrepancy has been answered, at least in part, by the discovery that GDAP1 is a nuclear encoded protein necessary for mitochondrial fragmentation, or "fission" to occur [54, 55]. Mitochondria supply energy to both neurons and Schwann cells. It is likely that abnormalities in mitochondria disrupt both cell types that may contribute to confusion about the causal cell in the neuropathy. CMT4A is an example of how mutations in ubiquitously expressed genes can disrupt axons and myelin by affecting both cell types.

## 11.9 *Fig4* Mutations Alter Neurons and Schwann Cells Independently

Recessive neuropathy CMT4J in humans and in pale tremor mice (*plt)* is caused by mutations in the *Fig4* gene that encodes a $PI(3,5)P_2$ 5-phosphatase. We have evaluated two patients with CMT4J and examined *plt* mice. The patients presented with an adult-onset (>35 years old), asymmetric, and rapidly progressive paralysis (from ambulation to wheel-chair bound within 5 years). In contrast sensory symptoms were absent despite of sensory axonal loss in the sural nerve biopsy. This novel phenotype clinically imitates motor neuron disease, and is also consistent with prominent neuronal vacuoles and severe neuronal loss in *plt* mice. Time-lapse imaging on fibroblasts isolated from CMT4J-patients' skin biopsies demonstrated an impairment of intracellular organelle trafficking, a potential mechanism for neuronal degeneration [56].

However, both CMT4J and plt mice demonstrate PNS demyelination in sensory as well as motor nerves [56, 57]. Despite these findings patients do not demonstrate evidence of length dependent sensory or motor abnormalities as are usually detected in demyelinating neuropathies. We hypothesize that there are cell autonomous abnormalities in neurons and Schwann cells that are responsible for the neuronal loss and demyelination respectively in CMT4J. This disorder serves as a reminder that axonal loss in demyelinating neuropathies cannot always be presumed to be secondary to demyelination, particularly when the causal gene is ubiquitously expressed.

## 11.10 Summary

Interactions between neurons and Schwann cells are necessary for the normal development and maintenance of myelinated peripheral nerves. If axonal contact is lost Schwann cells alter their phenotype and cease making myelin. Alternatively, mutations in Schwann cell genes cause axonal as well as myelin damage. In many cases axonal damage is responsible for more clinical impairment than demyelination in patients with various forms of CMT1. Comprehension of molecular pathways involved in Schwann cell-axonal interactions is likely to not only increase the understanding of PNS biology but also identify therapeutic targets for inherited neuropathies.

## References


1. Charcot J, Marie P. Sue une forme particulaire d'atrophie musculaire progressive souvent familial debutant par les pieds et les jamber et atteingnant plus tard les mains. Re Med 1886;6:97–138.
2. Tooth H. The peroneal type of progressive muscular atrophy. London: Lewis, 1886.
3. Skre H. Genetic and clinical aspects of Charcot-Marie-Tooth's disease. Clin Genet 1974;6: 98–118.
4. Harrison RG. Neuroblast versus sheath cell in the development of peripheral nerves. J Comp Neurol 1924;37: 123–245.
5. Le Douarin NM, Dupin E. Cell lineage analysis in neural crest ontogeny. J Neurobiol 1993;24:146–161.
6. Webster, H, De F. Development of peripheral nerve fibres. In: Peripheral Neuropathy In: P.J. Dyck, P.K. Thomas, J.W. Griffin, P.A. Low, and J.F. Poduslo, eds. 3rd ed. Philadelphia: W. B. Saunders, 1993: 243–266.
7. Scherer SS. The biology and pathobiology of Schwann cells. Curr Opin Neurol 1997;10: 386–397.
8. Mirsky R, Jessen KR. Schwann cell development, differentiation and myelination. Curr Opin Neurobiol 1996;6:89–96.
9. Falls DL. Neuregulins: functions, forms, and signaling strategies. Exp Cell Res 2003;284: 14–30.
10. Meyer D, Yamaai T, Garratt A, et al. Isoform-specific expression and function of neuregulin. Development 1997;124:3575–3586.
11. Wang JY, Miller SJ, Falls DL. The N-terminal region of neuregulin isoforms determines the accumulation of cell surface and released neuregulin ectodomain. J Biol Chem 2001;276:2841–2851.
12. Citri A, Skaria KB, Yarden Y. The deaf and the dumb: the biology of ErbB-2 and ErbB-3. Exp Cell Res 2003;284:54–65.
13. Nave KA, Salzer JL. Axonal regulation of myelination by neuregulin 1. Curr Opin Neurobiol 2006;16:492–500.
14. Taveggia C, Zanazzi G, Petrylak A, et al. Neuregulin-1 type III determines the ensheathment fate of axons. Neuron 2005;47:681–694.
15. Michailov GV, Sereda MW, Brinkmann BG, et al. Axonal neuregulin-1 regulates myelin sheath thickness. Science 2004;304:700–703.
16. Shy ME. Therapeutic strategies for the inherited neuropathies. Neuromolecular Med 2006;8:255–278.
17. Warner LE, Mancias P, Butler IJ, et al. Mutations in the early growth response 2 (EGR2) gene are associated with hereditary myelinopathies. Nat Genet 1998;18: 382–384.
18. Inoue K, Shilo K, Boerkoel CF, et al. Congenital hypomyelinating neuropathy, central dysmyelination, and Waardenburg-Hirschsprung disease: phenotypes linked by SOX10 mutation. Ann Neurol 2002;52:836–842.
19. Arroyo EJ, Scherer SS. On the molecular architecture of myelinated fibers. Histochem Cell Biol 2000;113:1–18.
20. Salzer JL. Polarized domains of myelinated axons. Neuron 2003;40:297–318.

21. Lupski JR, de Oca-Luna RM, Slaugenhaupt S, et al. DNA duplication associated with Charcot-Marie-Tooth disease type 1A. Cell 1991;66:219–232.
22. Raeymaekers P, Timmerman V, Nelis E, et al. Duplication in chromosome 17p11.2 in Charcot-Marie-Tooth neuropathy type 1a (CMT 1a). The HMSN Collaborative Research Group. Neuromuscul Disord 1991;1:93–97.
23. Lupski JR, de Oca-Luna RM, Slaugenhaupt S, et al. DNA duplication associated with Charcot-Marie-Tooth disease type 1A. Cell 1991;66:219–232.
24. Magyar JP, Martini R, Ruelicke T, et al. Impaired differentiation of Schwann cells in transgenic mice with increased PMP22 gene dosage. J Neurosci 1996;16: 5351–5360.
25. Sereda M, Griffiths I, Puhlhofer A, et al. A transgenic rat model of Charcot-Marie-Tooth disease. Neuron 1996;16:1049–1060.
26. Huxley C, Passage E, Manson A, et al. Construction of a mouse model of Charcot-Marie-Tooth disease type 1A by pronuclear injection of human YAC DNA. Hum Mol Genet 1996;5:563–569.
27. Roa BB, Garcia CA, Pentao L, et al. Evidence for a recessive PMP22 point mutation in Charcot-Marie-Tooth disease type 1A. Nat Genet 1993;5:189–194.
28. Roa BB, Garcia CA, Suter U, et al. Charcot-Marie-Tooth disease type 1A. Association with a spontaneous point mutation in the PMP22 gene. N Engl J Med 1993;329:96–101.
29. Vallat JM, Sindou P, Preux PM, et al. Ultrastructural PMP22 expression in inherited demyelinating neuropathies. Ann Neurol 1996;39:813–817.
30. Krajewski KM, Lewis RA, Fuerst DR, et al. Neurological dysfunction and axonal degeneration in Charcot-Marie-Tooth disease type 1A. Brain 2000;123:1516–1527.
31. Berciano J, Garcia A, Calleja J, Combarros O. Clinico-electrophysiological correlation of extensor digitorum brevis muscle atrophy in children with charcot-marie-tooth disease 1A duplication. Neuromuscul Disord 2000;10:419–424.
32. Suter U, Nave KA. Transgenic mouse models of CMT1A and HNPP. Ann N Y Acad Sci 1999;883:247–253.
33. de Waegh S, Brady ST. Altered slow axonal transport and regeneration in a myelin-deficient mutant mouse: the trembler as an in vivo model for Schwann cell-axon interactions. J Neurosci 1990;10:1855–1865.
34. Sahenk Z, Chen L, Mendell JR. Effects of PMP22 duplication and deletions on the axonal cytoskeleton. Ann Neurol 1999;45:16–24.
35. Massicotte C, Scherer SS. Neuropathies-translating causes into treatments. In: Neuroscience, Molecular Medicine, and the Therapeutic Transformation of Neurology. In: S.G. Waxman, ed. London: Elsevier Science, 2004: 401-414.
36. Eylar EH, Uyemura K, Brostoff SW, Kitamura K, Ishaque A, Greenfield S. Proposed nomenclature for PNS myelin proteins. Neurochem Res 1979;4:289–293.
37. Greenfield S, Brostoff S, Eylar EH, Morell P. Protein composition of myelin of the peripheral nervous system. J Neurochem 1973;20:1207–1216.
38. Lemke G, Axel R. Isolation and sequence of a cDNA encoding the major structural protein of peripheral myelin. Cell 1985;40:501–508.
39. Uyemura K, Asou H, Takeda Y. Structure and function of peripheral nerve myelin proteins. Prog Brain Res 1995;105:311–318.
40. D'Urso D, Brophy PJ, Staugaitis SM, et al. Protein zero of peripheral nerve myelin: biosynthesis, membrane insertion, and evidence for homotypic interaction. Neuron 1990;4: 449–460.
41. Eichberg J, Iyer S. Phosphorylation of myelin protein: recent advances. Neurochem Res 1996;21:527–535.
42. Filbin MT, Walsh FS, Trapp BD, Pizzey JA, Tennekoon GI. Role of myelin P0 protein as a homophilic adhesion molecule. Nature 1990;344:871–872.
43. Xu W, Shy M, Kamholz J, et al. Mutations in the cytoplasmic domain of P0 reveal a role for PKC-mediated phosphorylation in adhesion and myelination. J Cell Biol 2001;155: 439–446.

44. Giese KP, Martini R, Lemke G, Soriano P, Schachner M. Mouse P0 gene disruption leads to hypomyelination, abnormal expression of recognition molecules, and degeneration of myelin and axons. Cell 1992;71:565–576.
45. Wrabetz L, Feltri ML, Quattrini A, et al. P(0) glycoprotein overexpression causes congenital hypomyelination of peripheral nerves. J Cell Biol 2000;148:1021–1034.
46. Shy ME, Jani A, Krajewski KM, et al. Phenotypic Clustering in MPZ mutations. Brain 2004;127:371–384.
47. Li J, Bai Y, Ianakova E, Grandis M, Uchwat F, Trostinskaia A, Krajewski KM, Garbern J, Kupsky WJ, Shy ME. Major myelin protein gene (P0) mutation causes a novel form of axonal degeneration. J Comp Neurol 2006 Sep 10;498(2):252–265.
48. Bai Y, Ianokova E, Pu Q, Ghandour K, Levinson R, Martin JJ, Ceuterick-de Groote C, Mazanec R, Seeman P, Shy ME, Li J. Effect of an R69C mutation in the myelin protein zero gene on myelination and ion channel subtypes. Arch Neurol 2006 Dec;63(12):1787–1794.
49. Warner LE, Hilz MJ, Appel SH, et al. Clinical phenotypes of different MPZ (P0) mutations may include Charcot-Marie-Tooth type 1B, Dejerine-Sottas, and congenital hypomyelination. Neuron 1996;17:451–460.
50. Komiyama A, Ohnishi A, Izawa K, Yamamori S, Ohashi H, Hasegawa O. De novo mutation (Arg98–>Cys) of the myelin P0 gene and uncompaction of the major dense line of the myelin sheath in a severe variant of Charcot-Marie-Tooth disease type 1B. J Neurol Sci 1997;149:103–109.
51. Gabreels-Festen AA, Hoogendijk JE, Meijerink PH, et al. Two divergent types of nerve pathology in patients with different P0 mutations in Charcot-Marie-Tooth disease. Neurology 1996;47:761–765.
52. Baxter RV, Ben Othmane K, Rochelle JM, et al. Ganglioside-induced differentiation-associated protein-1 is mutant in Charcot-Marie-Tooth disease type 4A/8q21. Nat Genet 2002;30:21–22.
53. Cuesta A, Pedrola L, Sevilla T, et al. The gene encoding ganglioside-induced differentiation-associated protein 1 is mutated in axonal Charcot-Marie-Tooth type 4A disease. Nat Genet 2002;30:22–25.
54. Pedrola L, Espert A, Valdés-Sánchez T, Sánchez-Piris M, Sirkowski EE, Scherer SS, Fariñas I, Palau F. Cell expression of GDAP1 in the nervous system and pathogenesis of Charcot-Marie-Tooth type 4A disease. J Cell Mol Med. 2008 Apr;12(2):679–689.
55. Niemann A, Ruegg M, La Padula V, Schenone A, Suter U. Ganglioside-induced differentiation associated protein 1 is a regulator of the mitochondrial network: new implications for Charcot-Marie-Tooth disease. J Cell Biol 2005;170:1067–1078.
56. Zhang X, Chow CY, Sahenk Z, Shy ME, Meisler MH, Li J. Mutation of FIG4 causes a rapidly progressive, asymmetric neuronal degeneration. Brain 2008;131:1990–2001.
57. Chow CY, Zhang Y, Dowling JJ, et al. Mutation of FIG4 causes neurodegeneration in the pale tremor mouse and patients with CMT4J. Nature 2007;448:68–72.

# Chapter 12
# Phenotype and Clinical Evolution of Charcot-Marie-Tooth Disease Type 1A Duplication

**José Berciano, Antonio García, Elena Gallardo, César Ramón, and Onofre Combarros**

**Abstract** In this paper we revise the phenotype and clinical evolution of Charcot-Marie-Tooth disease type 1A duplication (CMT1A). We mainly focus on four phenotypic hallmarks: (i) "classic" phenotype, as currently observed in proband patients; (ii) evolution of mild phenotype of secondary cases in infancy and early childhood; (iii) proximal lower-limb musculature involvement as a late phenotypic feature; and (iv) minimal adult phenotype. We also briefly revise genetic, electrophysiological, pathological and neuroimaging data of the disease.

**Keywords:** Axon · Axonal degeneration · Charcot-Marie-Tooth disease · Demyelination · Magnetic resonance imaging · Myelin · Nerve conduction study · Phenotype · 17p11.2 duplication

## 12.1 Introduction

Charcot-Marie-Tooth disease type 1A (CMT1A) is an autosomal dominant demyelinating polyneuropathy usually associated with a large DNA duplication on the short arm of chromosome 17 [14, 19, 22]. Essentially the disease is characterized by a peroneal muscular atrophy syndrome of variable severity and marked and diffuse slowing of nerve conduction velocity [1, 9, 13, 15, 21] .

When CMT1A patients come to physicians, they present in four ways [24]: (i) with symptoms of muscle weakness or atrophy (this is the usual case of proband patients); (ii) with symptoms related to structural foot abnormalities or soft tissue complications usually attended by a traumatologist (again this is the usual case of probands); (iii) with incidental discovery or a telltale indicator of neuropathy (e.g., slowing nerve conduction velocity investigated for other causes); and (iv) with findings from evaluation of family members (secondary cases).

J. Berciano (✉)
Service of Neurology, Clinical Neurophysiology and Radiology, University Hospital "Marqués de Valdecilla", University of Cantabria, CIBERNED and IFIMAV, Santander, Spain
e-mail: jaberciano@humv.es

C. Espinós et al. (eds.), *Inherited Neuromuscular Diseases*, Advances in Experimental Medicine and Biology 652, DOI 10.1007/978-90-481-2813-6_12,

Based upon our reported descriptions, the aim of this paper is to analyze different CMT1A clinical hallmarks: (i) "classic" phenotype, as currently observed in proband patients; (ii) evolution of mild phenotype of secondary cases in infancy and early childhood; (iii) proximal lower-limb musculature involvement as a late phenotypic feature; and (iv) minimal adult phenotype. But before doing so, we will briefly revise genetic, electrophysiological, pathological and neuroimaging data of the disease.

## 12.2 Genetic, Electrophysiologic, Pathologic and Neuroimaging Data

The genetic locus for CMT1A was mapped to chromosome 17 and is most often associated with a 1.4-Mb duplication in band 17p12 containing the *PMP22* gene. CMT1A duplication, which may also occur as a *de novo* event, is responsible of 60% to 68% of CMT1 patients in North America, and 70.7% of 819 unrelated CMT1 patients in a European collaborative study (for review, see [24]). Demyelination probably is caused by increased gene dose (trisomic over-expression). Point mutations of *PMP22* gene account for a minority of CMT1A phenotypes.

Independently of clinical severity and patients' age, there is marked and uniform slowing of nerve conduction velocity. In our reported CMT1 series comprising 11 families (17p duplication was eventually demonstrated in all 11 families) with 73 electrophysiologically examined patients, mean motor conduction velocities (MCV) were 19.2 m/s (range, 9–35) for median nerve and 18.1 m/s (range, 11–32) for peroneal nerve (Fig. 12.1) [1]. Conversely to acquired demyelinating polyneuropathies, the morphology of compound muscle action potentials (CMAP) was preserved. Muscle weakness correlated with CMAP amplitude but not with the degree of MCV slowing, a fact suggesting that motor deficit is accounted for by secondary axonal loss to primary demyelination (see below).

As illustrated in Fig. 12.2, nerve biopsies demonstrate reduced numbers of myelinated fibers, remyelination, and Schwann cell proliferation resulting in onion-bulbs; active demyelination is extremely rare excepting in early childhood [11]. Histograms of myelinated fibers show a unimodal distribution with a broad middle peak due to predominant loss of large demyelinated fibers [8].

By using magnetic resonance imaging (MRI) study of lower limb musculature, we have reported the patterns of amyotrophy in different clinical stages of 11 CMT1A duplication patients [12] (Figs. 12.3–12.6). In all six CMT1A mild cases not referring to gait disturbances, MRI showed muscle fatty infiltration of intrinsic foot muscles mainly involving the lumbricals, all four leg muscle compartments being preserved (Figs. 12.4 and 12.5). Conversely, the remaining five patients exhibiting moderate gait difficulties, MRI showed, besides intrinsic foot muscle involvement, variable and distally accentuated fatty infiltration of the lateral, anterior and superficial posterior leg muscle compartments and, to a lesser degree, of the deep posterior compartment (Figs. 12.5 and 12.6).

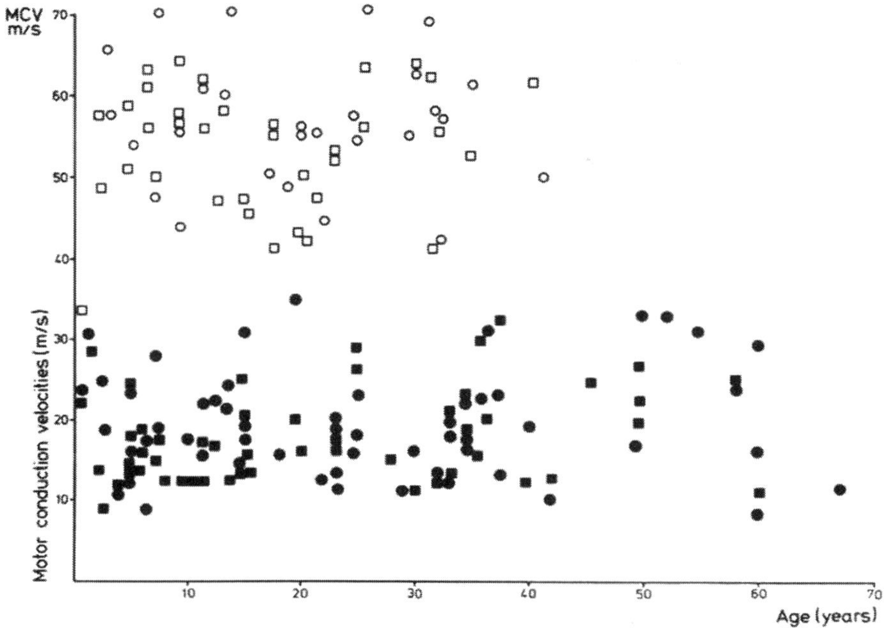


**Fig. 12.1** MCV of median (circles) and peroneal (*squares*) nerves in 99 CMT1 patients coming from 11 pedigrees reported by Berciano et al. [1]; 17p 11.2 duplication was detected afterwards in all 11 pedigrees. Results for 73 subjects with clinical signs of CMT1 are indicated with solid symbols, and results from their clinically normal relatives (26 subjects) are indicated with open symbols. Note that MCV is normal in clinically normal subjects, whereas it is systematically slowed in patients. Note also that the degree of MCV slowing is similar from the first to the seventh decade of life

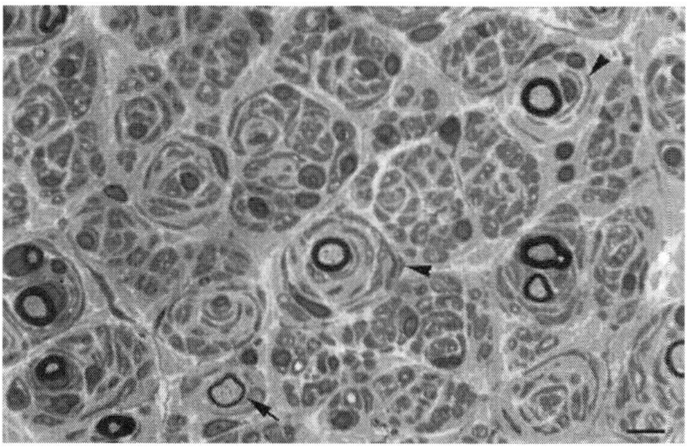

**Fig. 12.2** Semithin section of sural nerve from a CMT1A duplication patient. There is severe loss of myelinated fibers, well developed onion bulbs (*arrowheads*), and a remyelinated fiber (*arrow*). Bar, 8 μm

These findings led to conclude that clinical-MRI patterns of lower limb muscle atrophy vary with evolution of semeiology. Selective involvement of intrinsic foot muscles is the characteristic pattern of CMT1A cases with minimal disease signs. Afterwards this pattern usually combines variable involvement of leg muscles.

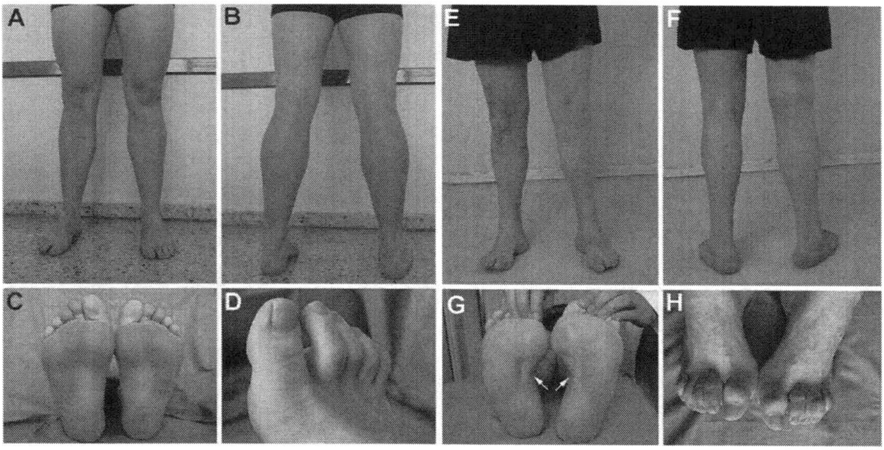


**Fig. 12.3** Pictures from two CMT1A patients aged 17 years (**A–D**; FDS=0 [normal]; CMTNS=2 [mild disease]) and 41 years (**E–H**; FDS=2 [inability to run]; CMTNS=12 [moderate disease]). (**A, B**) Note the absence of leg amyotrophy. (**C, D**) Close-up pictures of the feet illustrating moderate pes cavus, and toe clawing that almost disappears on standing up (**A**) indicating its reducible nature. (**E, F**) Note the presence of peroneal muscular atrophy, toe clawing and marked varus deformity of the ankles. (**G, H**) Close-up pictures of the feet showing severe pes cavus and varus, toe clawing and atrophy of abductor hallucis muscles (*arrows*). Reproduced with permission from [12]

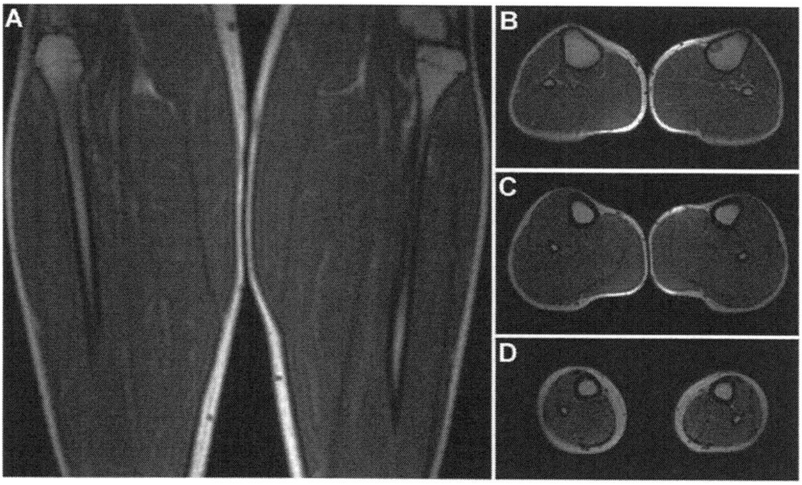


**Fig. 12.4** Coronal (**A**) and axial (**B**, proximal calves; **C**, mid calves; and **D**, lower calves) T1-weighted images from the patient illustrated in Fig. 3A-D showing that all four leg muscle compartments are preserved. Reproduced with permission from [12]

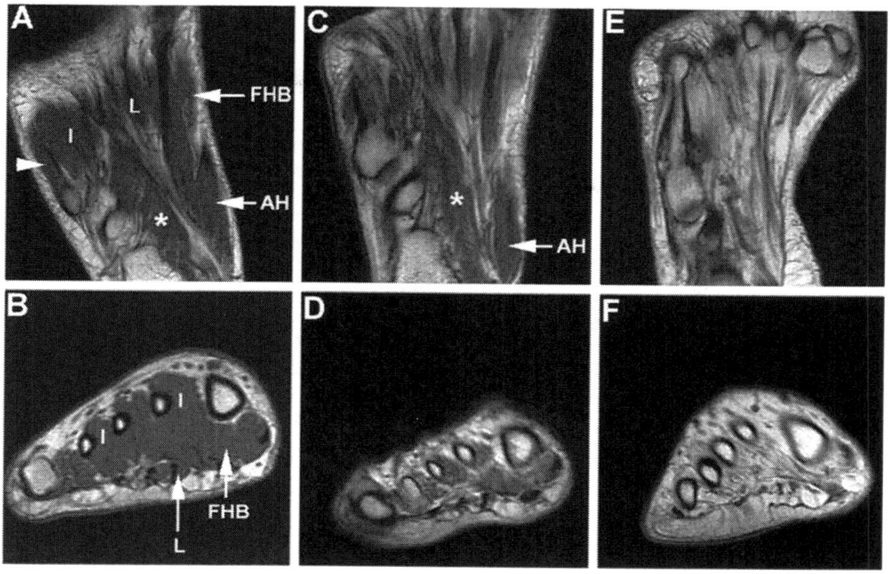


**Fig. 12.5** T1-weighted images of intrinsic foot muscles in a control subject (**A, B**), and in cases illustrated in Fig. 12.3 (patient aged 17 years, **C** and **D**; patient aged 41 years, **E** and **F**) obtained at the same levels in the long and short foot axes. (**A**) Normal identified muscles in this axial section include lumbricals (L), abductor hallucis (AH), and flexor digitorum accesorius (*asterisk*); interossei (I) and flexor digiti minimi (arrowhead) are partially visualized. (**B**) Normal muscles clearly identified in this coronal section include interossei (I), lumbricals (L) and flexor hallucis brevis (FHB). (**C, D**) Note evident but not complete fatty infiltration of the lumbricals, FHB, flexor digiti minimi brevis and interossei; AH and flexor digitorum accesorius (*asterisk*) are relatively preserved. (**E, F**) Compared with the previous case, note massive fatty infiltration of all mentioned foot muscles. Reproduced with permission from [12]

Our findings help to clarify the pathogenesis of pes cavus in the initial stages of the disease. We proposed that forefoot cavus is initiated by selective weakness of intrinsic foot muscles [2, 3, 13, 23]. The first event would be atrophy and weakness of the lumbricals and other intrinsic foot muscles causing dorsiflexion of metatarsophalangeal joints, initially manifested as clawing of the toes and flattening of the transverse arcus plantaris. During gait, prior to toe-off, as metarsophalangeal joints extend, the plantar aponeurosis is wrapped around the metatarsal heads and the short flexors contract, approximating the pillars of the longitudinal arch and shortening the Achilles tendon that limits ankle dorsiflexion. Therefore initial foot deformities and walking difficulties correlate with abnormal foot architecture due to selective denervation of the intrinsic foot muscles [2, 3]. In a recent study, Chung et al. [7] corroborated that foot MRI studies show that fatty infiltration and atrophy may be limited to intrinsic foot muscles with preservation of leg muscles in mildly affected CMT1A (36%) and CMT2A patients (5%). As whole, the MRI findings so far described provide a useful guideline for planning the correction of foot deformities in early and mild cases.

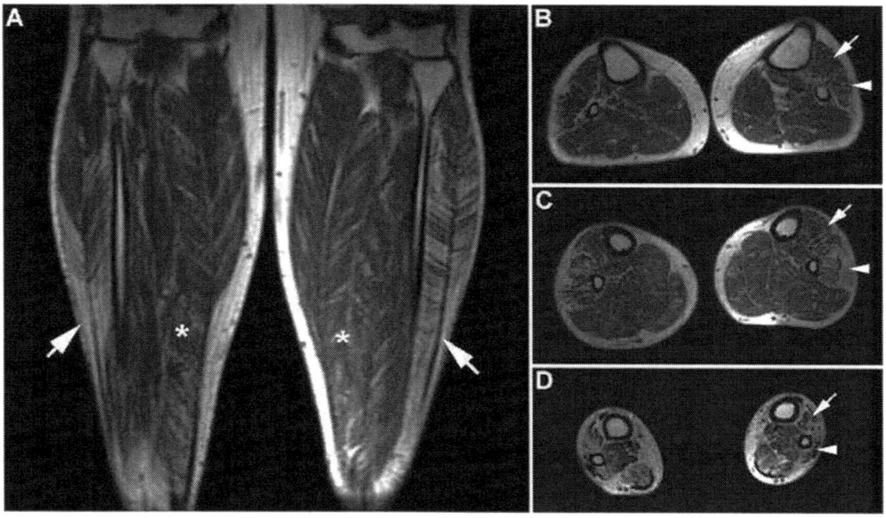


**Fig. 12.6** MRI study from the patient illustrated in Figure 3E–H. (**A**) Coronal T1 weighted image of the legs showing extensive and striped fatty infiltration of peroneal muscles, which is distally accentuated (*arrows*). Distal soleus muscles (*asterisks*) are involved to a lesser degree. Axial T1-weighted images (**B**, proximal calves; **C**, mid calves; and **D**, lower calves) showing distally accentuated fatty infiltration of the lateral (*arrowheads*) and anterior (*arrows*) muscle compartments. There is also fatty infiltration of the soleus muscles (**D**), but other muscles of the superficial and deep posterior compartments are preserved. Reproduced with permission from [12]

## 12.3 "Classic" CMT1A Phenotype

The description of "classic" phenotype has been based upon clinical semeiology observed in symptomatic CMT1A patients seeking medical attention, and usually being the first cases of the corresponding pedigrees in revealing the familial nature of the disease (proband cases). Occasionally, however, the proband may be an asymptomatic patient in whom the disease was identified either because of incidental detection of clinical signs of the disease (e.g., pes cavus, toe clawing or areflexia) or nerve conduction slowing (see above).

### *12.3.1 Initial Symptoms*

The disease has an insidious onset over several years and a slowly progressive course. As reported in CMT1 seminal series by Harding and Thomas [15], symptoms usually appear at the end of the first decade or early in the second decade and may be preceded by slight delayed milestones [16]. In the majority of cases initial symptoms are referred to lower limbs and include foot deformities with shoes not fitting, ankle weakness manifested as difficulty in walking or running, and cramps. Conversely to acquired demyelinating polyneuropathies, in only a small percentage of patients are negative or positive neuropathic sensory symptoms recorded. Symptomatic onset with hand weakness is exceptional.

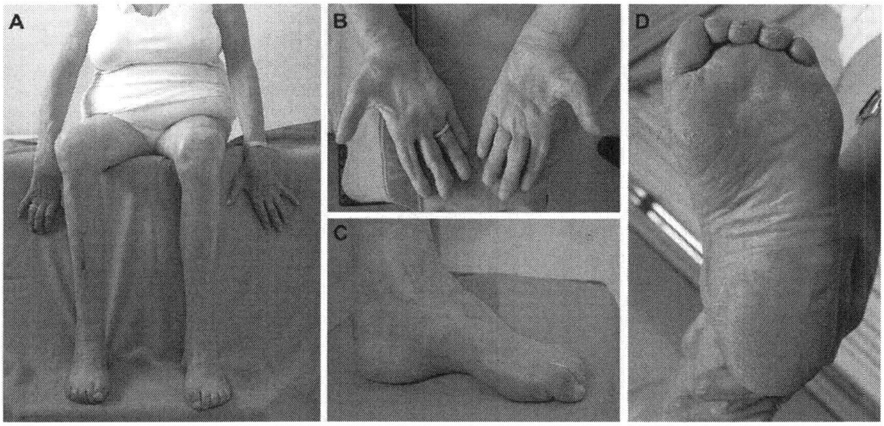


**Fig. 12.7** Severe phenotype in a proband woman aged 75 years. (**A**) Note marked lower leg amyotrophy predominating in peroneal musculature and clawing of the right hand. (**B**) Note atrophy of thenar, hypothenar and interosseus muscles. (**C** and **D**) Hollow left foot with toe clawing, flattening of tranverse arcus plantaris and callosity over the first and fifth metatarsals

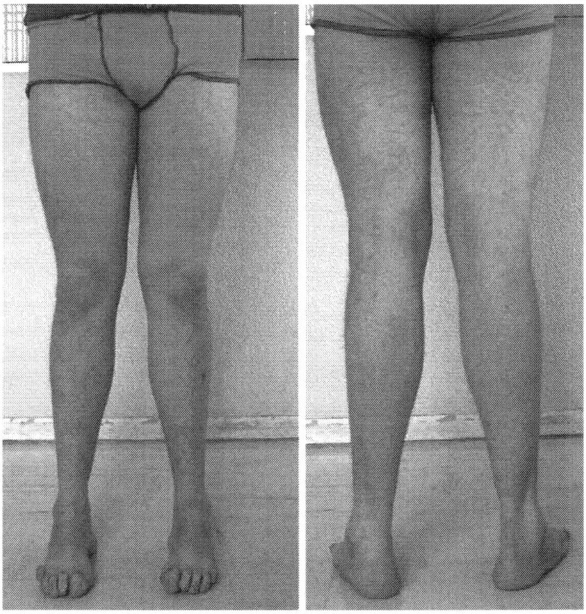

**Fig. 12.8** "Classic" phenotype in a CMT1A patient aged 17 years. There is lower leg amyotrophy predominating in peroneal musculature and in the left leg, toe clawing, and prominence of Achilles' tendons with cavus-varus deformity of the feet

Figures 12.7 and 12.8 illustrate the clinical phenotype in two proband patients aged 75 and 17 years. Established clinical picture comprises a variable constellation of signs including pes cavus or cavus-varus (rarely is pes planus observed), wasting of intrinsic foot muscles (extensor digitorum brevis and abductor hallucis),

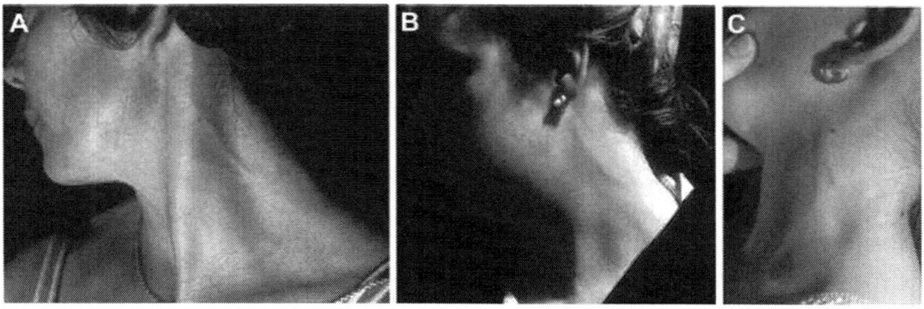


**Fig. 12.9** Great auricular nerve enlargement in the three patients from pedigree MO [4, 8]. (**A**) Picture taken in 1980 from a patient of the second generation, when she was aged 40 years. (**B**) Picture taken in 1982 from a patient of the third generation, then aged 20. (**C**) Picture taken in 1996 from a patient of the fourth generation, when she was aged 8

toe clawing, shortening of tendo Achillis, flattening of transverse arcus plantaris with callosity over metatarsals, lower-leg amyotrophy and weakness predominating in peroneal musculature eventually resulting in steppage gait, areflexia (usually generalized areflexia), nerve enlargement (Fig. 12.9), and stocking hypoesthesia. Ascending amyotrophy to involve vastus medialis muscle without apparent thigh weakness may occur. If subcutaneous fat is not abundant, the combination of lower leg and thigh amyotrophy may give the impression of "storklegs". Severe involvement of thigh and pelvic musculature may occur in the long run, in which case a waddling component is added during gait (see below). Hand amyotrophy and weakness is a relatively late event. Scoliosis may occur in a small proportion of cases.

CMT1A duplication has been associated with other phenotypes including Dejerine- Sottas disease, Roussy-Léwy syndrome, calf hypertrophy (Fig. 12.10), and Davidenkow syndrome (scapuloperoneal syndrome).

### *12.3.2 Progression of Disease*

Evolution of the disease during childhood is analyzed in the next paragraph. The clinical course is quiescent in adult patients [17], though significant age-dependent increase of either mean weakness score [1, 15] or neuropathic deficit [10] has been reported in cross-sectional studies. Functional disability increases with disease duration [6]; moreover progression of CMT1A can be detected by using the CMT Neuropathy Score (CMTNS) [26].

## 12.4 Evolution of Mild Phenotype of Secondary Cases in Infancy and Early Childhood

We addressed this question performing a longitudinal study over up two decades in 12 secondary CMT1A children, aged between 1 month and 5 years (mean, 2 years) at first initial examination, and final ages (in 2002) between 6 and 23

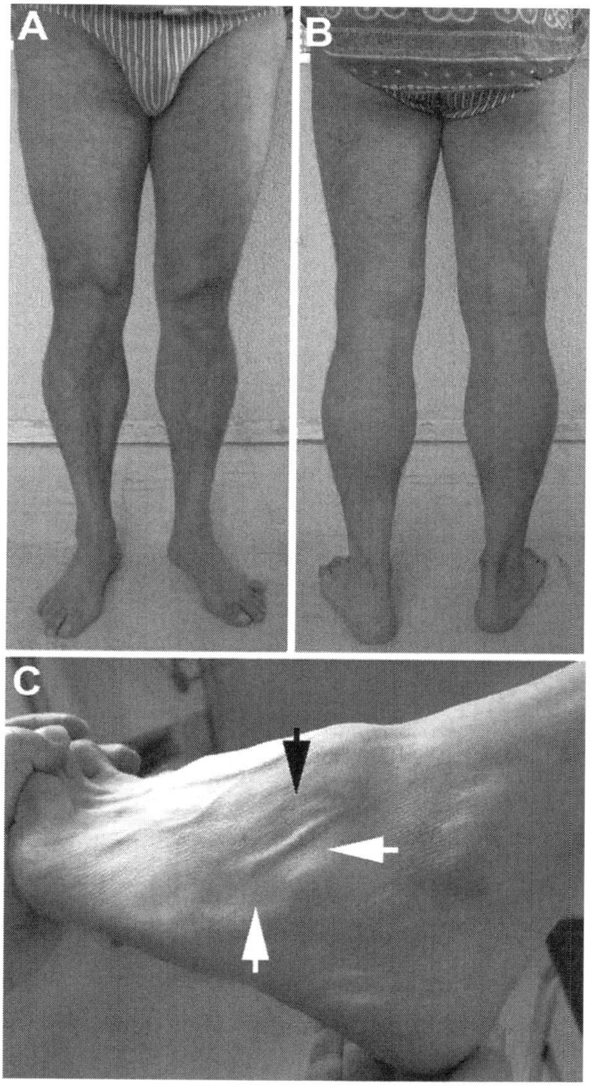


**Fig. 12.10** (**A**, **B**) Hypertrophy of lower-legmusculature in a CMT1A duplication patient aged 38 years. (**C**) This hypertrophy contrast with extensive wasting of extensor digitorum brevis muscle (*arrows*)

years (mean, 13 years) [2, 3, 13] (Fig. 12.11). Initially just 2 (17%) patients had developed symptoms, whereas at the end of the study 5 (42%) patients were symptomatic. Symptomatic children were at most only slightly disabled, the cardinal symptoms being some difficulty in running or walking, or foot deformity; in fact, they were able to do school gymnastics with no apparent trouble. This mild clinical presentation differs from that of autosomal recessive patients, either axonal or demyelinating, usually exhibiting prominent leg or even hand amyotrophy. Be that

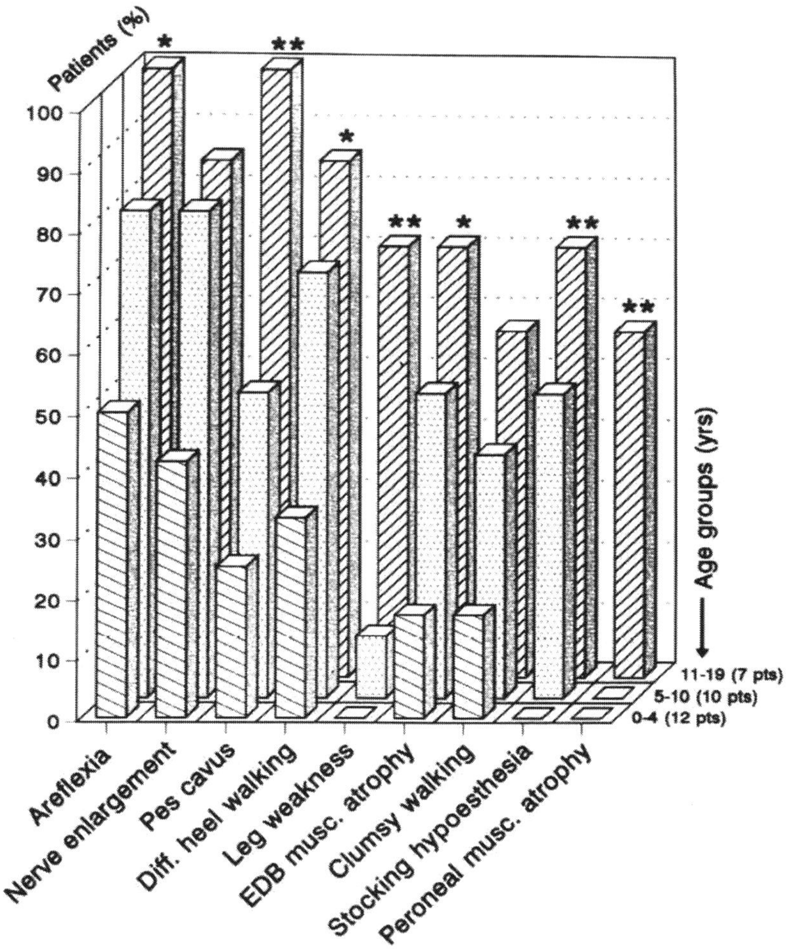


**Fig. 12.11** Frecuency of clinical signs and symptoms in CMT1A children according to age groups. Asterisks indicate significant differences between older and younger groups (∗p<0.05; ∗∗p<0.01). Reproduced with permission from [13]

as it may, the first two decades of life are an exceptional period of observation since CMT1A clinical picture evolves appreciably more quickly than in adult patients, where very slow progression of symptoms makes it difficult to assess the clinical course of the disease.

The clinical picture in children with any type of CMT is similar and basically consists of a peroneal muscular atrophy syndrome of variable severity. It is classical to consider that clinical signs are more severe in CMT1 than in CMT2, but exceptions to this rule are not rare. Figure 12.11 illustrates frequency of clinical signs in our referred CMT1A children splitting the series into three arbitrary age groups: 0–4 years (inclusion period), 5–10 years and 11–19 years (in 1998). Note that by definition all 12 affected children were present in the youngest age group, but at the end of the study only 10 and 7 patients reached the required age to be included in the remaining groups. During the inclusion period, the most frequent manifestations were areflexia (six cases), nerve enlargement (four cases), and difficulty in heel walking sometimes accompanied by shortening of tendo Achillis and incipient clawing of toes (Fig. 12.12). Other signs observed include pes cavus (three cases), pes planus (three cases), and extensor digitorum brevis (EDB) muscle atrophy. In no case was leg muscle atrophy observed. Because of the difficulty in assessing sensation in small children, stocking hypoesthesia was considered to be absent (see Fig. 12.11). In the second age group (5–10 years), frequencies of all previous signs increased, areflexia, nerve enlargement (Fig. 12.9C) and difficulty in heel walking again being most frequent. Stocking hypoesthesia was now noted in half the patients. Once again, leg muscle atrophy was not observed. The above-mentioned triad of signs was almost constant in patients aged 11 years or more (see Fig. 12.11). EDB muscle atrophy was an outstanding sign in some patients, and for the first time incipient to moderate atrophy of antero-lateral leg compartment was observed (four cases). Significant percentages of clinical signs of this and the youngest group appear in Fig. 12.11. Two children with pes planus at first examination then exhibited pes cavus.

From 2 years of age, all affected children had abnormal MCV, sensory conduction velocity, F-waves and distal motor latencies [13]. Furthermore, prolonged distal latency was already present in the first months of life and preceded slowing of MCV in three cases. It is worth noting that electrophysiologic studies were concordant with the presence or absence of the CMT1A duplication. We therefore concluded that in most CMT1A patients, symptoms appear early in childhood, although the florid clinical picture does not occur until the second decade of life. Serial electrophysiological studies can detect the CMT1A gene carrier in infancy.

### *12.4.1 Pathophysiology*

In the referred longitudinal study in CMT1A children, we investigated the electrophysiologic abnormalities accounting for the appearance and progression of EDB atrophy [2, 13]. All patients had two or more electrophysiological studies of peroneal nerve. EDB atrophy was observed in two out of 12 (17%) by age 5, in eight out of 10 (80%) examined between 5 and 9 years, and in all eight patients (100%) who had reached the second decade at the end. Nerve conduction maturation was systematically abnormal, but by age of 5 the mean values of nerve conduction parameters of peroneal nerve did not significantly differ from those in older patients. CMAP amplitudes of EDB were reduced in 42% of cases initially and 100% upon last exam. Furthermore, a constant finding throughout the study was progressive attenuation of CMAPs, these becoming unobtainable in four cases. As in adult

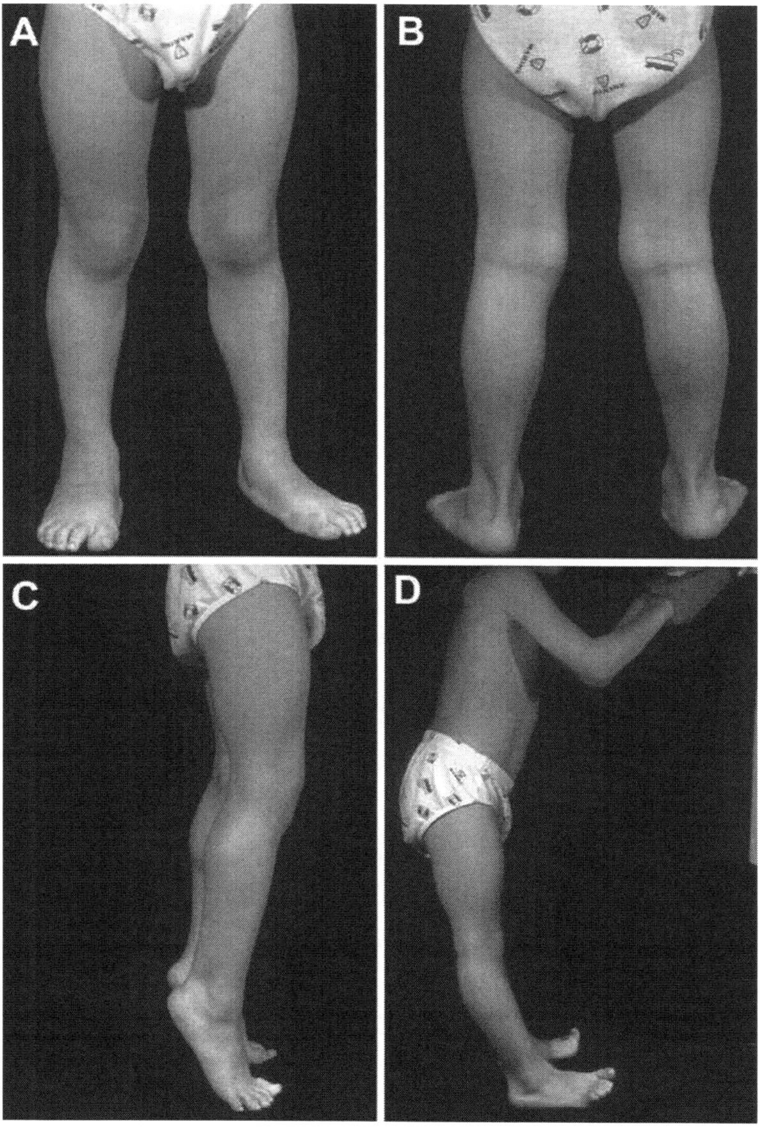


**Fig. 12.12** Affected CMT1A boy aged 5 years. (**A**) Neither peroneal muscular atrophy nor hammertoes is observed. (**B**) Note prominence of Achilles' tendons. (**C**) There is no difficulty in standing on tiptoe. (**D**) Conversely, standing on the heels is not possible; not dynamic clawing of toes. Reproduced with permission from [3]

series reported by Krajewski et al. [18], we concluded that EDB muscle atrophy in CMT1A children is an age-dependent sign which is accounted for by gradual reduction of the distal peroneal nerve CMAP amplitudes, not the degree of slowing of motor conduction velocity, that is, to the degree of secondary axonal degeneration to demyelination.

## 12.5 Proximal Lower-Limb Involvement

As indicated in previous paragraphs, the predominant clinical signs are distally accentuated muscle weakness and wasting in the lower limbs usually associated with loss of tendon reflexes and foot deformity. With the advance of the disease, more proximal muscles may become weak [27], though clinically perceptible paresis of proximal limb muscles is extremely rare in CMT1 [16]. Recently we have described a longitudinal study over three decades of a large CMT1A duplication pedigree, where severe pelvic and thigh musculature weakness occurred in two cases analyzing the MRI pattern of proximal lower limb muscle atrophy in one of them [4].

The pedigree comprised 18 examined patients, aged between 15 and 85 years (median, 46), serially evaluated for up to three decades. All 18 patients and 13 non-affected at-risk individuals had electrophysiological examination. MRI study of lower limb musculature was performed in four patients. Three patients had sural nerve biopsy. Genetic testing was performed in 17 patients and all 13 at-risk normal subjects.

Fourteen patients were asymptomatic or slightly disabled. The two oldest patients, aged 85 and 80, showed a moderate phenotype. The remaining two patients, aged 70 and 53, showed late onset and gradually progressive peroneal paresis extending up to thigh and pelvic musculature rending gait waddling (Fig. 12.13). In all three scanned patients with mild phenotype, MRI showed subtle and subclinical fatty infiltration of calf antero-lateral muscle compartments with thigh muscle involvement in one, and extensive atrophy of intrinsic foot muscles. In the youngest patient with proximal leg weakness, MRI showed massive fatty atrophy of all calf muscles, postero-medial thigh muscle compartments, and internal and external hip rotator muscles (Fig. 12.14). Sural nerve biopsies showed hypertrophic neuropathy with no superimposed inflammation. There was good correlation between electrophysiological and genetic testing.

We concluded that late in the clinical course a small proportion of CMT1A patients develop severe proximal leg weakness, long-term follow-up being essential for its detection. MRI may show subclinical involvement of thigh musculature.

## 12.6 Minimal CMT1A-Adult Phenotype

Whereas absence of signs on physical examination is relatively common in CMT1A duplication patients under 5 years of age (see above), this is an extremely rare and not well characterized situation in adult patients. We have addressed this question in a recent study describing clinico-electrophysiological features and MRI patterns of lower limb amyotrophy in two sub-clinical CMT1A adult patients [5]. After reviewing the clinical records of our cohort of 126 CMT1A cases coming from 25 unrelated pedigrees, we found two clinically normal adult individuals (Patient 1 and Patient 2) studied as secondary cases. Disease severity in terms of ability to walk and run was established using a nine-point functional disability scale (FDS) [6]. We administered the CMTNS, based on patients' symptoms, neurological examination

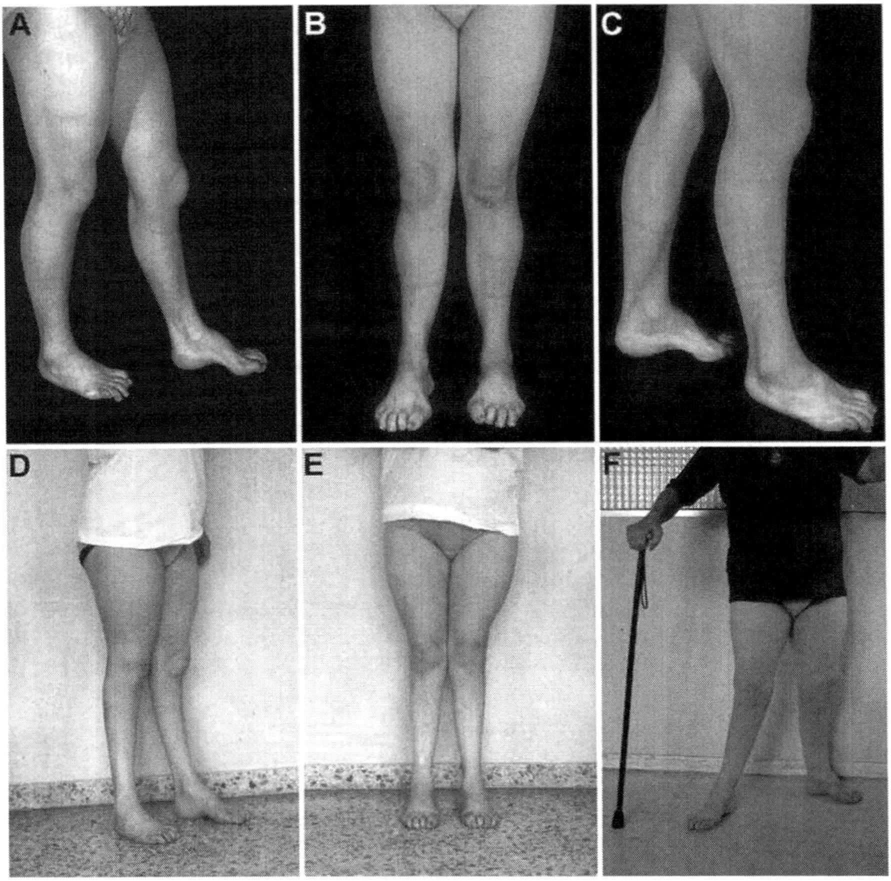


**Fig. 12.13** Serial photographs of the youngest patient with proximal weakness cited in the text. (**A**) This picture taken in 1976, at age 23, shows pes cavus and toe clawing, but preservation of leg musculature. (**B**, **C**). Eight years later, note the appearance of incipient bilateral peroneal muscular atrophy. (**D**, **E**) At age 38, marked and bilateral leg muscle atrophy is evident. (**F**) At age 53, note wide-based stance, which is only possible with support; lower limb amyotrophy and foot deformities are now masked by fattening up. Reproduced with permission from [4]

and neurophysiological testing [25]. MRI study included T1- and T2-weighted spin-echo sequences of thigh, lower-leg and foot musculature.

Patient 1 is aged 48 years and has been serially evaluated since age 17, her repeated physical examinations revealing just pes cavus with normal FDS score and mild CMTNS. Patient 2 is aged 58 and has been serially evaluated since age 48, her repeated physical examinations also revealing pes cavus with normal FDS score and mild CMTNS (Fig. 12.15). There was marked and uniform slowing of nerve conduction velocities with preservation or attenuation of foot CMAPs (see Fig. 12.15). Myoelectric deep tendon reflex (T-reflex) responses from biceps and gastrocnemius

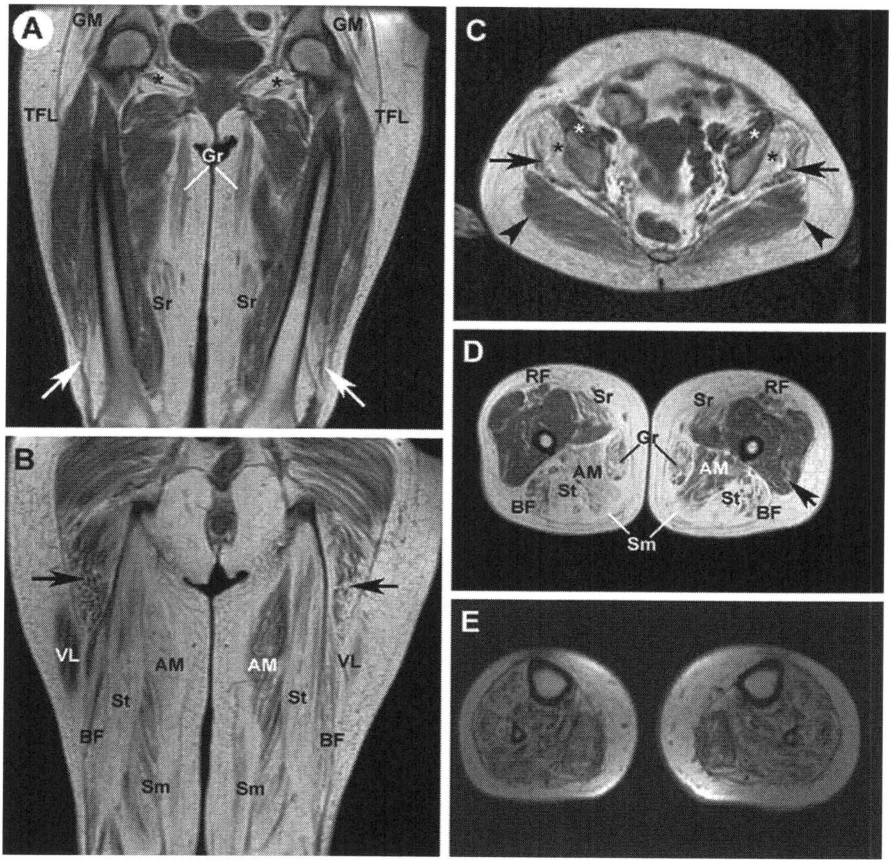


**Fig. 12.14** T1-weighted MR images of pelvis, thighs and legs from the patient illustrated in the previous figure. (**A**) Anterior coronal thigh image through femoral heads showing marked and symmetric hypersignal, indicative of fatty atrophy, of gluteus medius and minimus (GM), tensor fasciae latae (TFL), obturator externus (asterisks), gracilis (Gr), and sartorius (Sr) muscles. Note also distal fatty atrophy of both vastus lateralis muscles (*arrows*). (**B**) Posterior coronal thigh image through ischiatic tuberosities showing severe and symmetric fatty atrophy of the posterior femoral muscles (BF, biceps femoris; Sm, semimembranosus; and St, semitendinosus). Note also fatty infiltration of the lower portion of gluteus maximus muscles (*arrows*), and asymmetric involvement of adductor magnus (AM) and vastus lateralis (VL) muscles. (**C**) Pelvic axial image through acetabular roof showing severe fatty atrophy of gluteus medius (*arrows*) and gluteus minimus (*black asterisks*) muscles; note preservation at this level of gluteus maximus (*arrowheads*) and iliopsoas (*white asterisks*) muscles. (**D**) Middle third axial thigh image showing extensive fatty atrophy of the posterior and medial muscle compartments (AM, adductor magnus; BF, biceps femoris; Gr, garcilis; Sm, semimembranosus; and St, semitendinosus) and sartorius (Sr) and rectus femoris (RF) muscles. There is slight fatty infiltration of the left vastus lateralis muscle (*arrow*); note also asymmetry of adductor magnus (AM) and rectus femoris (RF) muscle involvement. (**E**) Middle third axial calf image showing extensive fatty infiltration of all four muscle compartments. Reproduced with permission from [4]

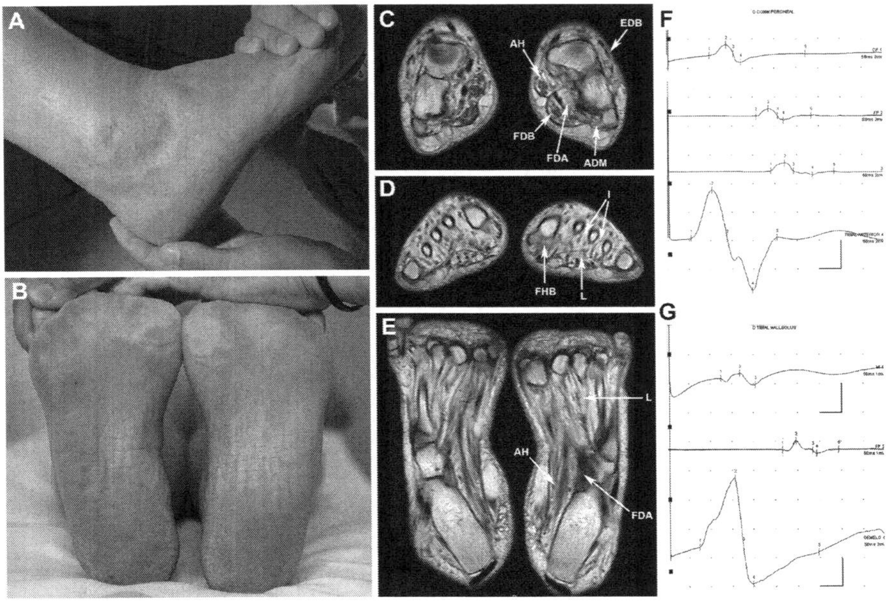


**Fig. 12.15** Lower-leg MRI study in Patient 2. (**A**) Coronal T1-weighted image of the legs showing a clear and striped fatty infiltration of both gastrocnemius medialis (GM) muscles. To a lesser degree note also fatty infiltration of soleus (So) and peroneus longus (PL), which is distally accentuated. (**B-D**) Axial T1-images (**B**, proximal calves; **C**, mid calves; **D**, lower calves) showing subtle and distally accentuated fatty infiltration of the anterior (*arrows*) and lateral (*arrowheads*) muscle compartments; in the superficial posterior compartment note outstanding fatty infiltration of GM and to a lesser extent of So, whereas the deep posterior muscle compartment is preserved

muscles were preserved though delayed. MRI showed a variable pattern of intrinsic foot musculature involvement: in Patient 1 fatty infiltration mainly involved the lumbricals, whereas Patient 2 showed extensive fatty atrophy of all foot muscles (Fig. 12.15). There was subtle and predominantly distal fatty infiltration of lower leg anterolateral muscle compartments in both patients. Patient 2 showed subclinical but evident fatty infiltration of gastrocnemius muscles. In no muscle was edema or contrast enhancement visualized.

We concluded that outstanding features of the minimal CMT1A adult phenotype include pes cavus with no other clinical signs of the disease, uniform slowing of nerve conduction velocity with either preservation or attenuation of foot CMAPs, delayed T-reflex, and fatty atrophy of intrinsic foot musculature on MRI study.

**Acknowledgements** Supported by CIBERNED and FIS Grant PI07/1323E.

# References

1. Berciano J, Combarros O, Calleja, Polo JM, Leno C. The application of nerve conduction and clinical studies to genetic counseling in hereditary motor and sensory neuropathy. Muscle Nerve 1989; 12: 302–306.

2. Berciano J, García A, Calleja J, Combarros O. Clinico-electrophysiological correlation of extensor digitorum brevis muscle atrophy in children with Charcot-Marie-Tooth disease 1A duplication. Neuromuscul Disord 2000; 10: 419–424.
3. Berciano J, García A, Combarros O. Initial semeiology in children with Charcot-Marie-Tooth disease. Muscle Nerve 2003; 27: 34–39.
4. Berciano J, Gallardo E, García, Infante J, Mateo I, Combarros O. Charcot-Marie-Tooth disease type 1A duplication with severe paresis of the proximal lower limb muscles: a long-term follow-up study. J Neurol Neurosurg Psychiatry 2006; 77: 1169–1176.
5. Berciano J, Gallardo E, García A, et al. Charcot-Marie-Tooth disease type 1A duplication: refining the minimal adult phenotype. J Peripher Nerv Syst 2008; 13: 310–2.
6. Birouk N, Gouider R, Le Guern E, et al. Charcot-Marie-Tooth disease type 1A with 17p11.2 duplication. Clinical and electrophysiological phenotype study and factors influencing disease severity in 119 cases. Brain 1997; 120: 813–823.
7. Chung KW, Suh BC, Shy ME, et al. Different clinical and magnetic resonance imaging features between Charcot-Marie-Tooth disease type 1A and 2A. Neuromuscul Disord (2008), doi: 10.1016/j.nmd.2008.05.012.
8. Combarros O, Calleja J, Figols J, Cabello A, Berciano J. Dominantly inherited motor and sensory neuropathy type I. Genetic, clinical, electrophysiological and pathological features in four families. J Neurol Sci 1983; 61: 181–191.
9. Dyck PJ, Lambert EH. Lower motor and primary sensory neuron diseases with peroneal muscular atrophy. I. Neurologic, genetic, and electrophysiologic findings in hereditary polyneuropathies. Arch Neurol 1968; 39: 1032–1038.
10. Dyck PJ, Karnes JL, Lambert EH. Longitudinal study of neuropathic deficits and nerve conduction abnormalities in hereditary motor and sensory neuropathy type 1. Neurology 1989; 39: 1302–1308.
11. Gabreëls-Festen AAWM, Joosten EMG, Gabreëls FJM, Jennekens FGI, Janssen-van Kempen TW. Early morphological features in dominantly inherited demyelinating motor and sensory neuropathy (HMSN type I). J Neurol Sci 1992; 107: 145–154.
12. Gallardo E, García A, Combarros O, Berciano J. Charcot-Marie-Tooth type 1A duplication: spectrum of clinical and magnetic resonance imaging features in leg and foot muscles. Brain 2006; 129: 426–437.
13. García A, Combarros O, Calleja J, Berciano J. Charcot-Marie-Tooth disease type 1A with 17p duplication in early infancy and childhood. A longitudinal clinical and electrophysiological study. Neurology 1998; 50: 1061–1067.
14. Hallam PJ, Harding AE, Berciano J, Barker DF, Malcolm S. Duplication of part of chromosome 17 is commonly associated with hereditary motor and sensory neuropathy type I (Charcot-Marie-Tooth disease type I). Ann Neurol 1992; 31: 570–572.
15. Harding AE, Thomas PK. The clinical features of hereditary motor and sensory neuropathy types I and II. Brain 1980; 103: 259–280.
16. Hoogendijk JE, De Visser M. Hereditary motor and sensory neuropathy types I and II (Charcot-Marie-Tooth disease). In: de Jong JMBV, ed. Handbook of clinical neurology, Vol. 16 (60), Amsterdam: Elsevier Science, 1991: 185–211.
17. Hoogendijk J, de Visser M, Bolhuis PA, Hart AAM, Ongerboer de Visser BM. Hereditary motor and sensory neuropathy type I: clinical and neurographical features of the 17p duplication subtype. Muscle Nerve 1994; 17: 85–90.
18. Krajewski KM, Lewis RA, Fuerst DR, et al. Neuronal dysfunction and axonal degeneration in Charcot-Marie-Tooth disease type 1A. Brain 2000; 123: 1516–1527.
19. Lupski JR, de Oca-Luna RM, Slaugenhaupt S, et al. DNA duplication associated with Charcot-Marie-Tooth disease type 1A. Cell 1991; 66: 219–232.
20. Lupski JR, García C, Parry GJ, Patel PI. Charcot-Marie-Tooth polyneuropathy syndrome: clinical, electrophysiologic and genetic aspects. In: Appel S, ed. Current neurology. Chicago: Mosby-Year Book, 1991:1–25.
21. Nicholson GA. Penetrance of the hereditary motor and sensory neuropathy Ia mutation: assessment by nerve conduction studies. Neurology 1991; 41: 547–552.

22. Raeymaekers P, Timmerman V, Nelis E, et al. Duplication in chromosome 17p11.2 in Charcot-Marie-Tooth neuropathy type Ia (CMT Ia). Neuromuscul Disord 1991; 1: 93–97.
23. Sabir M, Lyttle D. Pathogenesis of pes cavus in Charcot-Marie-Tooth disease. Clin Orthop Relat Res 1983; 175: 173–178.
24. Shy ME, Lupski JR, Chance PH, Klein CJ, Dyck PJ. Hereditary motor and sensory neuropathies. In: Dyck PJ, Thomas PK, eds.. Peripheral neuropathy. Philadelphia: Elsevier Saunders, 2005: 1623–1658.
25. Shy ME, Blake J, Krajewski K, et al. Reliability and validity of the CMT neuropathy score as a measure of disability. Neurology 2005; 64: 1209–1214.
26. Shy ME, Chen L, Swan ER, et al. Neuropathy progression in Charcot-Marie-Tooth disease type 1A. Neurology 2008; 70: 378–383.
27. Thomas PK, Marques W, Davis MB, et al. The phenotypic manifestations of chromosome 17p11.2 duplication. Brain 1997; 120: 465–478.

# Chapter 13
# Genotypes & Sensory Phenotypes in 2 New X-Linked Neuropathies (CMTX3 and dSMAX) and Dominant CMT/HMN Overlap Syndromes

**Garth Nicholson, Marina Kennerson, Megan Brewer, James Garbern, and Michael Shy**

**Abstract** Classification of neuropathies into Charcot-Marie-Tooth syndrome (CMT, hereditary motor and sensory neuropathy) or purely motor neuropathies is relatively easy in single patients but subtle sensory findings can vary in different affected individuals in a family. We examined the extent of sensory involvement in different individuals in two new X-linked neuropathy syndromes (CMTX3 and dSMAX) and in some dominantly inherited mainly motor neuropathies. CMTX3 is a mild X- linked recessive CMT phenotype linked to Xq26-28. dSMAX (distal spinal muscular atrophy linked to Xq13-21). We describe a new family linked to this locus that has some sensory findings which could also be described as a motor and sensory neuropathy i.e. a form of CMT.

In our dominant distal hereditary motor neuropathy (HMN) family linked to chromosome 7 (dHMN1) we also found some affected individuals with sensory signs as well as reduced sensory action potentials. In reported HMN families with known mutations in *GARS, SETX, HSPB1* and *HSPB8* genes and in many of our HMN families with unknown gene mutations, there is sensory involvement producing a CMT phenotype in some individuals. These disorders do not easily fit into traditional hereditary neuropathy classifications and should be recognised as CMT/HMN overlap syndromes. Recognition of overlap syndromes may assist development of more accurate gene screening paradigms.

**Keywords:** Hereditary neuropathy · Sensory signs · CMT · dSMAX · Motor neuropathy

## 13.1 Introduction

CMT is a common group of hereditary neuropathies caused by a large number of individually rare gene mutations and the common CMT1A duplication. At the present time even with a comprehensive gene screening program only about 50%

---

G. Nicholson (✉)
University of Sydney at the ANZAC Research Institute, Concord Hospital, NSW 2139, Australia
e-mail: garthn@med.usyd.edu.au

C. Espinós et al. (eds.), *Inherited Neuromuscular Diseases*, Advances in Experimental Medicine and Biology 652, DOI 10.1007/978-90-481-2813-6_13,

of families with hereditary neuropathies can be diagnosed at a molecular level. This suggests that there may be many more neuropathy genes to be found. As there are now a large number of genes to test, if phenotype recognition could be improved, then gene testing would become more affordable. However over selective phenotype definitions would exclude some individuals with gene mutations when the phenotype range is broad. One possible way to improve molecular diagnosis might be to better separate CMT syndromes from purely motor neuropathies. This review examines sensory involvement in X-linked and dominantly inherited motor neuropathies to see whether such distinctions are possible.

## 13.2 X-Linked Syndromes

### *13.2.1 CMTX3*

In our clinic less than half on our families with possible X-linked inheritance have *connexin32* mutations. Similar figures were reported from a Spanish clinic [2]. We recently reported a large Australian and New Zealand family of English origin linked to the X-chromosomal region reported as CMTX3 by Ionasescu [6]. Since then we have found a number of other families at the same locus. These families were selected from families with possible X-linked inheritance defined by no male to male inheritance of a CMT1A duplication negative CMT syndrome, with no *connexin32* mutations. Families large enough to yield a haplotype were screened with probe DXS453 to determine if any families might have *connexin32* promoter mutations. No such families were found. 6 families had a haplotype closely linked to the CMTX3 region.

**Table 13.1** Sensory action potentials and sensory exam in CMTX3 males

| Sex | Age | Median SNAP uV | Clinical |
|---|---|---|---|
| M | 9 | 0.0 | N |
| M | 11 | 0.1 | N |
| M | 14 | 0.0 | N |
| M | 14 | 5.0 | N |
| M | 14 | 15.0 | N |
| M | 17 | 13.0 | N |
| M | 19 | 1.0 | N |
| M | 27 | 0.0 | N |
| M | 28 | 0.0 | N |
| M | 32 | 0.7 | N |
| M | 32 | 14 | Pin & vibration not felt below ankles |
| M | 33 | 0.7 | N |
| M | 52 | 2.0 | N |
| M | 54 | 1.0 | Vibration absent at ankles |
| M | 56 | 2.0 | N |
| M | 60 | 1.8 | Vibration absent at ankles |

CMTX3 appears to be a truly X-linked recessive syndrome as no clinically affected obligate carrier females were found. Table 13.1 shows the sensory nerve action potential (SNAP) data in males. A HMN phenotype was found in 3 individuals with normal sensory action potentials (Table 13.1). These results show that CMTX3 males can have a CMT or HMN phenotype.

### *13.2.2 Distal X-Linked HMN (Distal X-Linked SMA, dSMAX)*

dSMAX (distal spinal muscular atrophy linked to the Xq13-21 chromosomal locus) was first described in a Brazilian family [12]. In a collaborative study we mapped a family with a HMN/CMT phenotype to the same locus [8]. Limited data from our dSMAX family shows individuals with borderline normal SNAPs. Two affected males had antidromic SNAPs of 17 and 27uV (normal >25uV). In some individuals there is minimal sensory involvement with subjective loss of sharp sensation. This syndrome therefore appears to be a HMN/CMT overlap syndrome (Table 13.3). Other *connexin32* negative families in our clinic, not linked to the *connexin32* locus were also screened for linkage to the dSMAX region. No new families were found suggesting that this may be a rare syndrome.

### *13.2.3 Dominant Neuropathies*

#### 13.2.3.1 HMN1, OMIM 182960

In our chromosome 7 linked distal motor neuropathy family (HMN1, OMIM 182960) there was minimal sensory involvement with reduced SNAPs in two older affected individuals (Table 13.2) producing a CMT phenotype. This syndrome is therefore another HMN/CMT overlap syndrome.

**Table 13.2** Sensory action potentials and sensory examination in chromosome 7 linked dominant HMN (HMN1, OMIM 182960)

| Sex | Age | Median SNAP | Sural SAP uV | Decreased sensation |
|---|---|---|---|---|
| M | 20 | 7 | 10 | Cold, sharp below knee |
| F | 20 | 34 | 7 | Pin, cold below knees |
| M | 23 | 19 | 2 | Cold, vibration below knee |
| M | 28 | 18 | 5 | Nil |
| F | 44 | 15 | ND | Nil |
| M | 50 | 8 | 5 | Cold, sharp below ankles |
| M | 54 | 5 | 2.8 | Sharp below ankle |
| M | 60 | 4 | ND | Temp, sharp below ankles |

Neurophysiology: Median sensory nerve action potentials (SNAPs) were tested by standard orthodromic methods. Reference values are shown in Table 13.1. The lower normal limit was 7uV.

**Table 13.3** Extent of sensory involvement in motor neuropathies

| Disease | Phenotype | SNAPs | Genotype |
|---|---|---|---|
| DSMAX | dHMN/CMT2 | Borderline & normal | unknown |
| CMTX3 | dHMN/CMT2 | Borderline & normal | unknown |
| CMT2A | CMT2 | Reduced | MFN2 |
| CMT2F | CMT2 | Reduced | HSPB1 |
| CMT2L | CMT2 | Reduced | HSPB8 |
| CMT2 | CMT2 | Reduced | SETX |
| CMT2D | CMT2 | Reduced | BSCL2 |
| HMN | dHMN | Normal | BSCL2 GARS |
| dHMN V | dHMN | Normal | GARS |
| dHMN II | dHMN | Normal | HSPB1 & B2 |
| dHMNP | dHMN & | Normal | BSCL2 |
| ALS4 | Pyramidal signs | Normal | SETX |
| | | Normal | HSPB1 |
| Silver | No sensory spasticity dMN | Normal | BSCL2 |
| HSP17 | No sensory spasticity +/− dMN | Normal | BSCL2 |

This table does not include infantile neuropathies or complicated forms with e.g. vocal cord paralysis. SN sensory neuropathy, dMN distal motor neuropathy, dHMNP is distal. HMN with pyramidal signs. dHMN II and V are childhood and adult onset distal spinal muscular atrophy (dSMA) syndromes.

#### 13.2.3.2 Other Dominant Motor Neuropathy Syndromes

In a chromosome 4 linked dominant HMN there was one individual with sensory signs but SNAPs were not reduced [9].

#### 13.2.3.3 Known Genes Causing Motor Neuropathies

A list of genes that have been reported to produce both CMT and HMN syndromes is shown in Table 13.4. The family in which *SETX* mutations were first discovered was initially classified as CMT [10]. There is one report of reduced median SNAPs in a *SETX* mutation [3] so this syndrome is a CMT/HMN overlap syndrome.

*GARS* mutations can occur in families with either CMT2D or HMN phenotypes. *MFN2* mutations can cause both CMT2 and a pyramidal CMT (CMTP) syndrome [14, 16].

Some individuals with sensory involvement have been reported in *HSPB8* [3, 13]. HSPB1 families were initially reported as CMT2F (see Table 13.1). These genes also can produce CMT/HMN overlap syndromes and as well as HMNP in the case of *HSPB1* [3].

There was no sensory involvement in individuals with *BSCL2* mutations causing dSMA or in dSMA with pyramidal signs including Silver's syndrome [7, 15], but 20% of affected individuals in a large Austrian family with a *BSCL2* mutation had a CMT2 phenotype [1].

**Table 13.4** List of genes with mutations producing mutigenerational hereditary, mainly motor neuropathy, overlap syndromes

| Gene/disease | Overlap syndrome |
|---|---|
| MFN2 | CMT2/CMTP |
| CMTX3 | CMT2/HMN |
| dSMAX | CMT2/HMN |
| GARS | CMT2D/HMN |
| HSPB8 | CMT2/HMN |
| SETX | CMT/HMN |
| HSPB1 | CMT2F/HMN/HMNP |
| BSCL2 | HMN/HMNP/SS/HSP17 |

SS, Silver syndrome; HSP, hereditary spastic paraplegia

#### 13.2.3.4 Unknown Genes Causing Motor Neuropathies

In our clinic there are 60 families out of approximately 1000 families with HMN phenotypes. In about 50% of these families, mutations in exon 3 of *BSCL2, GARS* and *MFN2* have been excluded indicating that these genes are rare causes of HMN. In our family (CMT105) with no known gene mutation there was a range of SNAPs from absent to normal [14]. In the majority of other HMN and HMNP families there was a range of median SNAPs from normal to reduced. *HSPB8* and *HSPB1* have not been tested in our series. These gene mutations are a rare cause of HMN especially *HSPB8* [5, 11]. In many of our families there are individuals with absent, reduced or normal SNAPs suggesting that such HMN/CMT overlap syndromes are not uncommon and may be caused by a number of as yet unknown gene mutations.

## 13.3 Discussion

Documentation of sensory involvement at clinical and neurophysiological levels is important to separate classical hereditary neuropathy syndromes. In some neuropathies SNAPs are extremely useful to define a phenotype e.g. CMT1A has minimal few sensory symptoms and minimal clinically detectable sensory loss but SNAPs are not detectable. Motor involvement was initially unrecognised in HSN1 but is present in all adult affected individuals. HSN1 is therefore a sensory and motor neuropathy, i.e. a form of CMT.

Our data on HMNs shows the variability of median SNAPs in different affected members of a family. More than one family is needed in order to reliably describe the phenotype range associated with mutations in a particular gene. Such families have CMT/HMN overlap syndromes caused by known genes (see Table 13.1) and more commonly by as yet unknown genes. Pure distal motor syndromes (dHMN) may be distinguished only after a number of individuals in a family consistently have no sensory signs. Attention to these clinical details and separation of phenotypes into true CMTs and broader CMT/dHMN overlap syndromes and pure HMN should help in selection of appropriate genes to screen. At present the yield of screening is low. Screening distal HMN syndromes produced a 7% yield of *BSCL2* mutations [4]. Screening *HSPB1* yielded mutations in 4/115 familial and in 2/30 sporadic cases [5].

## References

1. Auer-Grumbach M, Schlotter-Weigel B, et al. (2005). Phenotypes of the N88S Berardinelli-Seip congenital lipodystrophy 2 mutation. Ann Neurol 57(3): 415–424.
2. Casasnovas C, Banchs I, et al. (2006). Clinical and molecular analysis of X-linked Charcot-Marie-Tooth disease type 1 in Spanish population. Clin Genet 70(6): 516–523.
3. Dierick I, Baets J, et al. (2008). Relative contribution of mutations in genes for autosomal dominant distal hereditary motor neuropathies: a genotype-phenotype correlation study. Brain 131(Pt 5): 1217–1227.
4. Dierick I, Irobi J, et al. (2007). Genetic variant in the HSPB1 promoter region impairs the HSP27 stress response. Hum Mutat 28(8): 830.
5. Houlden H, Laura M, et al. (2008). Mutations in the HSP27 (HSPB1) gene cause dominant, recessive, and sporadic distal HMN/CMT type 2. Neurology 71(21): 1660–8.
6. Ionasescu VV, Trofatter J, et al. (1991). Heterogeneity in X-linked recessive Charcot-Marie-Tooth neuropathy. Am J Hum Genet 48(6): 1075–1083.
7. Irobi J, Van den Bergh P, et al. (2004). The phenotype of motor neuropathies associated with BSCL2 mutations is broader than Silver syndrome and distal HMN type V. Brain 127(Pt 9): 2124–2130.
8. Kennerson M, Nicholson G, et al. (2009). X-linked distal hereditary motor neuropathy maps to the DSMAX locus on chromosome Xq13.1-q21. Neurology 72(3): 246–52.
9. Muglia M, Magariello A, et al. (2008). A novel locus for dHMN with pyramidal features maps to chromosome 4q34.3-q35.2. Clin Genet 73(5): 486–491.
10. Myrianthopoulos NC, Lane MH, et al. (1964). Nerve Conduction and Other Studies in Families with Charcot-Marie-Tooth Disease. Brain 87: 589–608.
11. Rohkamm B, Reilly MM, et al. (2007). Further evidence for genetic heterogeneity of distal HMN type V, CMT2 with predominant hand involvement and Silver syndrome. J Neurol Sci 263(1–2): 100–106.
12. Takata RI, Speck Martins CE, et al. (2004). A new locus for recessive distal spinal muscular atrophy at Xq13.1-q21. J Med Genet 41(3): 224–229.
13. Tang B, Liu X, et al. (2005). Mutation analysis of the small heat shock protein 27 gene in chinese patients with Charcot-Marie-Tooth disease. Arch Neurol 62(8): 1201–1207.
14. Vucic S, Kennerson M, et al. (2003). CMT with pyramidal features. Charcot-Marie-Tooth. Neurology 60(4): 696–699.
15. Windpassinger C, Auer-Grumbach M, et al. (2004). Heterozygous missense mutations in BSCL2 are associated with distal hereditary motor neuropathy and Silver syndrome. Nat Genet 36(3): 271–276.
16. Zhu D, Kennerson ML, et al. (2005). Charcot-Marie-Tooth with pyramidal signs is genetically heterogeneous: families with and without MFN2 mutations. Neurology 65(3): 496–497.

# Chapter 14
# Natural History and Treatment of Peripheral Inherited Neuropathies

**Davide Pareyson and Chiara Marchesi**

**Abstract** Charcot-Marie-Tooth disease (CMT) is genetically highly heterogeneous. Disease course and severity vary according to CMT type, causative gene, and mutation type, but considerable phenotypic variability may occur also for the same CMT type. Research is focused on possible modifier factors particularly in CMT1A associated with Peripheral Myelin Protein 22 (PMP22) overexpression. Natural history studies are important to define disease course in different CMT types and to allow better assessment of intervention efficacy. Only a few such studies have been carried out, mainly on CMT1A, and described impairment and disability progression. Motor potential amplitudes seem to correlate with disease severity and progression, suggesting that axonal loss is the basis of disability in CMT. There is need to develop suitable and reproducible outcome measures: the CMT Neuropathy Score is the only validated outcome score specific for CMT, but others have been tested during the last few years. Currently there is no effective drug therapy for CMT and supportive treatment is limited to physical therapy, orthotics, surgical treatment of skeletal deformities and soft tissue abnormalities, and symptomatic drug treatment. Research is focused on developing new treatment strategies and approaches. The progesterone antagonist onapristone proved to be effective in a rat model of CMT1A; unfortunately, currently available progesterone antagonists are too toxic to be safely administered to patients. Neurotrophin-3 (NT3), a neurotrophic factor known to promote axonal growth, was tested with favourable results in two animal models and in a pilot study involving eight CMT1A patients. Ascorbic acid (AA) administration to CMT1A mice improved clinical and neuropathological findings, possibly by down-regulating PMP22 through a cAMP mediated mechanism. Clinical trials of AA in the human disease are currently being performed. Curcumin stimulates translocation of misfolded protein from the endoplasmic reticulum and proved useful for selected myelin protein zero and PMP22 mutants *in vitro* and in the animal models Trembler and Trembler$^{J}$.

---

D. Pareyson (✉)
IRCCS Foundation, "C. Besta" Neurological Institute, Via Celoria 11, 20133, Milan, Italy
e-mail: dpareys@istituto-besta.it

C. Espinós et al. (eds.), *Inherited Neuromuscular Diseases*, Advances in Experimental Medicine and Biology 652, DOI 10.1007/978-90-481-2813-6_14,

**Keywords:** Charcot-Marie-Tooth disease · Natural history · Outcome measures · Therapy

## 14.1 Introduction

This review deals with natural history and treatment of Charcot-Marie-Tooth disease (CMT), which is the most frequent hereditary neuropathy with an estimated prevalence of 17–40 affected person every 100,000 [64]. CMT is a heterogeneous group of diseases sharing the same clinical phenotype, usually characterized by wasting and weakness of distal limb muscles, decreased to absent deep tendon reflexes (DTRs), distal sensory loss, and frequent skeletal deformities [76]. The clinical phenotype is the consequence of a progressive axonal loss affecting preferentially the longest axons whatever the underlying primary pathogenic mechanism, either a myelinopathy or an axonopathy [51, 76].

## 14.2 CMT Subtype Frequency

CMT is usually subdivided into a demyelinating form (CMT1 if autosomal dominant, CMT4 if autosomal recessive), an axonal type (CMT2), and a pure motor form (distal Hereditary Motor Neuronopathy). Intermediate forms between CMT1 and CMT2 are recognised, the main one being X-linked CMT (CMTX), but rare dominant intermediate CMT types have also been identified. Pyramidal involvement (Hereditary Motor and Sensory Neuropathy, HMSN, type V) or optic atrophy (HMSN VI) rarely occur. Further subdivision is based mainly on causative genes and assigned loci [39, 76, 84].

CMT type 1A (CMT1A) associated with the 1.4 Mb 17p11.2-p12 duplication including the Peripheral Myelin Protein 22 (*PMP22*) gene is the most common form, accounting for 40–50% of all CMT cases; rarely, CMT1A is caused by *PMP22* point mutations (1% of CMT1 cases). CMTX associated with mutations in the Gap Junction B1 (*GJB1*) gene encoding Connexin-32 (Cx32) is the second most common type (about 10% of all CMT patients). Mutations in Myelin Protein Zero (*MPZ*) and Mitofusin 2 (*MFN2*) genes are less frequent but important causes of CMT: *MPZ* mutations usually cause the demyelinating subtype CMT1B (4–5% of all CMT1 cases) and less frequently axonal CMT2 with late onset; mutations in *MFN2* account for 10–20% of axonal CMT2 cases (CMT2A). All the other genetically defined forms thus far identified are much less common [5, 23, 39, 72, 75, 78, 84].

## 14.3 Phenotypes

### *14.3.1 Disease Severity Depends on CMT Type*

The typical CMT patients have symptom onset in the first or second decade of life and show slow progression of disease thereafter. However, age at onset, clinical course, progression rate, and overall severity vary according to CMT type, causative gene, and type of mutation.

CMT1A associated with the duplication is the most studied form, as several series of patients have been reported and followed up [4, 13, 35, 37, 45, 61, 103, 105]. In most cases onset occurs within the first two decades of life, the subsequent course is usually very slowly progressive, and almost all patients remain ambulatory throughout their life. *PMP22* point mutations are usually associated with more severe CMT1 [61].

Most *MPZ* mutations give rise to early onset and sometimes quite severe demyelinating CMT1B, whereas others are associated with late onset axonal CMT2, which, despite later onset, usually runs a severe course and some patients eventually become chairbound [20, 52, 93]. *MFN2* mutations originate CMT2A, which, depending on mutation type, may start early in life and show rapid progression, or contrariwise have later onset and milder course [16, 106]. In affected males, CMTX usually starts in the first or second decade of life and runs a restlessly progressive course, often leading to relevant impairment later in life; heterozygous female carriers may be completely asymptomatic or show mild clinical and electrophysiological evidence of CMT, but rare cases with severe CMT have been reported and attributed to skewed X-chromosome inactivation [54, 55, 96]. Autosomal recessive (AR) forms of CMT (AR-CMT2 and demyelinating CMT4) are usually more severe than the dominant forms, have early onset and more severe course [76, 104].

### *14.3.2 Phenotypic Variability*

CMT, however, is characterized by considerable phenotypic variability even within the same CMT type. This is particularly true for CMT1A, which is considered on average a rather "mild" form of CMT, but is characterised by a wide range of disease severity. CMT1A patients may have delayed motor milestones, severe skeletal deformities including scoliosis, develop relevant proximal weakness, require walking aids or even exceptionally become chairbound [2, 4, 35, 61, 103, 105]. On the other side, some CMT1A patients are almost or completely asymptomatic, are unaware of being affected, and conduct a normal or near-normal life. Such considerable disease severity variability may occur also within the same family; different disease severity has been recorded even between monozygous twins [29]. The research is now focused on the search for modifier factors, either genetic and environmental, which influence disease severity. The pathogenic mechanism of CMT1A is a gene dosage effect, related to soprannumerary *PMP22* gene copy number which leads to protein overexpression. The interest of such studies is self-explanatory, as the identification of factors modifying PMP22 expression levels may help developing therapeutic agents. Different disease expression in patients carrying the same mutation in other CMT-related gene has been reported [93, 100].

Comorbidity also appears to be relevant by worsening CMT. Concomitant diabetes, vincristine administration, superimposition of chronic or acute inflammatory demyelinating polyradiculoneuropathy (CIDP or Guillain-Barré syndrome), have been reported to result in more severe disease course or sudden worsening [1, 31,

32, 75]. Rare instances of "double trouble", i.e. co-occurrence of mutations in two different CMT-related genes in the same patient, have been described and have led to more severe disease [36].

## 14.4 Natural History Studies

Natural history studies are important to define more precisely the disease course of the different CMT types. Knowledge of the natural course of the disease allows accurate prognosis, appropriate genetic counselling, and better assessment of efficacy of intervention studies. However, a few studies on natural history of CMT have been carried out, and were mainly focused on CMT1A, as it is the most frequent type [74, 97, 101, 105]. There are important descriptive cross-sectional studies of patients' series, reporting age of onset, frequency of clinical signs and symptoms, nerve conduction study values, nerve biopsy changes [3, 4, 13, 24, 35, 103, 105]. Some studies tried also to employ or develop scores and scales to evaluate disease severity, and described the series accordingly [4, 51, 105].

Only a few prospective studies, however, were aimed at following disease course, and these researches described progression of impairment and disability in CMT1A [24, 74, 97] and CMT2 [101]. A few patients with the rare CMT types associated with mutations in the *EGR2* gene were shown to have important disease progression, and some of them died during the follow-up period [100, 102].

What is not known is whether disease progression rate is constant or dependent on age. Possible slower progression during adolescence was reported by Dyck et al. [24] in CMT1, and Shy et al. [97] provided data suggestive of slightly faster progression in older subjects with CMT1A. There is some evidence that early onset CMT1A patients show slower nerve conduction velocities (NCVs) and develop more severe functional impairment [4].

Analysis of electrophysiological data such as NCV and amplitude of compound muscle action potentials (CMAPs) have shown that in CMT1A NCVs do not correlate with disease severity, but CMAP amplitudes do [24, 51, 105]. Krajewski et al. [51] in a cross-sectional studies of 42 CMT1A patients showed a good correlation between upper limb CMAP amplitude and hand strength, and between sensory loss and sensory action potential (SAP) amplitude decrease. Longitudinal studies of CMT1A NCV and CMAP have confirmed that NCV does not significantly change over time, while the change of CMAP amplitude is less well established, with evidence of lower values in older subjects and possible decrease as time passes [45, 30, 51]. These observations are important and leads to the conclusion that axonal loss is the basis of disease severity and progression in CMT1A.

### *14.4.1 Outcome Measures*

As stated above, many studies were retrospective, and only a few prospective studies have been performed [74, 97, 101]. There is need to perform such prospective studies and to develop suitable outcome measures which are reproducible and can be diffusely employed to define natural history of CMT and to monitor its course.

A good clinical rating scale should be simple (easy to perform), valid (really computing what it is intended to measure), reliable (its results are reproducible in the same conditions by the same examiner or by a different examiner), and sensitive to change (able to detect significant modifications). Outcome measures may assess impairment, activity limitations (formerly disability), and handicap with reference to participation and context (WHO-ICH2/ICF [116]).

The definition of outcome measures for CMT is a particularly hard task, as CMT is a slowly progressive disease with subtle changes over time, the main characteristic is weakness of distal muscles that is traditionally difficult to measure, foot deformities are relevant and interfere with assessment of walking and strength, and the resulting disability is the product of impairment of different functions. During the last few years, different outcome measures have been used for CMT assessment. Some measures have been derived from other diseases such as multiple sclerosis or acquired peripheral neuropathies and have been tested and validated in CMT, others have been developed specifically for CMT. Two ENMC workshops dealt with outcome measures either in peripheral neuropathy (131th) and particularly in CMT (136th) [67, 85]. Participants to the latter workshop agreed on a series of recommended outcome measures for evaluation of CMT in clinical trial design.

#### 14.4.1.1 CMT Neuropathy Score (CMTNS)

The first and main outcome score specifically developed for CMT is the CMT Neuropathy Score [94], which is a composite scale with 9 items and a score ranging from 0, normal, to 36, worst score. It takes into account sensory symptoms in the legs, motor symptoms in upper and lower limbs (disability component), motor signs in upper and lower limbs, loss of pinprick and vibration sense in lower limbs (clinical impairment), values of CMAP and SAP amplitude in upper limbs (instrumental impairment). The CMTNS is derived from the Total Neuropathy Score [19], which has been developed for diabetic and toxic neuropathies and has introduced the important concept of measuring the length–dependent progression of sensory and motor limb involvement. The CMTNS has shown excellent intra- and inter-examiner reliability, and good correlation with other outcome measures of peripheral neuropathy including the Neuropathy Impairment Score (NIS) [94], and it has gained popularity among CMT clinical researchers. Moreover, it has been shown to change over time, with a 0.686 point increase per year in CMT1A [97]. These are all important observations, allowing a shared method of assessment of disease severity in different CMT populations and calculation of sample size necessary for clinical trials' design. Shy et al. [96] evaluated a large series of CMTX patients using the CMTNS and concluded that the scores positively correlated with the age of the patients, and both the motor and the sensory components worsened with advancing age, indicating that the disease is overall progressive.

#### 14.4.1.2 Other Outcome Measures in CMT

The NIS is a composite impairment scale measuring weakness, sensory loss, and DTRs changes and has been widely used in the evaluation of peripheral neuropathy. It has never been validated in CMT, but Dyck and coauthors performed a longitudinal studies of 31 CMT1 patients followed for up to 30 years and showed a worsening of the score of about 1 point/year in adults, whereas patients younger than age 15 at first examination tended to have slower progression [24]. Recently, Shy and coworkers found a 1.386 point per year worsening of the NIS in adults CMT1A subjects [97].

Our group [98] assessed intra- and inter-rater reliability of different outcome measures in a series of 40 CMT patients. We evaluated measures of impairment such as quantitative evaluation of muscle strength (maximal voluntary isometric contraction, MVIC) employing a hand-held myometer; reliability was excellent for hand movement (hand-grip and three point pinch), but unsatisfactory for proximal muscle and foot movements. We therefore developed a device to immobilise the foot and obtained good to excellent reliability for foot dorsiflexion and plantar flexion, whereas we chose to exclude proximal muscle MVIC evaluation from candidate measures for CMT. We also assessed reliability of disability outcome measures such as the 10-m timed walking test (10MTW, a quantitative mobility and leg function performance test), the 9-hole peg test (9HPT, a brief quantitative test of upper extremity function), and the Overall Neuropathy Limitations Scale (ONLS, an arm and leg disability scale evaluating the patient's ability to accomplish different tasks involving the upper limbs and the walking ability). All these measures demonstrated good to excellent reliability and were therefore recommended as outcome measures for CMT trials [84] and were adopted in a clinical trial of ascorbic acid in CMT1A [77].

Different methods to assess muscle strength have been tested and validated in subject with CMT.

Kilmer et al. [47] tested intra- and inter-examiner reliability of strength assessment in 10 CMT1 patients using a hand-held-myometer and obtained good to excellent results for most muscles; however, results were not so satisfactory for ankle dorsiflexion. Videler et al. [107] found excellent reproducibility of handgrip peak force in 20 CMT subjects. Sellers et al. [90] tested the Rotterdam Intrinsic Hand Myometer (RIHM), a hand-held dynamometer that directly measures intrinsic hand muscle strength, in 41 CMT patients. Both intra- and inter-rater reliability were excellent. Burns et al. [8] found excellent intrarater reliability of foot movement strength using a hand-held-myometer in 11 CMT patients with foot deformities); inter-examiner reliability was not assessed, however. Kilmer et al. [48] evaluated reliability of simulated work performance tasks including isokinetic peak torque in nine CMT1 patients, and found excellent reproducibility.

Svensson et al. [99] evaluated test-retest reliability of different measures of hand function in 20 CMT patients. They performed measurement of dexterity (Box and Block Test and 9HPT), strength (grip and pinch using the Grippit instrument), and tactile sensation (Shape Texture Identification Test). Reliability was good to excellent for all tests, which proved valuable to evaluate hand function in CMT. However, authors underlined that certain factors must be taken into account, such as limited time aspects for the Nine-Hole Peg Test and the number of trials used, and standard error of measurement, which is fairly large for strength evaluation, and may limit measure sensitivity.

Recently, Videler et al. [108] found excellent test-retest reliability of two measures of hand dexterity, namely the Sollerman test and the Functional Dexterity test, in 49 CMT1A patients.

Other outcome measures have been employed, but they have never been validated for CMT. Krajewski et al. [51] used the Modified Ambulation Index, and Verhamme et al. [105] used the physical dimension score of the Sickness Impact Profile (SIP), and the arm and leg subsets of the Guy's neurological disability scale. The walk-12 is another disability scale which was validated in a series of 50 neuropathic patients including 3 subjects with CMT [33].

A few studies assessed Quality of Life in CMT patients and all of them found that it is reduced in CMT patients with respect to the normal population [74, 82, 105, 110]; the SF36 questionnaire is the most frequently used scale, although it is not disease specific. Older age, female sex, and nonworking status were all associated with lower scores [74, 110], but an intriguing finding is that QoL did not appear to worsen during the follow up period [74, 105], suggesting that either QoL scales are not enough sensitive to change, or that the chronic course of the disease allows compensation.

The number of different tested measures and some difficulties in obtaining reproducible results underline the need for standardization of outcome measures. Particularly in multicentre clinical trial (which will be the great majority in the future for CMT) it is fundamental to define the procedure in great detail, agree on the procedures, perform session for instruction and standardization between examiners, and provide written or video/DVD material for instruction.

There is also need to develop specific outcome measures for children, as the CMTNS cannot be used for children under age 6, and motor development and growth may hamper accurate evaluation of disease course. Another major issue is the need of biological markers in CMT. One option is skin biopsy, that can provide an easy access to nerve tissue for assessing mRNA and protein content, and for immunological studies [53, 60, 66]. Another interesting approach is the study of muscle involvement with magnetic resonance imaging, which demonstrated that the disease process involves at first the intrinsic foot muscles and then progresses to leg muscles; in CMT1A denervation involves at first the anterolateral compartment and then the posterior compartment, whereas in CMT2A the superficial posterior leg compartment seems to be the first to be involved [17, 28].

## 14.5 Treatment of Charcot-Marie-Tooth Disease

Currently there is no effective drug therapy for CMT. Supportive treatment is limited to physical therapy, use of orthotics, surgical treatment of skeletal deformities and soft tissue abnormalities, and symptomatic drug treatment for pain (and possibly fatigue). A multidisciplinary approach to treatment of CMT is needed, where the neurologist collaborates with the physiatrist, the physical therapist, the orthopaedic surgeon, the genetic counsellor, and sometimes other professional figures such as the psychologist, the occupational therapist, the polmunary specialist [39].

Ongoing research is focused on developing new treatment strategies and approaches. Animal models are being useful for such purposes, and clinical trials are underway.

### 14.5.1 Rehabilitative Treatment

Several different approaches to rehabilitation have been proposed and employed for CMT treatment. Unfortunately, however, there are only a few clinical randomised trials and two Cochrane reviews [87, 118] considered suitable for review only two studies, which provided evidence for benefit of rehabilitation therapy in CMT. Lindeman et al. [56–58] demonstrated that moderate exercise is effective and safe for CMT patients, and results in significant improvement as assessed through the 6-m timed walking and isokinetic knee extension torques. In this trial, 29 CMT participants were randomly assigned to receive a training of 3 times a week for 24 weeks with weights adapted to their force. Training and most measurement tools involved the proximal lower extremity muscles.

Chetlin and coauthors also provided evidence that a moderate-intensity 12-week home-based resistance training program produced improvement in quantitative muscle assessment of isometric strength and in activities of daily living (ADLs) [14, 15].

Recently, El Mhandi et al. [26] treated 4 CMT1A and 4 CMT2 patients with a 24-week interval-training exercise program performed three times per week. This was well tolerated and of benefit to their subjective perception of pain/fatigue, and improved the functional capacity, dynamic strength and physiological capacities.

In other studies, a few CMT patients underwent home-based aerobic exercise training and showed minimal benefits in walking ability and aerobic capacity [50, 117]. These results suggest that aerobic exercise may be of help in the CMT rehabilitation [49].

However, in a study performed by Kilmer et al. [46] a high resistance exercise program was not superior to moderate exercise training but caused some deleterious effects to some patients. Vinci et al. [109] indirectly furnished elements in support of overwork weakness, by showing that the dominant hand is weaker in CMT subjects.

Complex approaches have also been proposed and in one study 16 CMT patients were randomly assigned to either an experimental protocol or control group. The intervention session consisted of passive stretching, muscle strengthening and dynamic balance training during standing and stepping. Benefit was reported but the intervention lasted only 12 days [65].

### 14.5.2 Orthotics

Plantars, orthoses and assistive devices may be of help in CMT patients. Plantars are commonly used to correct foot position and to avoid pressure and contact sores and callosities. Ankle-foot orthoses (AFOs) are the most commonly prescribed tools to overcome foot drop and facilitate gait. Several different types made of different materials exist. They are, however, often poorly tolerated [111]. Custom-fitted AFOs are more comfortable, provide better compliance, and may give relief to painful pes cavus [9, 12]. Refshauge et al. [83] observed that stretching by wearing night splints (6 weeks/leg) did not increase ankle range of motion or strength in 14 CMT1A patients. Bracing orthotics are also of help for patients with severe upper limb involvement [12, 39].

### *14.5.3 Respiratory Failure*

Respiratory involvement is exceedingly rare in CMT, but it has been reported in CMT1A [103], CMT2C [70], and other CMT types [34]. Diaphragm weakness, vocal cord palsy in adduction, sleep apnoeas, and restless-leg syndrome may be associated with CMT and require proper treatment [1].

### *14.5.4 Surgical Treatment*

Many different approaches have been used along the years to treat skeletal deformities, and particularly foot deformities. Again, there is lack of prospective studies and treatment efficacy has been evaluated only in retrospective studies, often conducted on limited series of patients, and treated with non-homogeneous approaches. Therefore, evidence for preferring one approach rather than another is scanty, and indications and timing of surgery are based at the best on experts' opinion. Triple artrhodesis has long been employed to treat the most severe foot deformities [63, 89, 112], but long-term results have been poorly investigated and probably are of limited reward. Predominant opinion is to reserve this approach to selected severe cases when other more conservative approaches are not feasible and surgery cannot be avoided [112]. More limited osteotomies are frequently employed to correct equino-cavo-varus deformities. Soft tissue surgery includes plantar fasciotomy to reduce cavus deformity and achilles tendon lengthening to treat retractions. Posterior tibial tendon transfer is also used to partly restore foot dorsiflexion, as the anterolateral leg muscle compartment is usually more severely affected than the posterior leg compartment; however, posterior leg compartment muscles are also weak to some extent, tend to become weaker over time, and loss of efficacy may occur with time [69]. Tendon transfers can also be of help to treat upper limb motor deficits and to recover thumb opposition or wrist extension [7, 68, 115].

Scoliosis is not infrequent in CMT and in the most severe cases requires surgical treatment. Karol et al. [41] found significant scoliosis in 45 out of 298 CMT patients. They found differences with idiopathic scoliosis, observed frequent thoracic hyperkyphosis, and concluded that bracing is usually unsuccessful. Surgical fusion was necessary in 14 cases and was not associated with a high rate of complications, although intraoperative monitoring was hampered by the neuropathy in three quarter of the cases in which it was attempted.

### 14.5.5 Symptomatic Drug Therapy

Pain is an emerging feature of CMT. Seventy-one percent of 617 CMT patients complained of some pain in the series reported by Carter et al. [10] and two-thirds used drugs to manage pain. Different series of CMT patients have been evaluated with the SF36 QoL questionnaire which has an item specific for body pain and in all studies body pain scores were significantly worse than in the normal population [73, 110]. We assessed pain in a series of 271 CMT1A adult patients using a visual analogue scale (VAS, score 0–10) and found a mean score of $3.68 \pm 3.02$, confirming that pain is a major issue in CMT [79]. Pain appears to be mainly of the ostheo-artrhropathic type related to skeletal deformities and posture abnormalities, partly related to muscle fatigue, and sometimes of true neuropathic type [10, 73]. However, further investigations to characterise frequency and characteristics of pain in CMT are needed and are underway. Therapy includes physical therapy and plantars to correct posture, surgery when deemed necessary, and drugs for non-neuropathic and neuropathic pain, depending on the single cases.

CMT patients complain also of fatigue. Mean VAS values for fatigue in the same 271 CMT1A patient series was $4.87 \pm 2.81$. It is probably related to different factors, including reduced muscle strength in different muscle groups and possibly impaired cardiopulmonary performance. Carter et al. [11] tested modafinil in four CMT1A patients, obtaining some benefit but also some side effects. Modafinil is a central nervous system acting analeptic drug, which has been employed in the treatment of somnolence and fatigue in several neurological disorders such as multiple sclerosis and myotonic dystrophy. Side effects included nervousness or anxiety, insomnia, headache, and diarrhoea [11]. Risk-benefit ratio is probably unfavourable and the central acting mechanism is non-specific.

### 14.5.6 Specific Drug Therapy

During the last 20 years some drugs have been tested in different studies but none was of benefit. Tested drugs included TRH [18], Coenzyme Q10 [27], GH replacement therapy [59], 3,4-diaminopyridine [86], Levodopa [40], Vitamin E combined with Cis-linolein, gamma-linoleic, oleic acid [114], parenteral gangliosides [6], and oral creatine monohydrate (5 g/day for 12 weeks) [14]. Most reports were anecdotal or included also patients affected by disease other than CMT. Only the two latter studies were specifically considered in the Cochrane review [118].

#### 14.5.6.1 Animal Models and Clinical Trials

Animal Models

Strategy for developing new treatments include now the study of pathogenic mechanisms and of treatment effects in animal models. Several spontaneous and transgenic animal models are available. The main models are related to CMT1A. Trembler (*Tr*) and Trembler$^J$ (*Tr*$^J$) mice are spontaneous *pmp22* mutants (carrying the G150D and L16P mutations, respectively) [43]. Transgenic rats [71, 91] and mice [38] carrying additional *PMP22* gene copy numbers have been constructed and have been used to test possible treatments aimed at down-regulating *PMP22* expression.

Progesterone and Onapristone

It is known that progesterone and derivates are able to increase *MPZ* and *PMP22* gene expression in vitro [21, 62]. Sereda et al. [92] tested the effect of progesterone administration (20 mg/kg/day for 7 weeks) to transgenic CMT1A rats overexpressing PMP22 and demonstrated a detrimental effect on clinical and neuropathological findings, whereas the progesterone antagonist onapristone (20 mg/kg/day for 7 weeks) proved to be effective in the same model, resulting in clinical and neuropathological improvement. A subsequent study with long-term onapristone treatment confirmed its beneficial effect and showed that it exerted a protective effect on axonal loss [66]. Unfortunately, onapristone and currently available progesterone antagonists are too toxic to be safely administered to CMT1A patients.

Neurotrophin-3

Neurotrophin-3 (NT3), a neurotrophic factor known to promote axonal growth, was tested in two animal models and in a pilot study involving eight CMT1A patients [88]. NT3 favoured axonal regeneration in a xenograft model based on axons of nude mice ensheathed by human CMT1A Schwann cells, and it increased nerve regeneration after sciatic nerve crush in Trembler$^J$ mice.

It also appeared to improve sensory loss (as assessed through the NIS) and increase the number of solitary small myelinated fibres (considered an index of axonal regeneration) in sural nerves in a double-blind placebo controlled small pilot trial with eight patients (four treated and four untreated) receiving intradermal NT3 (150 μg/kg three times a week) for 6 months. No further studies have been designed thus far.

Ascorbic Acid

Ascorbic acid has long been known to favour myelination in vitro, and scurvy has exceptionally been associated with peripheral neuropathy. The group of Fontés and colleagues [80] tested the efficacy of ascorbic acid administration to the CMT1A model mice (C22 line). The animals were force-fed with ascorbic acid (1.12 mg once a week for a 20 g mouse) for months, in three different sets of experiments.

Treated rats performed considerably better than the control group and some clinical measures improved as compared to the beginning of the treatment; histological examination showed that sciatic nerve of treated mice contained more myelinated fibres and myelin was thicker. The mechanism of AA effect is not entirely understood, but the authors provided evidence for a role of AA in reducing PMP22 mRNA levels through a cAMP mediated pathway [42, 80]. It has been debated whether the CMT1A mouse is a proper model for the human disease, as life span of the animals is reduced and the PMP22 copy number is high (up to seven) as compared to the three copies of the human disease. However, AA is easily available and almost devoid of side effects and these factors prompted the design of clinical trials that are currently being performed. An ENMC workshop was extremely useful to decide a protocol with common core features so that subsequent prospective metanalysis can be performed. A trial is being conducted by the French group itself [118]. Another large multicentric trial is ongoing in Italy [77] (CMT-TRI*AA*L) in conjunction with a parallel trial (CMT-TRAUK) in London, and 272 patients have been recruited overall in these parallel trials [79]. Both are placebo-controlled randomised trials with a placebo group and a treated group assuming 1.5 g/day of oral ascorbic acid for a 2-year period. A third large trial is being conducted in the US. This is a futility trial with higher doses (4 g of AA, for 2 years). Other trials are underway in the Czech Republic and in Germany, the latter involving children [118]. In all trials the CMTNS is the primary outcome measure and there are secondary outcome measures.

### Curcumin

Another interesting and promising molecule is curcumin, which is derived from the curry spice turmeric. As suggested by experiments in different disease models, it plays a role in stimulating the translocation of misfolded proteins from the endoplasmic reticulum (ER) to the plasma membrane, thereby reducing cytotoxicity of the mutant proteins. This mechanism is particularly intriguing in inherited neuropathies, in which various *MPZ* and *PMP22* mutations are known to cause intracellular accumulation of mutant proteins, primarily within the ER. Khajavi and colleagues [44] demonstrated that curcumin releases selected *MPZ* mutants from the ER into the cytoplasm and reduces the toxic effect of mutant aggregation-induced apoptosis *"in vitro"*. In a second set of experiments [43], they tested the effects of curcumin in the *Tr*$^{J}$ and *Tr pmp22* mutants. They provided evidence that curcumin can lessen the toxicity of PMP22 mutant proteins, as evidenced by a decreased percentage of apoptotic cells *"in vitro"* and *"in vivo"*, and can mitigate the clinical and morphological phenotype of *Tr*$^{J}$ mouse, with no observed side effects.

### Immunomodulatory Treatment

There is evidence for a role of the immune system in the pathophysiology of CMT, deriving from animal model studies and from some reports of inflammation superimposed on CMT [22, 25, 32, 75]. Some of these patients responded to steroids or intravenous immunoglobulin treatment. However, it seems unlikely that immunomodulatory treatment may be beneficial in the majority of CMT cases.

### Other Issues and Future Perspectives

Proper genetic counselling should be given. Prenatal and preimplantation genetic diagnosis may be offered taking into account the ethic issues and the national regulations and laws [95]. There is a list of medications, potentially causing peripheral nerve toxicity *per se*, that should be avoided or administered with caution in CMT after consideration of risk-benefit ratios [113]. Vinka alkaloids can precipitate acute neuropathy resembling Guillain-Barré syndrome in patients with unrecognised CMT. Cis-platin and taxol derivatives should also be avoided if possible.

Novel treatments are under study. Disease severity can be modified by a tetracycline responsive element in the PMP22 promoter in the mouse model of CMT1A [81].

Post-transcriptional gene silencing may theoretically be used to reduce PMP22 overexpression in CMT1A and expression of other mutated genes causing gain-of-function toxicity: small double-stranded RNAs (dsRNA), small interfering RNAs (siRNAs), ribozymes and antisense oligonucleotides can down-regulate mRNAs in a sequence-specific way [95].

**Acknowledgements** Supported by Telethon Italy (Grants N° GUP04002 and GUP05007), and by the Italian Medicines Agency (AIFA) within the independent drug research program (contract no. FARM53APAH).

## References

1. Aboussouan LS, Lewis RA, Shy ME. Disorders of pulmonary function, sleep, and the upper airway in Charcot-Marie-Tooth Disease. Lung 2007; 185:1–7.
2. Berciano J, Gallardo E, García A, et al. Charcot-Marie-Tooth disease type 1A duplication with severe paresis of the proximal lower limb muscles: a long-term follow-up study. J Neurol Neurosurg Psychiatry 2006; 77:1169–1176.
3. Bienfait HM, Baas F, Koelman JH, et al. Phenotype of Charcot-Marie-Tooth disease Type 2. Neurology 2007; 68:1658–1667.
4. Birouk N, Gouider R, Le Guern E, et al. Charcot-Marie-Tooth disease type 1A with 17p11.2 duplication. Clinical and electrophysiological phenotype study and factors influencing disease severity in 119 cases. Brain 1997; 120:813–823.
5. Boerkoel CF, Takashima H, Garcia CA, et al. Charcot-Marie-Tooth disease and related neuropathies: mutation distribution and genotype-phenotype correlation. Ann Neurol 2000; 51:190–201.
6. Bradley WG, Badger GJ, Tandan R, Fillyaw MJ, Young J, Fries TJ, et al. Double-blind controlled trials of Cronassial in chronic neuromuscular diseases and ataxia. Neurology 1988; 38:1731–1739.
7. Brown RE, Zamboni WA, Zook EG, et al. Evaluation and management of upper extremity neuropathies in Charcot-Marie-Tooth-disease. J Hand Surg 1992; 17:523–530.
8. Burns J, Redmond A, Ouvrier R, et al. Quantification of muscle strength and imbalance in neurogenic pes cavus, compared to health controls, using hand-held dynamometry. Foot Ankle Int 2005; 26:540–544.

9. Burns J, Crosbie J, Ouvrier R et al. Effective orthotic therapy for the painful cavus foot: a randomized controlled trial. J Am Podiatr Med Assoc 2006; 96:205–211.
10. Carter GT, Jensen MP, Galer BS, et al. Neuropathic pain in Charcot-Marie-Tooth disease. Arch Phys Med Rehabil 1998; 79:1560–1564.
11. Carter GT, Han JJ, Mayadev A, Weiss MD. Modafinil reduces fatigue in Charcot-Marie-Tooth disease type 1A: a case series. Am J Hosp Palliat Care 2006; 23:412–416.
12. Carter GT, Weiss MD, Han JJ, et al. Charcot-Marie-Tooth disease. Curr Treat Options Neurol 2008; 10:94–102.
13. Carvalho AA, Vital A, Ferrer X, et al. Charcot-Marie-Tooth disease type 1A: clinicopathological correlations in 24 patients. J Peripher Nerv Syst 2005; 10:85–92.
14. Chetlin RD, Gutmann L, Tarnopolsky MA, et al. Resistance training exercise and creatine in patients with Charcot-Marie-Tooth disease. Muscle Nerve 2004; 30:69–76.
15. Chetlin RD, Gutmann L, Tarnopolsky M, et al. Resistance training effectiveness in patients with Charcot-Marie-Tooth disease: recommendations for exercise prescription. Arch Phys Med Rehabil 2004; 85:1217–1223.
16. Chung KW, Kim SB, Park KD, et al. Early onset severe and late-onset mild Charcot-Marie-Tooth disease with mitofusin 2 (MFN2) mutations. Brain 2006; 129:2103–2118.
17. Chung KW, Suh BC, Shy ME, et al. Different clinical and magnetic resonance imaging features between Charcot-Marie-Tooth disease type 1A and 2A. Neuromuscul Disord 2008; 18:610–618.
18. Congia S, Tronci S, Ledda M, et al. Low doses of TRH in amyotrophic lateral sclerosis and in other neurological diseases. Italian Journal of Neurological Sciences 1991; 12:193–198.
19. Cornblath DR, Chaudhry V, Carter K, et al. Total neuropathy score: validation and reliability study. Neurology 1999; 53:1660–1664.
20. De Jonghe P, Timmerman V, Ceuterick C, et al. The Thr124Met mutation in the peripheral myelin protein zero (MPZ) gene is associated with a clinically distinct Charcot-Marie-Tooth phenotype. Brain 1999; 122:281–290.
21. Désarnaud F, Do Thi AN, Brown AM, et al. Progesterone stimulates the activity of the promoters of peripheral myelin protein 22 and protein zero genes in Schwann cells. J Neurochem 1998; 71:1765–1768.
22. Donaghy M, Sisodiya SM, Kennett R, et al. Steroid responsive polyneuropathy in a family with a novel myelin protein zero mutation. J Neurol Neurosurg Psychiatry 2000; 69: 799–805.
23. Dubourg O, Tardieu S, Birouk N, et al. Clinical, electrophysiological and molecular genetic characteristics of 93 patients with X-linked Charcot-Marie-Tooth disease. Brain 2001; 124:1958–1967.
24. Dyck PJ, Swanson CJ, Low PA, et al. Prednisone-responsive hereditary motor and sensory neuropathy. Mayo Clin Proc 1982; 57:239–246.
25. Dyck PJ, Karnes JL, Lambert EH. Longitudinal study of neuropathic deficits and nerve conduction abnormalities in hereditary motor and sensory neuropathy type 1. Neurology 1989; 39:1302–1308.
26. El Mhandi L, Millet GY, Calmels P, et al. Benefits of interval-training on fatigue and functional capacities in Charcot-Marie-Tooth disease. Muscle Nerve 2008; 37: 601–610.
27. Folkers K, Simonsen R. Two successful double-blind trials with coenzyme Q10 (vitamin Q10) on muscular dystrophies and neurogenic atrophies. Biochimica et Biophysica Acta 1995; 127:281–286.
28. Gallardo E, García A, Combarros O, et al. Charcot-Marie-Tooth disease type 1A duplication: spectrum of clinical and magnetic resonance imaging features in leg and foot muscles. Brain 2006; 129:426–437.
29. Garcia CA, Malamut RE, England JD, et al. Clinical variability in two pairs of identical twins with the Charcot-Marie-Tooth disease type 1A duplication. 1: Neurology 1995; 45: 2090–2093.

30. Garcia A, Combarros O, Calleja J, et al. Charcot-Marie-Tooth disease type 1A with 17p duplication in infancy and early childhood: a longitudinal clinical and electrophysiologic study. Neurology 1998; 50:1061–1067.
31. Gemignani F, Marbini A. Disease course of Charcot-Marie-Tooth disease type 2 and comorbidity. Arch Neurol 2004; 61:1470.
32. Ginsberg L, Malik O, Kenton AR, et al. Coexistent hereditary and inflammatory neuropathy. Brain 2004; 127:193–202.
33. Graham RC, Huges RAC. Clinimetric properties of a walking scale in peripheral neuropathy. J Neurol Neurosurg Psychiatry 2006; 77:977–979.
34. Hardie R, Harding AE, Hirsh N, et al. Diaphragmatic weakness in hereditary moto rand sensory neuropathy. J Neurol Neurosurg Psychiatry 1990; 53:348–350.
35. Hattori N, Yamamoto M, Yoshihara T, et al. Demyelinating and axonal features of Charcot-Marie-Tooth disease with mutations of myelin-related proteins (PMP22, MPZ and Cx32): a clinicopathological study of 205 Japanese patients. Brain 2003; 126:134–151.
36. Hodapp JA, Carter GT, Lipe HP, et al. Double trouble in hereditary neuropathy: concomitant mutations in the PMP-22 gene and another gene produce novel phenotypes. Arch Neurol 2006; 63:112–117.
37. Hoogendijk JE, De Visser M, Bolhuis PA, et al. Hereditary motor and sensory neuropathy type I: clinical and neurographical features of the 17p duplication subtype. Muscle Nerve 1994; 17:85–90.
38. Huxley C, Passage E, Manson A, et al. Construction of a mouse model of Charcot-Marie-Tooth disease type 1A by pronuclear injection of human YAC DNA. Hum Mol Genet 1996; 5:563–569.
39. Jani-Acsadi A, Krajewski K, Shy ME. Charcot-Marie-Tooth neuropathies: diagnosis and management. Semin Neurol 2008; 28:185–194.
40. Jaradeh S, Dyck PJ. Hereditary motor and sensory neuropathy with treatable extrapyramidal features. Arch Neurol 1992; 49:175–178.
41. Karol LA, Elerson E. Scoliosis in patients with Charcot-Marie-Tooth disease. J Bone Joint Surg Am 2007; 89:1504–1510.
42. Kaya F, Belin S, Bourgeois P, et al. Ascorbic acid inhibits PMP22 expression by reducing cAMP levels. Neuromuscul Disord 2007; 17:248–253.
43. Khajavi M, Inoue K, Wiszniewski W, et al. Curcumin treatment abrogates endoplasmic reticulum retention and aggregation-induced apoptosis associated with neuropathy-causing myelin protein zero-truncating mutants. Am J Hum Genet 2005; 77:841–850.
44. Khajavi M, Shiga K, Wiszniewski W, et al. Oral curcumin mitigates the clinical and neuropathologic phenotype of the Trembler-J mouse: a potential therapy for inheritedneuropathy. Am J Hum Genet 2007; 81:438–453.
45. Killian JM, Tiwari PS, Jacobson S, et al. Longitudinal studies of the duplication form of Charcot-Marie-Tooth polyneuropathy. Muscle Nerve 1996; 19:74–78.
46. Kilmer DD, McCrory MA, Wright NC, et al. The effect of a high resistance exercise program in slowly progressive neuromuscular disease. Arch Phys Med Rehabil 1994; 75:560–563.
47. Kilmer DD, McCrory MA, Wright NC, et al. Hand-held dynamometry reliability in persons with neuropathic weakness. Arch Phys Med Rehabil 1997; 78:1364–1368.
48. Kilmer DD, Aitkens SG, Wright NC, et al. Simulated work performance tasks in persons with neuropathic and myopathic weakness. Arch Phys Med Rehabil. 2000; 81: 938–943.
49. Kilmer DD. Response to aerobic exercise training in humans with neuromuscular disease. Am J Phys Med Rehabil 2002; 81:S148–150.
50. Kilmer DD, Wright NC, Aitkens S. Impact of a home-based activity and dietary intervention in people with slowly progressive neuromuscular diseases. Arch Phys Med Rehabil 2005; 86:2150–2156.
51. Krajewski KM, Lewis RA, Fuerst DR, et al. Neurological dysfunction and axonal degeneration in Charcot-Marie-Tooth disease type 1A. Brain 2000; 123:1516–1527.

52. Laurà M, Milani M, Morbin M, et al. Rapid progression of late onset axonal Charcot-Marie-Tooth disease associated with a novel MPZ mutation in the extracellular domain. J Neurol Neurosurg Psychiatry 2007; 78:1263–1266.
53. Li J, Bai Y, Ghandour K, Qin P, et al. Skin biopsies in myelin-related neuropathies: bringing molecular pathology to the bedside. Brain 2005; 128:1168–1177.
54. Liang GSL, de Miguel M, Gomez-Hernandez JM, et al. Severe neuropathy with leaky connexin32 hemichannels. Ann Neurol 2005; 57:749–754.
55. Lin GS, Glass JD, Shumas S, et al. A unique mutation in connexin32 associated with severe, early onset CMTX in a heterozygous female. Ann NY Acad Sci 1999; 883:481–484.
56. Lindeman E, Leffers P, Reulen J, et al. Reduction of knee torques and leg–related functional abilities in hereditary motor and sensory neuropathy. Arch Phys Med Rehabil 1994; 75:1201–1205.
57. Lindeman E, Leffers P, Spaans F, et al. Strength training in patients with myotonic dystrophy and hereditary motor and sensory neuropathy: a randomized clinical trial. Arch Phys Med Rehabil 1995; 76:612–620.
58. Lindeman E, Spaans F, Reulen J, et al. Progressive resistance training in neuromuscular patients. Effects on force and surface EMG. J Electromyogr Kinesiol 1999; 9: 379–384.
59. Lissett CA, Toogood AA, Didi M. Growth hormone replacement in an adult with mild growth hormone deficiency and hereditary motor and sensory neuropathy: growth hormone restores independent mobility. Hormone Research 1998; 50:232–236.
60. Lombardi R, Erne B, Lauria G, et al. IgM deposits on skin nerves in anti-myelin-associated glycoprotein neuropathy. Ann Neurol 2005; 57:180–187.
61. Lupski JR and Chance PFK. Hereditary Motor and Sensory Neuropathies involving altered dosage or mutation of PMP22: the CMT1A duplication and HNPP deletion. In: Dyck PJ, Thomas PK (eds.) Peripheral neuropathy. 4th ed. Elsevier Saunders, Philadelphia, 2005; pp. 1659–1680.
62. Magnaghi V, Cavarretta I, Galbiati M, et al. Neuroactive steroids and peripheral myelin proteins. Brain Res Brain Res Rev 2001; 37:360–371.
63. Mann DC, Hsu JD. Triple arthrodesis in the treatment of fixed cavovarus deformity in adolescent patients with Charcot-Marie-Tooth disease. Foot Ankle 1992;13:1–6.
64. Martyn CN, Hughes RAC. Epidemiology of peripheral neuropathy. J Neurol Neurosurg Psychiatry 1997; 62: 310–318.
65. Matjacić Z, Zupan A. Effects of dynamic balance training during standing and stepping in patients with hereditary sensory motor neuropathy. Disabil Rehabil 2006; 28: 1455–1459.
66. Meyer zu Horste G, Prukop T, Liebetanz D, et al. Antiprogesterone therapy uncouples axonal loss from demyelination in a transgenic rat model of CMT1A neuropathy. Ann Neurol 2007; 61:61–72.
67. Merkies IS, Lauria G. 131st ENMC international workshop: selection of outcome measures for peripheral neuropathy clinical trials 10-12 December 2004, Naarden, The Netherlands. Neuromuscul Disord 2006; 16:149–156.
68. Miller MJ, Williams LL, Slack SL, et al. The hand in Charcot-Marie-Tooth disease. J Hand Surg 1991; 16:191–196.
69. Miller GM, Hsu JD, Hoffer MM, et al. Posterior tibial tendon transfer: a review of the literature and analysis of 74 procedures. J Pediatr Orthop 1982; 2:363–370.
70. Nagamatsu M, Jenkins RB, Schaid DJ, et al. Hereditary motor and sensory neuropathy type 2C is genetically distinct from types 2B and 2D. Arch Neurol 2000; 57:669–672.
71. Nave KA, Sereda MW, Ehrenreich H. Mechanisms of disease: inherited demyelinating neuropathies–from basic to clinical research. Nat Clin Pract Neurol 2007; 3:453–464.
72. Nelis E, Van Broeckhoven C, De Jonghe P, et al. Estimation of the mutation frequencies in Charcot-Marie-Tooth disease type 1 and hereditary neuropathy with liability to pressure palsies: a European collaborative study. Eur J Hum Genet 1996; 4:25–33.

73. Padua L, Cavallaro T, Pareyson D, et al; Italian CMT QoL Study Group. Charcot-Marie-Tooth and pain: correlations with neurophysiological, clinical, and disability findings. Neurol Sci 2008; 29:193–194.
74. Padua L, Pareyson D, Aprile I, et al. Natural history of CMT1A including QoL: a 2-year prospective study. Neuromuscul Disord 2008; 18:199–203.
75. Pareyson D. Differential diagnosis of Charcot-Marie-Tooth disease and related neuropathies. Neurol Sci 2004; 25:72–82.
76. Pareyson D, Scaioli V, Laurà M. Clinical and electrophysiological aspects of Charcot-Marie-Tooth disease. Neuromolecular Med 2006; 8:3–22.
77. Pareyson D, Schenone A, Fabrizi GM, et al. for the CMT-TRIAAL Group. A multicenter, randomized, double-blind, placebo-controlled trial of long-term ascorbic acid treatment in Charcot-Marie-Tooth disease type 1A (CMT-TRIAAL): the study protocol [EudraCT no.: 2006-000032-27]. Pharmacol Res 2006; 54:436–441.
78. Pareyson D. Axonal Charcot-Marie-Tooth disease: the fog is only slowly lifting. Neurology 2007; 68:1649–1650.
79. Pareyson D, Schenone A, Rizzuto N, et al. Clinical and electrophysiological evaluation of 222 patients with Charcot- Marie-Tooth disease type 1A recruited in the CMT-TRIAAL (ascorbic acid therapy for Charcot-Marie-Tooth 1A disease). J Neurol 2008; 255 (Suppl 2): 104 (P422).
80. Passage E, Norreel JC, Noack Fraissignes P, et al. Ascorbic acid treatment corrects the phenotype of a mouse model of Charcot-Marie-Tooth disease. Nat Med 2004; 10:396–401.
81. Perea J, Robertson A, Tolmachova T, et al. Induced myelination and demyelination in a conditional mouse model of Charcot-Marie-Tooth disease type 1A. Hum Mol Genet 2001; 10:1007–1018.
82. Pfeiffer G, Wicklein EM, Ratusinski T, et al. Disability and quality of life in Charcot-Marie-Tooth disease type 1. J Neurol Neurosurg Psychiatry 2001; 70:548–550.
83. Refshauge KM, Raymond J, Nicholson G, et al. Night splinting does not increase ankle range of motion in people with Charcot-Marie-Tooth disease: a randomised, cross-over trial. Aust J Physiother 2006; 52:193–199.
84. Reilly MM. Sorting out the inherited neuropathies. Pract Neurol 2007; 7:93–105.
85. Reilly MM, de Jonghe P, Pareyson D. 136th ENMC International Workshop: Charcot-Marie-Tooth disease type 1A (CMT1A). 8-10 April 2005, Naarden, The Netherlands. Neuromuscul Disord 2006; 16: 396–402.
86. Russell JW, Windebank AJ, Harper CM. Treatment of stable chronic demyelinating polyneuropathy with 3,4-diaminopyridine. Mayo Clinic Proceedings 1995; 70:532–539.
87. Sackley C, Disler PB, Turner-Stokes L, et al. Rehabilitation interventions for foot drop in neuromuscular disease. Cochrane Database Syst Rev 2007 18; (2):CD003908.
88. Sahenk Z, Nagaraja HN, McCracken BS, et al. NT-3 promotes nerve regeneration and sensory improvement in CMT1A mouse models and in patients. Neurology 2005; 6:681–689.
89. Saltzman CL, Fehrle MJ, Cooper RR, et al. Triple arthrodesis: twenty-five and forty-four-year average follow-up of the same patients. J Bone Joint Surg Am 1999; 81:1391–402.
90. Sellers RW, van Ginneken BTJ, Schreuders TAR, et al. Dynamometry of intrinsic hand muscles in patients with Charcot-Marie-Tooth disease. Neurology 2006; 67: 2022–2027.
91. Sereda M, Griffiths I, Pühlhofer A, et al. A transgenic rat model of Charcot-Marie-Tooth disease. Neuron 1996; 16:1049–1060.
92. Sereda MW, Meyer zu Hörste G, Suter U, et al. Therapeutic administration of progesterone antagonist in a model of Charcot-Marie-Tooth disease (CMT1A). Nat Med 2003; 9: 1533–1537.
93. Shy ME, Jáni A, Krajewski K, et al. Phenotypic clustering in MPZ mutations. Brain 2004; 127:371–384.
94. Shy ME, Blake J, Krajewski K, et al. Reliability and validity of the CMT neuropathy score as a measure of disability. Neurology 2005; 64:1209–1214.

95. Shy ME. Therapeutic strategies for the inherited neuropathies. Neuromolecular Med 2006; 8:255–278.
96. Shy ME, Siskind C, Swan ER, et al. CMT1X phenotypes represent loss of gene function GJB1. Neurology 2007; 68:849–855.
97. Shy ME, Chen L, Swan ER, et al. Neuropathy progression in Charcot-Marie-Tooth disease type 1A.Neurology 2008; 70:378–83.
98. Solari A, Laurà M, Salsano E, Radice D, Pareyson D; CMT–TRIAAL Study Group. Reliability of clinical outcome measures in Charcot–Marie–Tooth disease. Neuromuscul Disord 2008; 18:19–26.
99. Svensson E, Häger-Ross C. Hand function in Charcot-Marie-Tooth: test-retest reliability of some measurements. Clin Rehab 2006; 20:896–908.
100. Szigeti K, Wiszniewski W, Saifi GM, et al. Functional, histopathologic and natural history study of neuropathy associated with EGR2 mutations. Neurogenetics 2007; 8:257–262.
101. Teunissen LL, Notermans NC, Franssen H, et al. Disease course of Charcot-Marie-Tooth disease type 2: a 5-year follow-up study. Arch Neurol 2003; 60:823–828.
102. Timmerman V, De Jonghe P, Ceuterick C, et al. Novel missense mutation in the early growth response 2 gene associated with Dejerine-Sottas syndrome phenotype. Neurology 1999; 52:1827–1832.
103. Thomas PK, Marques W Jr, Davis MB et al. The phenotypic manifestations of chromosome 17p11.2 duplication. Brain 1997; 120:465–478.
104. Vallat JM, Tazir M, Magdelaine C, et al. Autosomal-recessive Charcot-Marie-Tooth diseases. J Neuropathol Exp Neurol 2005; 64:363–370.
105. Verhamme C, van Schaik IN, Koelman JH, et al. Clinical disease severity and axonal dysfunction in hereditary motor and sensory neuropathy IA. J Neurol 2004; 251:1491–1497.
106. Verhoeven K, Claeys KG, Zuchner S, et al. MFN2 mutation distribution and genotype/phenotype correlation in Charcot-Marie-Tooth type 2. Brain 2006; 129:2093–2102.
107. Videler AJ, Beelen A, Aufdemkampe G, et al. Hand strength and fatigue in patients with hereditary motor and sensory neuropathy (types I and II). Arch Phys Med Rehabil 2002; 83:1274–1278.
108. Videler AJ, Beelen A, van Schaik IN, et al. Manual dexterity in hereditary motor and sensory neuropathy type 1a: severity of limitations and feasibility and reliability of two assessment instruments. J Rehabil Med 2008; 40:132–136.
109. Vinci P, Esposito C, Perelli SL, et al. Overwork weakness in Charcot–Marie–Tooth disease. Arch Phys Med Rehabil 2003; 84:825–827.
110. Vinci P, Serrao M, Millul A, et al. Quality of life in patients with Charcot-Marie-Tooth disease. Neurology 2005; 65:922–924.
111. Vinci P, Gargiulo P. Poor compliance with ankle-foot-orthoses in Charcot-Marie-Tooth disease. Eur J Phys Rehabil Med 2008; 44:27–31.
112. Wetmore RS, Drennan JC. Long-term results of triple arthrodesis in Charcot-Marie-Tooth disease. J Bone Joint Surg Am 1989; 71:417–422.
113. Weimer LH, Podwall D. Medication-induced exacerbation of neuropathy in Charcot-Marie-Tooth disease. J Neurol Sci 2006; 242:47–54.
114. Williams LL, O'Dougherty MM, Wright FS, et al. Dietary essential fatty acids, vitamin E, and Charcot-Marie-Tooth disease. Neurology 1986; 36:1200–1205.
115. Wood VE, Huene D, Nguyen J. Treatment of the upper limb in Charcot-Marie-Tooth disease. J Hand Surg 1995; 20:511–518.
116. World Health Organization. International Classification of Functioning. Disability and Health. Geneva: WHO, 2001.
117. Wright NC, Kilmer DD, McCrory MA, et al. Aerobic walking in slowly progressive neuromuscular disease: effect of a 12-week program. Arch Phys Med Rehabil 1996; 77:64–69.
118. Young P, De Jonghe P, Stögbauer F, et al. Treatment for Charcot-Marie-Tooth disease. Cochrane Database Syst Rev. 2008; 23(1):CD006052.

# Chapter 15
# Spinal Muscular Atrophy During Human Development: Where Are the Early Pathogenic Findings?

**Eduardo Tizzano**

**Abstract** Spinal muscular atrophy (SMA) is an autosomal recessive disorder that affects motor neurons. It is caused by mutations in the survival motor neuron gene 1 (SMN1). The SMN2 gene, which is the highly homologous SMN1 copy that is present in all patients, is unable to prevent the disease. SMA patients can be classified into four groups based on age at onset and acquired milestones (type I or severe acute disease, with onset before 6 months; type II, before 18 months; type III, after 18 months and type IV, in adult life). The human developmental period is believed to play an essential role in SMA pathogenesis. However, the neuropathologic study of SMA comes largely from postnatal necropsy samples, which describe the end-stage of the disease. With the exception of severe congenital SMA (or Type 0 SMA), type I patients tend to present a short but variable presymptomatic period after birth. Our main interest lies in studying SMA during human development so as to gain insight into the mechanism of the disease in the prenatal – presymptomatic stage. In fetuses of 12–15 weeks' gestational age we systematically studied histology, cell death and gene expression in spinal cord and muscle, the key tissues involved in the disease. Furthermore, ultrasound parameters were investigated at these stages. These studies may help to delineate an early intervention in SMA, in particular during the potential therapeutic window.

**Keywords:** Spinal muscular atrophy · Prenatal diagnosis · Histology · Apoptosis · Gene and protein expression · Fetal movements

## 15.1 Introduction to SMA

Spinal muscular atrophy (SMA) is a common autosomal recessive neuromuscular disorder characterised by degeneration and loss of the alpha-motor neurons in the

---

E. Tizzano (✉)
Genetic and Research Institute Hospital de la Santa Creu i Sant Pau and CIBERER, Sant Antoni Ma. Claret 167, 08025 Barcelona, Spain
e-mail: etizzano@santpau.cat

Supported by FIS 05, 2416; 08-0729 and GENAME PROJECT

C. Espinós et al. (eds.), *Inherited Neuromuscular Diseases*, Advances in Experimental Medicine and Biology 652, DOI 10.1007/978-90-481-2813-6_15,

anterior horn of the spinal cord. The disease manifests as symmetrical weakness and atrophy of voluntary muscles of the limbs and trunk. SMA has an estimated incidence of 1/6000 to 1/10,000 live births and a carrier frequency of 1/35–1/60 [14, 40].

SMA has been classified into four clinical groups based on motor milestones and age of onset. Type I SMA (Werdnig-Hoffmann disease, MIM #253300) is the most severe form, with a generalised muscular hypotonia and disease onset within 6 months of birth. Affected children are unable to sit or walk, and usually die before the age of 2. Type II SMA (MIM # 253550) is the intermediate form of the disease with onset before 18 months. These patients are able to sit but are unable to walk unaided. Type III SMA (Kugelberg-Welander disease, MIM#253400) is the mild form with onset after 18 months. These patients are able to stand and walk, but often become wheelchair bound during youth or adulthood. Type IV SMA (MIM#271150) is the adult form and it is characterised by onset after the age of 30. The patients in this group have generally a normal life expectancy.

Mutations in the telomeric survival motor neuron gene (SMN1) are responsible for SMA. The SMN1 gene is duplicated owing to the existence of a highly homologous copy, SMN2, which act as a phenotypic modifier of the disease [8, 27]. In our experience, 87% of type I patients have one or two copies of SMN2 but none have four copies. On the other hand, 87% of type III–IV patients have three or four SMN2 copies, but none have one copy. Most type II patients have two or three copies [13]. Prognosis based on SMN2 copy number is not therefore reliable, but may be applicable for a subset of patients, such as those harboring four or more SMN2 copies [57]. It has been reported that type I SMA patients have fewer copies of SMN2 (because of true deletions in the SMN1) than type III-IV patients, who show the highest number (as a consequence of gene conversion events) [7]. Asymptomatic siblings with absence of the SMN1 gene have been described by several groups ([13] and references therein). Recently, transcriptional studies in cells from these type of asymptomatic siblings, led to the identification of a possible phenotype modifier gene [38]. The sequence variations found in the SMN1 and SMN2 promoters do not predict differences in expression between the two genes. However, most of the SMN1 transcripts are full length, and a single nucleotide change in exon 7 is involved in the skipping of this exon in approximately half the SMN2 transcripts [30, 35].

The SMN protein (38 kDa) appears to be ubiquitously expressed. It is located diffusely in the cytoplasm and concentrated in aggregates called gems in the nucleus [29]. SMN is part of a complex with various proteins that function in the biogenesis of small nuclear ribonucleoprotein assembly which are the essential components of pre-messenger RNA (reviewed by [41]). This apparent essential function for all cells appears critical for motor neurons where the levels of the SMN protein are reduced and the cells lack gems [10, 28]. SMA patients without SMN2 have not been reported to date, which suggests that SMN plays an essential role during embryonic development. In mice, the SMN gene (Smn) is not duplicated and homozygous inactivation of the Smn gene leads to massive cell death in the early blastocyst stage [44]. This embryonic lethality is reversed generating mice with a different number of transgenic copies of SMN2 [22, 36]. All these observations support the hypothesis that the SMN2 gene can be regarded as a phenotype modifier of the disease.

## 15.2 Therapeutic Prospective in SMA

There is as yet no effective therapy to mitigate the motor neuron degeneration and loss in SMA and a number of approaches should be taken into account when defining the targets for therapeutic development against the disease. These approaches can be divided into four categories: clinical, genetic, neurobiological and pharmacological.

The clinical approach deals with the natural history of the disease, the possibility of early intervention and the definition and validation of outcome measures. A second approach is the unravelling of genetic influences that directly affect SMN protein abundance or role in the motor neurons and neuromuscular system. The third approach considers neurobiology of the disease studying the animal models and comparing them with studies in humans. Regarding the fourth approach, SMA may be treated by classical pharmacological methods with compounds that upregulate SMN2 expression leading to a more abundant FL mRNA and protein. Furthermore, drugs promoting neuroprotection or muscle strength may also be strong candidates and have been tested in pilot or placebo trials (Fig. 15.1) (reviewed by [52]).

Activating the SMN2 gene and/or exon 7 inclusion have been delineated in order to increase the functional SMN protein. Histone deacetylase inhibitors are compounds that can modulate gene expression and they are thus candidate drugs for

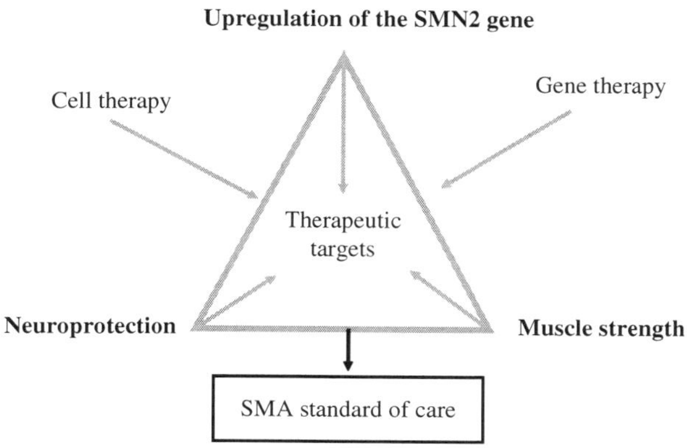


**Fig. 15.1** Scheme of the therapeutic alternatives in spinal muscular atrophy. The pharmacologic approach may include drugs that upregulate SMN2, that increase muscle strength and neuroprotectors. In a second level, gene and cell therapy emerges as possible therapies in the future. This combined approach for therapy may be incorporated to the standard of care of the disease (phisiotherapy, nutrition, orthopedic, surgical and respiratory care)

a therapeutic approach. Indeed, studies with aromatic fatty acids such as phenylbutyrate (PB) or valproic acid (VA) have been reported. PB and VA are effective in increasing SMN expression in SMA fibroblast cell lines [1, 5, 51]. These compounds were chosen given that it is well known that both can cross the blood-brain barrier, they were approved by the US Food and Drug Administration (FDA) for the treatment of urea cycle disorders (PB) and epilepsy (VA), and are generally well tolerated. In short pilot trials including SMA patients and carriers it has been shown that PB and VA also increase SMN expression in peripheral blood leukocytes [4, 6]. These data provide the first evidence that SMN transcript levels can be enhanced in vivo by the administration of a drug. Another drug under investigation is hydroxyurea (HU), used to treat patients with sickle cell anemia and thalassemias. HU activates fetal globin and appears to increase FL SMN transcripts, protein and gems in lymphoblasts from SMA patients [20]. Consequently, these drugs have been selected for early clinical trials in SMA patients, although efficacy in SMA animal models has not yet been totally proven. A randomized, double-blind, placebo-controlled trial with PB was not effective in 107 Italian patients [34] and results of clinical trials using VA and HU are still inconclusive [53]. Recently, salbutamol has been reported to increase the SMN2-full length mRNA and SMN protein in SMA fibroblasts, predominantly by promoting exon 7 inclusion in SMN2 transcripts [2]. Moreover, salbutamol administered for 12 months in non-ambulant young SMA patients may improve muscle function without significant adverse effects [39]. New therapeutic approaches include gene therapy using adeno-associated viruses (AAV) and lentivirus to incorporate the SMN1 gene, non-viral experiments to modify the SMN2 gene (reviewed in [52]) and cell therapy to restore the motor neuron population and its environment [11]. All are promising prospects, but further experimental data from animal models are needed. More immediate advances in treatment are likely to occur using the classic pharmacological approach and unravelling the basic function of SMN. The explanation for the neuromuscular phenotype in SMA is to assume that insufficient SMN protein causes motor neuron dysfunction and death, and that muscle atrophy is a consequence of denervation. However, investigation is ongoing to ascertain whether muscle, neuromuscular junctions, or motor neurons alone are the critical target tissue in SMA. Based on studies in mouse models, the hypothesis that SMN function may be critical in axonal transport and neuromuscular junctions is gaining support [24]. Indeed, one of the earliest detectable morphological defects in the SMA mice appears to be the loss of synapse occupation by motor axons [32].

The neuropathologic description of SMA comes largely from postnatal necropsy samples [12]. Autopsy studies of SMA, however, have the drawback that they demonstrate the end-stage of the disease. The human developmental period appears to play an essential role in SMA pathogenesis. With the exception of severe congenital SMA, varying age at onset in the four SMA types provides evidence of a latency period without clear manifestations in most SMA patients. Given that studies of patients' preclinical status are lacking, the main objective of our group is to study SMA during development to gain insight into the mechanism of disease in the prenatal and presymptomatic stage. These investigations may help to delineate an early intervention in SMA, in particular during the potential therapeutic window.

## 15.3 The Neuromuscular System During Development

During the fetal period, a first step of excess of neurogenesis is followed by a phase in which a large number of neurons degenerate and die as the result of a non-pathological process known as programmed cell death (PCD) [37]; this event is crucial to achieve the final organization of the nervous system. Dying cells display similar changes in morphology and nuclear DNA degradation, characterizing apoptosis. Neuronal death occurs naturally during development of the spinal cord. Even when a high proportion of neurons in a given region die in a few days, only isolated cells are visibly dying at a given moment. This is because neuron death and elimination appears to be very rapid. PCD in spinal cord motor neurons coincides with the stage of synaptogenesis, where myotubes and muscle fibers in formation express acetylcholine receptors, which stimulates axon growth and targeting. During skeletal muscle development, the myoblasts fuse into myotubes, centrally multinucleated cells that subsequently become muscle fibers. At this stage innervation in muscle fibers is polyaxonal, with several motor endplates and several axons per neuromuscular junction (Reviewed by [46]). Motor neuron survival in the spinal cord is regulated by signals derived mainly from skeletal muscle, glial cells and neurotrophic factors. Target contacts between motor neurons and muscle have already started around 7–8 weeks and motor neurons begin to express choline acetyltransferase (ChAT), the enzyme which catalyses the biosynthesis of acetylcholine. ChAT is one of the most specific markers for determining the functional state of cholinergic neurons in the central and peripheral nervous systems. A reduced expression of ChAT can be attributed to the loss of function in motor neurons [58]. The first spontaneous movements of the human fetus appear at 7 weeks post-menstrual age. The number and frequency of fetal movements increase from 10 weeks onward and are related to the development of neuronal connections through axonal in-growth, synaptogenesis and dendrite proliferation. Around 11–16 weeks, several fetal movements can be observed, these including general movements, isolated arm or leg movements, head flexion and rotation, startle and hiccup [26].

## 15.4 SMA During Human Development

### *15.4.1 Histologic Observations*

SMA motor neurons show no clear demonstrable alterations under light microscope examination at 12–15 weeks [48], although nuclei with abnormal shape and high nucleoplasm density have been observed in SMA motor neurons at 20 weeks under electronic microscopy [17]. Interestingly, in a conditional mouse model of the disease, indentation of the nuclear membrane of postnatal motor neurons has been observed [19]. Morphometric analysis has shown that nuclear diameter of motor neurons is similar in controls and SMA [48]. The typical neurodegeneration changes described in motor neurons from autopsy material has not been observed prenatally in SMA samples. It has been found that the number of motor neurons in the fetal and neonatal period is reduced by 20–57%, respectively, compared with controls [45, 48]. We investigated the expression of choline acetyl transferase (ChAT) in SMA spinal cords and observed reduced ChAT expression with respect to controls, probably due to the lower number of motor neurons in SMA. However, the proportion of motor neurons negative for ChAT immunostaining was similar in both groups, suggesting that the surviving motor neurons in SMA fetuses could have a cholinergic function akin to that of controls. On the other hand, the chromatolytic motor neurons detected in SMA after birth were ChAT negative. Therefore, ChAT does not identify early pathogenic events in fetal SMA spinal cord, but is, rather, a late marker for motor neuron degeneration [49].

Regarding the development of muscle, myotubes in SMA fetuses tend to be smaller and arranged in clusters close to isolated and larger myotubes [33]. This finding may represent a delay in growth and maturation of SMA muscle and supports previous findings of immature cells and fibers in postnatal SMA muscle [15, 21].

### 15.4.2 DNA Fragmentation

DNA fragmentation is an early marker of cell death and is usually studied by the TUNEL technique. A significant increase in DNA fragmentation has been observed in the spinal cord of fetuses predicted to have type I SMA, particularly around 12–15 weeks, a period that coincides with the establishment of functional neuromuscular synapses [17, 48]. This finding suggests supraphysiologic cell death as a possible disease mechanism during development. Analysing postnatal spinal cords, we did not detect apoptotic cells by morphologic appearance or TUNEL positive staining in the anterior horns. The sparse TUNEL positive cells identified were cells of glial origin [48]. Yet apoptotic cells have been reported in the spinal cord in children with Werdnig-Hoffmann disease [44]. It therefore appears that motor neuron death largely takes place prenatally.

In contrast to the spinal cord, fetal SMA muscle displays a rate of isolated TUNEL positive cells similar to that of controls [33]. However, in the postnatal period, an increase of TUNEL-positive cells in muscle fibers of SMA patients has been reported [16, 47]. It has been proposed that the elimination of muscle fibers could result in secondary motor neuron death in postnatal tissues [16, 43]. This increase may not only indicate apoptosis but also delayed development of myocytes and the presence of defective innervation [54].

### 15.4.3 SMN Expression

A significant expression of the SMN gene has not only been detected in the motor neurons (body, dendrites and axons) of the spinal cord but also in other neuronal populations such as sensory neurons of the dorsal root ganglia, neurons of the medulla oblongata, Purkinje cells of the cerebellum, and pyramidal cells of the cortex [3, 18, 55]. The expression of the SMN gene is reportedly higher in the prenatal period than in the postnatal period [9], suggesting that very high levels of SMN are necessary for normal development. In the control fetal spinal cord, there appears to be a high level of full-length (FL) SMN transcripts (including exon 7) with tiny amounts of Δ7 isoform. These transcripts are yielded from SMN1 with very little (if any) contribution from SMN2 [50]. SMN gene regulation in the SMA spinal cord therefore has to accommodate to the presence of SMN2 alone. Consequently, a decrease in the FL occurs with a substantial increase in Δ7 in SMA motor neurons that could contribute to the disease [25]. Furthermore, tissues that are unaffected by the disease, such as intestine, lung, adrenal gland, kidney and eye, seem to compensate for the absence of SMN1 [50].

### *15.4.4 Ultrasound Evaluation*

Although fetal movements are diminished in some SMA pregnancies [31], it is not known exactly how early such a reduction initiates. We recently evaluated fetal movements prior to chorionic villi sampling in couples at risk for SMA. Gestational ages ranged from 11 to 14 weeks. Cases were subsequently classified as healthy, carriers or affected, in relation to molecular results. The presence of a wide range of movements was documented. Movements in SMA fetuses were similar to those observed in controls and carriers of the same gestational age. Nuchal translucency (NT), an ultrasound marker used in the first and early second trimesters, has been associated with chromosomal anomalies, genetic syndromes, and structural malformations [23]. A number of authors have reported an increase in NT as an early sign of SMA albeit a proportion of these reported cases had hydrops fetalis [41, 56]). In our series, we found that the mean NT values did not differ between healthy, carrier or affected SMA fetuses (Parra et al., submitted).

## 15.5 SMA as a Developmental Disorder

The striking differences between the development of SMA in spinal cord and muscle suggest that pathogenic responses to lower levels of SMN protein differ in these key tissues at this stage (Table 15.1). It is thus reasonable to assume that the dynamic interaction between motor neurons and muscle could be disrupted early in SMA disease, lending support to the view of SMA as a developmental disorder. The observation in SMA fetuses of ChAT positive motor neurons and of a wide range of fetal movements provides indirect evidence that neuromuscular connections may be functionally sufficient or compensated at this time of development. As therapeutic advances emerge, future research should address the possible benefits of early intervention in SMA.

**Table 15. 1** Comparison of spinal cord and muscle observations in prenatal and postnatal SMA

| Stage | Tissue | Light microscope histology | DNA fragmentation | References |
|---|---|---|---|---|
| Prenatal | SMA Spinal Cord | No detectable alterations in SMA with respect to controls. Motor neuron nuclear size similar to controls | Significantly increased in SMA spinal cord | [17, 48] |
| | SMA muscle | SMA myotubes smaller with respect to control. Clusters of small myotubes with isolated bigger myotubes in SMA | No numerical differences with respect to control | [33] |
| Postnatal | SMA spinal cord | MNs with excentric nuclei, less Nissl corpuscles, chromatolysis | Isolated | [45, 47] |
| | SMA muscle | Groups of atrophic and hyperthrophic fibers | Increased | [16, 54] |

**Acknowledgements** I wish to thank the consenting parents and SMA families who made this study possible and Rebeca Martínez-Hernández, Carolina Soler-Botija, Laura Alias, Sara Bernal, Eva Also, María Jesús Barceló, Juan Parra, Ivón Cuscó, Carolyn Newey and Montserrat Baiget for their invaluable collaboration.

## References

1. Andreassi C, Angelozzi C, Tiziano FD, Vitali T, De Vincenzi E, Boninsegna A, Villanova M, Bertini E, Pini A, Neri G, Brahe C. Phenylbutyrate increases SMN expression in vitro: relevance for treatment of spinal muscular atrophy. Eur J Hum Genet 2004;12:59–65.
2. Angelozzi C, Borgo F, Tiziano FD, Martella A, Neri G, Brahe C. Salbutamol increases SMN mRNA and protein levels in spinal muscular atrophy cells. J Med Genet 2008;45:29–31.
3. Battaglia G, Princivalle A, Forti F, Lizier C, Zeviani M. Expression of the SMN gene, the spinal muscular atrophy determining gene, in the mammalian central nervous system. Hum Mol Genet 1997;6:1961–1971.
4. Brahe C, Vitali T, Tiziano FD, Angelozzi C, Pinto AM, Borgo F, et al. Phenylbutyrate increases SMN gene expression in spinal muscular atrophy patients. Eur J Hum Genet 2005;13:256–259.
5. Brichta L, Hofmann Y, Hahnen E, Siebzehnrubl FA, Raschke H, Blumcke I, et al. Valproic acid increases the SMN2 protein level: a well-known drug as a potential therapy for spinal muscular atrophy. Hum Mol Genet 2003;19:2481–2489.
6. Brichta L, Holker I, Haug K, Klockgether T, Wirth B. In vivo activation of SMN in spinal muscular atrophy carriers and patients treated with valproate. Ann Neurol 2006;59:970–975.
7. Burghes A. When is a deletion not a deletion? When it is converted. Am J Hum Genet 1997;61:9.
8. Bürglen L, Lefebvre S, Clermont O, Burlet P, Viollet L, Cruaud C, Munnich A, Melki J. Structure and organization of the human survival motor neurone (SMN) gene. Genomics 1996;32:479–482.
9. Burlet P, Huber C, Bertrandy S, et al. The distribution of SMN protein complex in human fetal tissues and its alteration in spinal muscular atrophy. Hum Mol Genet 1998;7: 1927–1933.

10. Coovert DD, Le TT, McAndrew PE, Strasswimmer J, Crawford TO, Mendell JR, et al. The survival motor neuron protein in spinal muscular atrophy. Hum Mol Genet 1997;6:1205–1214.
11. Corti S, Nizzardo M, Nardini M, Donadoni C, Salani S, Ronchi D, Saladino F, Bordoni A, Fortunato F, Del Bo R, Papadimitriou D, Locatelli F, Menozzi G, Strazzer S, Bresolin N, Comi GP. Neural stem cell transplantation can ameliorate the phenotype of a mouse model of spinal muscular atrophy. J Clin Invest 2008 Sep 2; [Epub ahead of print].
12. Crawford TO, Pardo CA. The neurobiology of childhood spinal muscular atrophy. Neurobiol Dis 1996;3:97–110.
13. Cuscó I, Barceló MJ, Rojas-García R, Illa I, Gámez J, Cervera C, Pou A, Izquierdo G, Baiget M, Tizzano EF. SMN2 copy number predicts acute or chronic spinal muscular atrophy but does not account for intrafamilial variability in siblings. J Neurol 2006;253:21–25.
14. Cusin V, Clermont O, Gérard B, Chantereau D, Elion J. Prevalence of SMN1 deletion and duplication in carrier and normal populations: implication for genetic counselling. J Med Genet 2003;40:e39.
15. Fidzianska A, Hausmanowa-Petrusewicz I. Morphology of the lower motor neuron and muscle. In: Gamstorp I, Sarnat HB editors. Progressive Spinal Muscular Atrophies. Raven Press, New York; 1984:55–89.
16. Fidzianska A, Goebel HH, Warlo I. Acute infantile spinal muscular atrophy. Muscle apoptosis as a proposed pathogenetic mechanism. Brain 1990;113 ( Pt 2):433–445.
17. Fidzianska A, Rafalowska J. Motoneuron death in normal and spinal muscular atrophy-affected human fetuses. Acta Neuropathol 2002;104:363–368.
18. Francis JW, Sandrock AW, Bhide PG, Vonsattel JP, Brown RH Jr. Heterogeneity of subcellular localization and electrophoretic mobility of survival motor neuron (SMN) protein in mammalian neural cells and tissues. Proc Natl Acad Sci USA 1998;95:6492–6497.
19. Frugier T, Tiziano FD, Cifuentes-Diaz C, Miniou P, Roblot N, Dierich A, et al. Nuclear targeting defect of SMN laking the C-terminus in a mouse model of spinal muscular atrophy. Hum Mol Genet 2000; 9:849–858.
20. Grzeschik SM, Ganta M, Prior TW, Heavlin WD, Wang CH. Hydroxyurea enhances SMN2 gene expression in spinal muscular atrophy cells. Ann Neurol 2005;58:194–202.
21. Hausmanowa-Petrusewicz I, Fidzianska A, Niebroj-Dobosz I, Strugalska MH. Is Kugelberg-Welander spinal muscular atrophy a fetal defect? Muscle Nerve 1980;3:389–402.
22. Hsieh-Li HM, Chang JG, Jong YJ, et al. A mouse model for spinal muscular atrophy. Nat Genet 2000;24:66–70.
23. Holzgreve W, Curry CJ, Golbus MS, Callen PW, Filly RA, Smith JC. Investigation of nonimmune hydrops fetalis. Am J Obstet Gynecol 1984;150:805–812.
24. Kariya S, Park GH, Maeno-Hikichi Y, Leykekhman O, Lutz C, Arkovitz MS, Landmesser LT, Monani UR. Reduced SMN protein impairs maturation of the neuromuscular junctions in mouse models of spinal muscular atrophy Hum Mol Genet 2008;17:2552–2569.
25. Kerr DA, Nery JP, Traystman RJ, Chau BN, Hardwick JM. Survival motor neuron protein modulates neuron-specific apoptosis. Proc Natl Acad Sci USA 2000;97:13312–13317.
26. Kurjak A.,Andonotopo W, Stanojevic M, Milenkovic D, Azumendi G., Hafner T, Ujevic B. Longitudinal study of fetal behaviour by 4D ultrasound. Ultras Rev Obst Gynecol 2005; 5:259–275.
27. Lefebvre S, Bürglen L, Reboullet S, Clermont O, Bulret P, Viollet L, et al. Identification and characterization of spinal muscular atrophy determining gene. Cell 1995; 80:155–165.
28. Lefebvre S, Burlet P, Liu Q, Bertrandy S, Clermont O, Munnich A, et al. Correlation between severity and SMN protein level in spinal muscular atrophy. Nat Genet 1997; 16:265–269.
29. Liu Q, Dreyfuss G. A novel nuclear structure containing the survival of motor neurons protein. EMBO J 1996;15:3555–3565.
30. Lorson CL, Hahnen E, Androphy EJ, Wirth B. A single nucleotide in the SMN gene regulates splicing and is responsible for spinal muscular atrophy. PNAS USA 1999; 96:6307–6311.
31. MacLeod MJ, Taylor JE, Lunt PW, Mathew CG, Robb SA. Prenatal onset spinal muscular atrophy. Eur J Paediatr Neurol 1999; 3:65–72.

32. McGovern VL, Gavrilina TO, Beattie CE, Burghes AH. Embryonic motor axon development in the severe SMA mouse. Hum Mol Genet 2008;17:2900–2909.
33. Martínez-Hernández R, Soler-Botija C, Also E, Alias E, Casellas L, Gich I, Bernal S, Tizzano EF. The developmental pattern of myotubes in spinal muscular atrophy indicates prenatal delay of muscle maturation. J Neuropathol Exp Neurol 2009;68:474–481.
34. Mercuri E, Bertini E, Messina S, Solari A, D'Amico A, Angelozzi C, Battini R, Berardinelli A, Boffi P, Bruno C, Cini C, Colitto F, Kinali M, Minetti C, Mongini T, Morandi L, Neri G, Orcesi S, Pane M, Pelliccioni M, Pini A, Tiziano FD, Villanova M, Vita G, Brahe C. Randomized, double-blind, placebo-controlled trial of phenylbutyrate in spinal muscular atrophy. Neurology 2007;68:51–55.
35. Monani UR, Lorson CL, Parsons DW, et al. A single nucleotide difference that alters splicing patterns distinguishes the SMA gene SMN1 from the copy gene SMN2. Hum Mol Genet 1999; 8:1177–1183.
36. Monani UR, Sendtner M, Coovert DD, Parsons DW, Andrassi C, Le TT, et al. The human centromeric survival motor neuron gene (SMN2) rescues embryonic lethality in Smn-/- mice and results in a mouse with spinal muscular atrophy. Hum Mol Genet 2000; 9: 333–339.
37. Oppenheim R. Cell death during development of the nervous system Annu Rev Neurosci 1991; 14:453–501.
38. Oprea GE, Kröber S, McWhorter ML, Rossoll W, Müller S, Krawczak M, Bassell GJ, Beattie CE, Wirth B. Plastin 3 is a protective modifier of autosomal recessive spinal muscular atrophy. Science 2008; 320(5875):524–527.
39. Pane M, Staccioli S, Messina S, D'Amico A, Pelliccioni M, Mazzone ES, Cuttini M, Alfieri P, Battini R, Main M, Muntoni F, Bertini E, Villanova M, Mercuri E. Daily salbutamol in young patients with SMA type II. Neuromuscul Disord 2008;18:536–540.
40. Pearn J. Incidence, prevalence, and gene frequency studies of chronic childhood spinal muscular atrophy. J Med Genet 1978; 15:409–413.
41. Pellizoni L. Chaperoning ribonucleoprotein biogenesis in health and disease. EMBO Report 2007; 8:340–345.
42. Rijhsinghani A, Yankowitz J, Howser D, Williamson R. Sonographic and maternal serum screening abnormalities in fetuses affected by spinal muscular atrophy. Prenat Diagn 1997; 17:166–169.
43. Sandri M, Carraro U. Apoptosis of skeletal muscles during development and disease. Int J Biochem Cell Biol 1999; 31:1373–1390.
44. Schrank B, Gotz R, Gunnersen JM, et al. Inactivation of the survival motor neuron gene, a candidate gene for human spinal muscular atrophy, leads to massive cell death in early mouse embryos. Proc Natl Acad Sci U S A 1997; 94:9920–9925.
45. Simic G, Seso-Simic D, Lucassen PL, Islam A, Krsnik Z, Cviko A, et al. Ultrastructural analysis and TUNEL demonstrate motor neuron apoptosis in Werdnig-Hoffmann disease. J Neuropathol Exp Neurol 2000; 59:398–407.
46. Sohal GS. Sixth Annual Stuart Reiner Memorial Lecture. Embryonic development of nerve and muscle. Muscle Nerve 1995;18:2–14.
47. Stathas D, Kalfakis N, Kararizou E, Manta P. Spinal muscular atrophy: DNA fragmentation and immaturity of muscle fibers. Acta Histochem 2008;110:53–58.
48. Soler-Botija C, Ferrer I, Gich I, Baiget M, Tizzano EF. Neuronal death is enhanced and begins during foetal development in type I spinal muscular atrophy spinal cord. Brain 2002; 125:1624–1634.
49. Soler-Botija C, Cusco I, Lopez E, Clua A, Gich I, Baiget M, et al. Choline acetyltransferase expression does not identify early pathogenic events in fetal SMA spinal cord. Neuromuscul Disord 2005a; 15:253–258.
50. Soler-Botija C, Cusco I, Caselles L, et al. Implication of fetal SMN2 expression in type I SMA pathogenesis: protection or pathological gain of function? J Neuropathol Exp Neurol 2005b; 64:215–223.

51. Sumner CJ, Huynh TN, Markowitz JA, Perhac JS, Hill B, Coovert DD, Schussler K, Chen X, Jarecki J, Burghes AH, Taylor JP, Fischbeck KH. Valproic acid increases SMN levels in spinal muscular atrophy patient cells. Ann Neurol 2003;54:647–654.
52. Sumner CJ. Therapeutics development for spinal muscular atrophy. NeuroRx. 2006; 3: 235–245.
53. Swoboda KJ, Scott CB, Reyna SP, Prior TW, LaSalle B, Sorenson SL, Wood J, Acsadi G, Crawford TO, Kissel JT, Krosschell KJ, D'Anjou G, Bromberg MB, Schroth MK, Chan GM, Elsheikh B, Simard LR. Phase II open label study of valproic acid in spinal muscular atrophy. PLoS One. 2009;4:e5268. Epub 2009 May 14.
54. Tews DS, Goebel HH. DNA fragmentation and BCL-2 expression in infantile spinal muscular atrophy. Neuromuscul Disord. 1996; 6:265–273.
55. Tizzano EF, Cabot C, Baiget M. Cell-specific survival motor neuron gene expression during human development of the central nervous system: implications for the pathogenesis of spinal muscular atrophy. Am J Pathol 1998; 153:355–361.
56. Van Vugt JM, Tinnemans BW, Van Zalen-Sprock RM. Outcome and early childhood follow-up of chromosomally normal fetuses with increased nuchal translucency at 10-14 weeks′ gestation. Ultrasound Obstet Gynecol 1998; 11:407–409.
57. Wirth B, Brichta L, Schrank B, Lochmüller H, Blick S, Baasner A, Heller R. Mildly affected patients with spinal muscular atrophy are partially protected by an increased SMN2 copy number. Hum Genet 2006;119:422–428.
58. Wu D, Hersh LB. Choline acetyltransferase: celebrating its fiftieth year. J Neurochem 1994; 5:1653–1663.

# Chapter 16
# Spinal Muscular Atrophy

**Jérémie Vitte, Ruben Attali, Nasim Warwar, Irena Gurt, and Judith Melki**

**Abstract** Spinal muscular atrophies (SMA) are frequent autosomal recessive disorders characterized by degeneration of lower motor neurons. SMA are caused by mutations of the survival of motor neuron gene (*SMN1*) leading to a reduction of the SMN protein amount. The identification of SMN interacting proteins involved in the formation of the spliceosome and splicing changes in SMN-deficient tissues of mutant mice strongly support the view that SMN is involved in the splicing reaction. However, the molecular pathway linking SMN defect to the SMA phenotype remains unclear. From a better knowledge of the genetic basis of SMA and the defects resulting from the mutations of *SMN1* in cellular or animal models, several therapeutics strategies have been selected aiming at targeting *SMN2*, a partially functional copy of *SMN1* gene which remains present in patients, or to prevent neurons from death. Refined characterization of the degenerative process in SMA and the identification of the defective molecular pathway downstream from the SMN defect will provide further exciting insight into this disease in the near future. They should contribute to clarify the pathophysiology of SMA, the function of SMN and should help in designing potential targeted or non-targeted therapeutic molecules.

**Keywords:** Spinal muscular atrophy · SMN · Motor neuron disease · Splicing

## 16.1 Clinical Aspects

Childhood SMA has been subdivided into three clinical types based on the age of onset of symptoms, achievement of motor milestones and age at death [36]. The acute form or Werdnig-Hoffmann disease (type I) is characterized by severe, generalized muscle weakness at birth or within the first 6 months. Feeding and breathing difficulties are usually responsible for death within 2 years of age. Type II SMA is a more slowly progressive disease with a variable prognosis. Infants are able to

---

J. Melki (✉)
Department of Human Genetics, Hadassah University Hospital, 91120, Jerusalem, Israel
e-mail: judith.melki@inserm.fr

C. Espinós et al. (eds.), *Inherited Neuromuscular Diseases*, Advances in Experimental Medicine and Biology 652, DOI 10.1007/978-90-481-2813-6_16,

sit unsupported but not able to stand or walk unaided. All children will develop, if untreated, severe scoliosis and respiratory ventilation defect. Life expectancy is highly variable, ranging up to adult life in some cases. In juvenile SMA or type III or Kugelberg-Welander disease, symptoms start from 18 months of life to the age of 20 years. All patients are able to walk unaided and proximal muscle weakness is progressive. Subsequently, the patients show difficulties in climbing stairs, gait become waddling. Finally, adult SMA is defined by an age of onset of symptoms beyond 20 years of age, the diagnostic criteria being identical to the other forms.

Whatever the age of onset, the clinical features consist of symmetrical muscle weakness (more proximal than distal) associated with muscle atrophy, absence or marked decrease of deep tendon reflexes, fasciculations of the tongue, and tremor of hands. There is no evidence of other symptoms. Electromyographic studies show pattern of denervation with neither sensory involvement nor marked decrease of motor nerve conduction velocities. Finally, muscle biopsy, which has been recently replaced by genetic testing for diagnosis, shows skeletal muscle denervation with groups of atrophic and hypertrophic fibers or fiber type grouping most often found in chronic cases (types II and III).

The most striking neuropathological feature found in autopsy material of SMA patients is a loss of the large anterior horn cells of the spinal cord. In the remaining surviving motor neurons, severe degree of central chromatolysis is visible. These cells appear as large ballooned cells without stored substances. Other anterior cells are pyknotic.

## 16.2 Genetic Basis of SMA

SMA is genetically heterogeneous. Most pedigrees with childhood proximal SMA show an autosomal recessive mode of inheritance with an incidence of 1 in 6,000–10,000 newborns. Severe SMA associated with arthrogryposis and bone fractures has been shown to be X-linked. Recessive and dominant autosomal types have also been described in the adult form. Finally, a recessive X-linked adult form known as Kennedy's syndrome or spinal and bulbar muscular atrophy has been described and is caused by a (CAG) triplet repeat expansion in the androgen receptor gene. Our review will focus on autosomal recessive proximal SMA, the most frequent form.

The SMA locus on chromosome 5q13 is characterized by an inverted duplication, each element (about 500 kb) containing several genes. Deletions or conversion events including exon 7 of the telomeric copy of the survival of motor neuron gene (*SMN* , renamed *SMN1* ) are found in 93% of SMA patients [24]. In the remaining patients, the same mutation is found on one allele, the other allele carrying intragenic mutations of *SMN1* including missense, non-sense or splice site mutations [14, 24].

*SMN1* is duplicated in a highly homologous gene called *SMN2* and both genes are transcribed [24]. The *SMN2* gene is present in all patients but is not able to compensate for the *SMN1* gene defects. At the genomic level, the presence of a gene dosage effect in type I but not in type III SMA has suggested that type I SMA was caused by deletion of *SMN1* whereas type III was associated with conversion event of *SMN1* into *SMN2* leading to an increased number of *SMN2* genes [4, 24]. This is in agreement with the tight inverted correlation between the amount of the protein encoded by the *SMN2* gene and the clinical severity of human SMA disease [10, 25].

Five nucleotides distinguish *SMN2* from *SMN1* genes without any effect on the amino acid sequence [24]. The critical difference between these two genes is a Cytosine (C) to Thymine (T) transition in exon 7 of *SMN2*. This silent mutation in *SMN2* exon 7 disrupts an exonic splicing enhancer [31] leading to an alternative splicing of exon 7 of *SMN2* transcripts only. Full-length transcripts are almost exclusively produced by *SMN1* , whereas the predominant form encoded by *SMN2* is lacking exon 7 ($SMN^{\triangle 7}$) [31, 24]. Full length transcript ($SMN^{FL}$) is also encoded by *SMN2* and translated into a functional protein but its abundance is much lower than that encoded by *SMN1* . The truncated transcript lacking exon 7 encodes a putative shorter protein in which the last 16 amino acids of $SMN^{FL}$ are replaced by 4 amino acids encoded by exon 8 ($SMN^{\triangle 7}$). In human and mice, the $SMN^{\triangle 7}$ protein encoded by the *SMN2* gene is highly unstable *in vivo* [46].

## 16.3 SMN: A Multifunctional Protein

SMN is an ubiquitously expressed protein of 294 amino acids with a molecular weight of 38 kDa. Sequence analysis showed that SMN contains a phylogenetically conserved sequence called the Tudor domain which has been found in many proteins involved in RNA metabolism [44]. The SMN complex is found both in the cytoplasm and in the nucleus where it is concentrated in a structure called gems (for "gemini of coiled bodies") most often associated with or identical to Cajal bodies (coiled bodies) depending on the cell type or tissue analyzed [28]. SMN forms a large complex of approximately 1 MDa. The interaction of SMN with components of this large complex is mediated directly by the Tudor domain and is enhanced by a SMN oligomerization domain [30]. The identification of SMN interacting proteins of known function strongly supports the view that SMN is involved and facilitates cytoplasmic assembly of snRNP into the spliceosome, a large RNA-protein complex that catalyzes the splicing reaction. In the nucleus, SMN appears to be directly involved in pre-mRNA splicing, transcription and metabolism of ribosomal RNA [11, 12, 29, 38, 39].

The link between defect in snRNP biogenesis mediated by SMN deficiency and motor axon defect has been strongly suggested in zebrafish. The injection of U snRNPs into SMN-deficient embryos of zebrafish is able to rescue motor axon outgrowth and pathfinding defects [49]. More recently, strong impairment of snRNP assembly has been observed in human SMA type I and in tissues of severe SMA mice [15, 48, 52]. In addition, alterations in splicing have been reported in various tissues of mutant mice indicating that splicing defects are not restricted to motor neurons [52]. Splicing pattern changes were observed in 1.3% of genes in SMA mutant spinal cord. However, the splicing pattern changes specific to motor neurons, the target cells in SMA and responsible for motor neuron degeneration remains unknown.

SMN has also been shown to interact directly with other proteins including Bcl-2 and p53, two proteins involved in apoptotic processes, although these interactions and the involvement of an apoptotic process in SMA remain to be clarified in vivo. Other direct or indirect partners of SMN include the FUSE binding protein, profilin II, a zinc-finger protein called ZRP1, RNA helicase A, RNA polymerase II, and RNA [14 for review]. Therefore, SMN appears to be a multifunctional protein.

## 16.4 SMA Pathogenesis

In various species including *M. musculus, D. melanogaster, C. elegans, SMN* ortholog is not duplicated and its deletion leads to early embryonic lethality [13, 18, 34, 35, 41]. To circumvent embryonic lethality of mice knocked out for the *Smn* gene, several transgenic approaches were undertaken to create mouse models. One strategy was based on the generation of mice carrying genomic organization similar to that of human SMA [18, 35]. It consisted of the creation of two mouse lines, one carrying a deletion of *Smn* through homologous recombination and the other line carrying a transgene expressing the human *SMN2* gene. Mice carrying both homozygous deletion of *Smn* and the human *SMN2* gene developed a phenotype depending on the number of *SMN2* copies. Mice carrying low copy number of *SMN2* and harbouring a homozygous null *Smn* allele developed a severe phenotype leading to death either *in utero* or in the first days after birth. Mutant mice displayed motor defects associated with a moderate loss of motor neurons of the spinal cord (up to 35%) at the latest stage of disease. The absence of abnormal phenotype in mutant mice carrying a high copy number of *SMN2* transgene indicated that *SMN2* was able to prevent the embryonic lethality of the *Smn* knock out mice and confirmed that an increased copy number of *SMN2* reduced the severity of the phenotype, feature in agreement with data observed in human SMA [18, 25, 35].

Another approach has been carried out by using the Cre-loxP recombination system. A mouse line carrying two loxP sequences flanking *Smn* exon 7 has been established through homologous recombination. Deletion of *Smn* exon 7, the most frequent mutation found in SMA patients, has been directed to neurons, skeletal muscle or liver ("neuronal", "muscular" or "liver" mutant) [7, 8, 13, 37, 45]. Neuronal mutant mice in which full length *Smn* transcripts are replaced by *Smn* $^{\Delta 7}$ displayed severe motor defect leading to complete paralysis and death at a mean age of 4 weeks. Analysis of skeletal muscle revealed a severe muscle denervation process contrasting with mild reduction of motor neuron cell bodies (29%). In addition, abnormal synaptic terminals of neuromuscular junctions filled with neurofilaments, including phosphorylated forms, were observed in mutant mice associated with defects in axonal sprouting [8, 13]. These findings are likely responsible for motor neuron dysfunction and suggest that loss of motor neurons occurs through a dying back axonopathy.

Interestingly, whatever the mouse models (*SMN2* copies on an *Smn* null background or homozygous *Smn* exon 7 deletion targeted to neurons), SMN deficiency causes moderate and late loss of motor neurons associated with severe defect in axonal outgrowth with poor terminal arborization and accumulation of neurofilament at the neuromuscular junctions [8, 20]. These data indicate that SMA is a cell autonomous process, mutation of *Smn* in neurons only being sufficient to recapitulate the disease phenotype [8].

Consistently, antisense morpholinos to reduce SMN levels in zebrafish cause defects in motor axon pathfinding suggesting that SMN has a role in motor axon development [33]. Moreover, point mutations in *SMN* similar to those found in SMA patients have been found in *Drosophila* resulting in defects of the neuromuscular junctions including disorganization of synaptic motor neuron boutons and reduction of post synaptic receptor subunits [5]. Therefore, defects in SMN in various organisms highlighted an essential role of SMN in motor axon and neuromuscular junction development or maintenance.

Using the same strategy, deletion of *Smn* exon 7 has been directed to murine skeletal muscle ("muscular mutant") [7, 37]. Unexpectedly, mutant mice displayed a myopathic phenotype characterized by muscle necrosis leading to muscle paralysis and death. These data suggest that SMN has an important role in skeletal muscle differentiation and/or regeneration. To know whether deletion of *Smn* exon 7 led to deleterious effect in all mammalian cell types *Smn* mutation has been directed to liver, a tissue non-affected in human SMA. Homozygous mutation leads to late embryonic lethality characterized by marked atrophy of liver associated with severe liver dysfunction and absence of regeneration [45].

How to explain the severe muscle or liver phenotype of mice lacking the full length SMN whereas the phenotype of the human disease is mainly characterized by the involvement of motor neurons? Murine *Smn* gene is present as a single copy while in human, *SMN1* is duplicated in a highly homologous gene *SMN2* . The residual amount of SMN produced by full length transcripts of the *SMN2* gene, which remains present in patients, is likely sufficient to ensure normal functions of various organs including liver but not motor neurons. Altogether, these data support the hypothesis that SMN plays an ubiquitous and essential role in all cells and indicate that $Smn^{\Delta7}$ is not able to compensate for the lack of full length SMN.

More recently, it has been suggested that SMN might have an additional function in neurons. SMN was shown to bind heterogeneous nuclear ribonucleoprotein-R (hnRNP-R), an mRNA binding protein that may associate with β-actin mRNA *in vitro* suggesting a role of SMN in the assembly and/or transport of β -actin mRNP complexes into growth cones [40]. Moreover, SMN was localized in granules and transported down axons of cultured neurons [51]. Altogether, these data suggested that SMN plays a role in metabolism of RNA in the nucleus and/or in the transport of some mRNA in axons.

In vitro, overexpression of $SMN^{\Delta7}$ appears to be pro-apoptotic suggesting a dominant effect of $SMN^{\Delta7}$ in motor neurons [21, 47]. In addition, in chick forebrain neurons, $SMN^{\Delta 7}$ was not able to be transported and overexpression of $SMN^{\Delta 7}$ resulted in shorter axons associated with a retention of $SMN^{\Delta 7}$ in the nucleus [51]. These data have suggested that in addition to the deleterious effect of a reduced dose of $SMN^{FL}$, $SMN^{\Delta 7}$ might have a dominant negative effect on neurons. Other studies using cell system or mouse model did not confirm these data [9, 26]. In addition, using new antibodies specific to $SMN^{FL}$ or $SMN^{\Delta 7}$, the $SMN^{\Delta 7}$ protein was not detected in either human neuronal (including spinal cord) or non-neuronal tissues excluding the hypothesis of a dominant effect of $SMN^{\Delta 7}$to explain the vulnerability of motor neurons in SMA [46]. These data indicate that motor neuron degeneration is caused uniquely by a reduction in $SMN^{FL}$ protein levels.

It has been hypothesized that the vulnerability of motor neurons in SMA might be related to the differential level of expression of the SMN protein in the spinal cord compared to other tissues. A dramatic reduction in $SMN^{FL}$ is observed in SMA samples, regardless of the tissue type examined indicating that the vulnerability of the spinal cord in SMA can not be ascribed to the differential expression, or a more dramatic reduction of $SMN^{FL}$, in spinal cord when compared to brain tissue [46].

In summary, the main hypothesis to explain the vulnerability of motor neurons in SMA is a role of SMN in splicing of some RNA specific to motor neurons. However, an additional function of SMN in these cells unrelated to snRNP biogenesis cannot be excluded. In spite of the generation and use of several model systems, the SMA pathogenesis remains unclear. This is partly due to the difficulty to generate model systems mimicking the situation of human SMA namely the presence of the *SMN2* locus in its genomic and cellular (motor neuron) contexts and the reduced dose of SMN which is likely regulated during development and postnatal period.

## 16.5 Therapeutic Strategies in SMA

From a better knowledge of genetic basis of SMA, upregulation of *SMN2* gene expression, preventing exon 7 skipping of *SMN2* transcripts or improving the stability of the $SMN^{\Delta 7}$ protein have been selected as therapeutic strategies in SMA. Several groups have generated cell systems and identified compounds able to induce these modifications [1–3, 6, 16, 19, 22, 32, 42, 50] (Table 16.1). Some of them should be validated *in vivo* on animal models carrying the human *SMN2* gene. In addition, validating their activity on deficient motor neurons in culture should be very helpful to undertake clinical trials in SMA. Several detection systems aiming at following SMN at either RNA or protein level in blood samples have been developed [23, 43]. Although we do not know whether the SMN level in blood correlates with SMN level in motor neurons, clinical trials aiming at acting on *SMN2* in SMA patients should take into account the circulating blood level of SMN (before and after the treatment) to interpret clinical data with respect to the presence or the absence of SMN response to a given drug.

**Table 16.1** Drugs able to activate *SMN2* gene promoter (transcription), modulate exon 7 skipping of *SMN2* transcripts (splicing) or acting on SMN protein level as determined by *in vitro* experiments. * indicates HDAC inhibitors

| Drug | Target | Reference |
|---|---|---|
| Interferon | Transcription | [2] |
| SAHA* | Transcription | [22] |
| Sodium butyrate* | Transcription, splicing | [6] |
| Valproic acid* | Transcription, splicing | [3, 42] |
| Indole | Transcription, splicing | [19] |
| Quinazoline | Transcription, splicing | [19] |
| Phenylbutyrate | Transcription, splicing | [1] |
| Hydroxyurea | Splicing | [16] |
| Aclarubicin | Splicing | [1] |
| Indoprofen | SMN protein | [32] |
| Aminoglycoside | SMN protein | [50] |

Alternatively, identifying molecules involved in or associated with the SMA degenerative process should help in designing targeted therapeutics in SMA. Cellular or animal models should contribute to elucidate the molecular events leading to SMA. Neuroprotective agents or neurotrophic factors may protect neurons against toxicity or promote axonal sprouting of motor neurons. The availability of animal models allowed us to test the efficiency, however non specific, of molecules with such biological properties. Cardiotrophin I and riluzole which exhibit neurotrophic activities have shown therapeutic benefits on neuronal mouse model of SMA indicating for the first time that SMA disease progression might be attenuated even after the onset of symptoms [17, 27].

## 16.6 Conclusion

Major advances have been made in the field of SMA, the gene has been identified, a genetic test has been developed for diagnostic purpose avoiding muscle biopsy. In addition, these developments have helped parents and family members in family planning options. The link between defect in RNA metabolism mediated by SMN deficiency and motor neuron defect has been strongly suggested. However, the splicing pattern changes specific to motor neurons, the target cells in SMA, and responsible for motor neuron degeneration remains unknown. A genetic approach aiming at identifying new genes responsible for lower motor neuron diseases should be very helpful to identify new proteins linking motor neuron degeneration and RNA metabolism.

Importantly, several therapeutic strategies have been developed aiming at acting on *SMN2* or at protecting motor neuron from death. Refined characterization of the degenerative process in SMA and the identification of the defective molecular pathway downstream from the SMN defect should help in designing novel targeted or non-targeted therapeutic molecules.

**Acknowledgements** Work made in our laboratory was supported by INSERM, the University of Evry, the Association Française contre les Myopathies, Families of SMA (U.S.A) and the Foundation Bettencourt Schueller. We gratefully thank Natacha Roblot, Bénédicte Desforges, Gaelle Millet, Vandana Joshi, and Vilayphone Vongsa for their invaluable assistance.

# References

1. Andreassi C, et al. Aclarubicin treatment restores SMN levels to cells derived from type I spinal muscular atrophy patients. Hum Mol Genet 2001; 10:2841–2849.
2. Baron-Delage S, et al. Interferons and IRF-1 induce expression of the survival motor neuron (SMN) genes. Mol Med 2000; 6:957–968.
3. Brichta L, et al. Valproic acid increases the SMN2 protein level: a well-known drug as a potential therapy for spinal muscular atrophy. Hum Mol Genet 2003; 12:2481–2489.
4. Campbell L, et al. Genomic variation and gene conversion in spinal muscular atrophy: implications for disease process and clinical phenotype. Am J Hum Genet 1997; 61:40–50.
5. Chan YB, et al. Neuromuscular defects in a Drosophila survival motor neuron gene mutant. Hum Mol Genet 2003; 12:1367–1376.
6. Chang JG, et al. Treatment of spinal muscular atrophy by sodium butyrate. Proc Natl Acad Sci USA 2001; 98:9808–9813.
7. Cifuentes-Diaz C, et al. Deletion of murine SMN exon 7 directed to skeletal muscle leads to severe muscular dystrophy. J Cell Biol 2001; 152:1107–1114.
8. Cifuentes-Diaz C, et al. Neurofilament accumulation at the motor endplate and lack of axonal sprouting in a spinal muscular atrophy mouse model. Hum Mol Genet 2002; 11:1439–1447.
9. Cisterni C, et al. Death of motoneurons induced by trophic deprivation or by excitotoxicity is not prevented by overexpression of SMN. Neurobiol Dis 2001; 8:240–251.
10. Coovert DD, et al. The survival motor neuron protein in spinal muscular atrophy. Hum Mol Genet 1997; 6:1205–1214.
11. Eggert C, et al. Spinal muscular atrophy: the RNP connection. Trends Mol Med 2006; 12: 113–121.
12. Fischer U, et al. The SMN-SIP1 complex has an essential role in spliceosomal snRNP biogenesis. Cell 1997; 90:1023–1029.
13. Frugier T, et al. Nuclear targeting defect of SMN lacking the C-terminus in a mouse model of spinal muscular atrophy. Hum Mol Genet 2000; 9:849–858.
14. Frugier T, et al. The molecular bases of spinal muscular atrophy. Curr Opin Genet Dev 2002; 12:294–298.
15. Gabanella F, et al. Ribonucleoprotein assembly defects correlate with spinal muscular atrophy severity and preferentially affect a subset of spliceosomal snRNPs. PLoS ONE. 2007 2:e921.
16. Grzeschik SM, et al. Hydroxyurea enhances SMN2 gene expression in spinal muscular atrophy cells. Ann Neurol 2005; 58:194–202.
17. Haddad H, et al. Riluzole attenuates spinal muscular atrophy disease progression in a mouse model. Muscle and Nerve 2003; 28:432–437.
18. Hsieh-Li HM, et al. A mouse model for spinal muscular atrophy. Nat Genet 2000; 24:66–70.
19. Jarecki J, et al. Diverse small-molecule modulators of SMN expression found by high-throughput compound screening: early leads towards a therapeutic for spinal muscular atrophy. Hum Mol Genet 2005; 14:2003–2018.
20. Kariya S, et al. Reduced SMN protein impairs maturation of the neuromuscular junctions in mouse models of spinal muscular atrophy. Hum Mol Genet. 2008 May 20 [Epub ahead of print].
21. Kerr DA, et al. Survival motor neuron protein modulates neuron-specific apoptosis. Proc Natl Acad Sci USA 2000; 97:13312–13317.
22. Kernochan LE, et al. The role of histone acetylation in SMN gene expression. Hum Mol Genet 2005; 14:1171–1182.

23. Kolb SJ, et al. A novel cell immunoassay to measure survival of motor neurons protein in blood cells. BMC Neurol 2006; 6:6.
24. Lefebvre S, et al. Identification and characterization of a spinal muscular atrophy-determining gene. Cell 1995; 80:155–165.
25. Lefebvre S, et al. Correlation between severity and SMN protein level in spinal muscular atrophy. Nat Genet 1997; 16:265–269.
26. Le TT, et al. SMNDelta7, the major product of the centromeric survival motor neuron (SMN2) gene, extends survival in mice with spinal muscular atrophy and associates with full-length SMN. Hum Mol Genet 2005; 14:845–857.
27. Lesbordes JC, et al. Therapeutic benefits of cardiotrophin-1 gene transfer in a mouse model of spinal muscular atrophy. Hum Mol Genet 2003; 12:1233–1239.
28. Liu Q, Dreyfuss G. A novel nuclear structure containing the survival of motor neurons protein. Embo J 1996; 15:3555–3565.
29. Liu Q, et al. The spinal muscular atrophy disease gene product, SMN, and its associated protein SIP1 are in a complex with spliceosomal snRNP proteins. Cell 1997; 90: 1013–1021.
30. Lorson CL, et al. SMN oligomerization defect correlates with spinal muscular atrophy severity. Nat Genet 1998; 19:63–66.
31. Lorson CL, et al. A single nucleotide in the SMN gene regulates splicing and is responsible for spinal muscular atrophy. Proc Natl Acad Sci USA 1999; 96:6307–6311.
32. Lunn MR, et al. Indoprofen upregulates the survival motor neuron protein through a cyclooxygenase-independent mechanism. Chem Biol 2004; 11:1489–1493.
33. McWhorter ML, et al. Knockdown of the survival motor neuron (Smn) protein in zebrafish causes defects in motor axon outgrowth and pathfinding. J Cell Biol 2003; 162:919–931.
34. Miguel-Aliaga I, et al. The Caenorhabditis elegans orthologue of the human gene responsible for spinal muscular atrophy is a maternal product critical for germline maturation and embryonic viability. Hum Mol Genet 1999; 8:2133–2143.
35. Monani UR, et al. The human centromeric survival motor neuron gene (SMN2) rescues embryonic lethality in Smn(-/-) mice and results in a mouse with spinal muscular atrophy. Hum Mol Genet 2000; 9:333–339.
36. Munsat TL. Workshop report: International SMA Collaboration. Neuromuscul Disord 1991; 1:81.
37. Nicole S, et al. Intact satellite cells lead to remarkable protection against Smn gene defect in differentiated skeletal muscle. J Cell Biol 2003; 161:571–582.
38. Paushkin S, et al. The SMN complex, an assemblyosome of ribonucleoproteins. Curr Opin Cell Biol 2002; 14:305–312.
39. Pellizzoni L, et al. A novel function for SMN, the spinal muscular atrophy disease gene product, in pre-mRNA splicing. Cell 1998; 95:615–624.
40. Rossoll W, et al. Smn, the spinal muscular atrophy-determining gene product, modulates axon growth and localization of beta-actin mRNA in growth cones of motoneurons. J Cell Biol 2003; 163:801–812.
41. Schrank B, et al. Inactivation of the survival motor neuron gene, a candidate gene for human spinal muscular atrophy, leads to massive cell death in early mouse embryos. Proc Natl Acad Sci USA 1997; 94:9920–9925.
42. Sumner CJ, et al. Valproic acid increases SMN levels in spinal muscular atrophy patient cells. Ann Neurol 2003; 54:647–654.
43. Sumner CJ, et al. SMN mRNA and protein levels in peripheral blood: biomarkers for SMA clinical trials. Neurology 2006; 66:1067–1073.
44. Talbot K, et al. Missense mutation clustering in the survival motor neuron gene: a role for a conserved tyrosine and glycine rich region of the protein in RNA metabolism? Hum Mol Genet 1997; 6:497–500.
45. Vitte JM, et al. Deletion of murine Smn exon 7 directed to liver leads to severe defect of liver development associated with iron overload. Am J Pathol 2004; 165:1731–1741.

46. Vitte J, et al. Refined characterization of the expression and stability of the SMN gene products. Am J Pathol 2007; 171:1269–1280.
47. Vyas S, et al. Involvement of survival motor neuron (SMN) protein in cell death. Hum Mol Genet 2002; 11:2751–2764.
48. Wan L, et al. The survival of motor neurons protein determines the capacity for snRNP assembly: biochemical deficiency in spinal muscular atrophy. Mol Cell Biol 2005; 25:5543–5551.
49. Winkler C, et al. Reduced U snRNP assembly causes motor axon degeneration in an animal model for spinal muscular atrophy. Genes Dev 2005; 19:2320–2330.
50. Wolstencroft EC, et al. A non-sequence-specific requirement for SMN protein activity: the role of aminoglycosides in inducing elevated SMN protein levels. Hum Mol Genet 2005; 14:1199–1210.
51. Zhang HL, et al. Active transport of the survival motor neuron protein and the role of exon-7 in cytoplasmic localization. J Neurosci 2003; 23:6627–6637.
52. Zhang Z, et al. SMN deficiency causes tissue-specific perturbations in the repertoire of snRNAs and widespread defects in splicing. Cell. 2008 133:585–600.

# Chapter 17
# Friedreich Ataxia: An Update on Animal Models, Frataxin Function and Therapies

**Pilar González-Cabo, José Vicente Llorens, Francesc Palau, and Maria Dolores Moltó**

**Abstract** Friedreich ataxia (FRDA) is an autosomal recessive progressively debilitating degenerative disease that principally affects the nervous system and the heart. Although FRDA is considered a rare disease, is the most common inherited ataxia. It is caused by loss-of-function mutations in the *FXN* gene, mainly an expanded GAA triplet repeat in the intron 1. The genetic defect results in the reduction of frataxin levels, a protein targeted to the mitochondria. Frataxin deficiency leads to mitochondrial dysfunction, oxidative damage and iron accumulation. Studies of the yeast and animal models of the disease have led to propose several different roles for frataxin. Animal models have also been important for dissecting the steps of pathogenesis in FRDA and they are essential for the development of effective therapies. Currently, antioxidant and iron chelation therapies are under evaluation in clinical trials. Gene reactivation, gene therapy and protein replacement strategies for FRDA are promising approaches.

This review focuses on the current models developed for FRDA, the different roles proposed for frataxin and the progress of potential treatment strategies for the disease.

**Keywords:** Friedreich ataxia · Frataxin · Mitochondria · Iron-sulfur clusters · Oxidative stress · Oxidative phosphorylation · Antioxidant therapy · Iron chelators · Recombinant human erythropoietin · Histone deacetylase inhibitors

## 17.1 Introduction

In 1863, Nicholaus Friedreich described a new disease that was given the name of Friedreich Ataxia (FRDA, MIM 229300) [1, 2]. FRDA is the most common hereditary ataxia in the Caucasian population. Specifically, in Spain the prevalence is about 4.7 cases per 100,000 habitants [3, 4]. FRDA is characterized by progressive

P. González-Cabo (✉)
Laboratory of Genetics and Molecular Medicine, Instituto de Biomedicina de Valencia, CSIC, C/Jaume Roig 11, Valencia, Spain
e-mail: pilargc@ibv.csic.es

C. Espinós et al. (eds.), *Inherited Neuromuscular Diseases*, Advances in Experimental Medicine and Biology 652, DOI 10.1007/978-90-481-2813-6_17,

sensory ataxia with onset typically before 25 years of age (cases with later onset are not uncommon), areflexia, dysarthria, extensor plantar responses and loss of position and vibration sense [5]. Patients have primary degeneration of dorsal root ganglia associated with axonal degeneration of the posterior columns, spinocerebellar tracts, and corticospinal tracts, and large myelinated fibers in peripheral nerves. Most individuals with FRDA develop a cardiomyopathy and diabetes mellitus.

## 17.2 Friedreich Ataxia Gene and Molecular Pathology

Friedreich ataxia is an autosomal recessive degenerative disorder caused by a GAA triplet repeat expansion or point mutations in the *FXN* gene on chromosome 9q13 [6, 7]. The *FXN* gene product, frataxin, is a widely expressed mitochondrial protein that is severely reduced in FRDA patients [6].

Most patients with FRDA (approximately 97%) have expansions of a GAA repeat in the first intron of both *FXN* alleles [6, 8, 9]. The GAA repeat expansions reduce the levels of messenger RNA by the formation of non-B DNA structures (triplexes or sticky DNA), the formation of a persistent DNA × RNA hybrid, or heterochromatin formation [10–14]. Individuals with FRDA have also been identified (aprox. 3%) who carry one allele with a trinucleotide repeat expansion and the other allele with a point mutation [8, 10, 11, 15–26].

The nuclear encoded human frataxin protein is synthesized as a 210 amino acid precursor with a N-terminal mitochondrial targeting sequence where it is proteolytically cleaved by the mitochondrial processing peptidase (MPP) to the mature form via a processing intermediate [27, 28]. The initial characterization demonstrated that the endogenous mature form (m-FXN) was encoded by amino acids 56-210. However, there are evidences that endogenous mature frataxin corresponds to amino acids 81-210 [29].

The structure of human frataxin reveals a novel protein fold. A five-stranded, antiparallel β sheets provides a flat platform, which supports a pair of parallel α-helix, to form a compact αβ sandwich [30]. A cluster of 12 acidic residues from the first helix and the first strand of the large sheet form a contiguous anionic surface on the protein. Many of these amino acids are necessary for stabilizing the structure and can not be replaced, being conserved in the different species and appear mutated in several patients. These structural findings predict potential modes of protein-protein and protein iron binding.

## 17.3 Animal Models in Friedreich Ataxia

Frataxin is a mitochondrial protein which shows high conservation throughout evolution, with orthologs in essentially all eukaryotes and some prokaryotes [31]. Many studies in model organisms as yeast, mice, *Drosophila* and *C. elegans* contribute extensively to our current understanding.

In yeast, the deletion of the *YFH1* gene, showed a strain unable to grow on a non-fermentable source of carbon suggesting that it could not carry out oxidative phosphorylation. The strain *yfh1Δ*, showed mitochondrial iron accumulation (up to 10 times more than the wild strain) and exhibited sensitivity to oxidative stress [32–36]. Frataxin can substitute for Yfh1p in yeast, indicating that the two proteins are functional as well as structural homologs [36, 37].

Generation of fly models of FRDA has been possible after the identification of the *Drosophila* frataxin homolog, *fh*, [38]. Systemic and tissue-specific down-regulation of *fh* has been carried out combining the UAS-GAL4 system and RNA interference (RNAi). Moderate systemic reduction of *fh* expression to levels of 30% of normal *fh*-mRNA, a situation which strikingly parallel with Friedreich ataxia, is compatible with normal embryonic development in fly [39]. However in such conditions, adults show a decrease of mean and maximum life span of 60 and 32% respectively, and a reduction in climbing abilities. Meanwhile, widespread *fh* silencing to undetectable levels results in lethality during development and larvae are incapable of reaching the adult phase [39, 40]. Silencing of *fh* in early mesoderm also provokes lethality in agreement with the essential role for frataxin during development, previously proposed in a mouse model [41]. At the biochemical level, almost completely depletion of *Drosophila* frataxin reduces activities of mitochondrial Fe-S containing enzymes, and confers hypersensitivity to iron and to hyperoxia [39, 40]. However aconitase is the only Fe-S enzyme which shows a dramatic reduction in its activity under hyperoxia in flies with moderate reduction of frataxin. Meanwhile the respiratory chain complexes operate normally in this condition. Therefore the decay in aconitase activity by reactive oxygen species (ROS) inactivation would be one of the primary events in frataxin deficient condition. In turn, aconitase inactivation would induce further oxidative stress by different mechanisms [42, 43] that might affect other macromolecules, such as the Fe-S enzymes of the respiratory chain. Frataxin would act as an aconitase chaperone protecting this enzyme against ROS-mediated inactivation supporting a causative role of oxidative stress in Friedreich ataxia. In addition, overexpression of the $H_2O_2$-scavenging enzymes (peroxisomal and mitochondrial catalases and a mitochondrial peroxiredoxin) rescues the shortened life span in frataxin deficient flies [44]. Expressions of these enzymes also alleviate the hypersensibility to $H_2O_2$ shown by flies showing deficiency of frataxin in the peripheral nervous system. These data suggest that $H_2O_2$ is an important pathogenic substrate in FRDA with promising therapeutic implications.

As expected overexpression of *Drosophila* frataxin improve antioxidant capability, resistance to oxidative stress insults, and longevity in flies [45]. However a 9-fold of normal level of *Drosophila* frataxin is lethal in flies [39]. Moreover, the excess of frataxin induces defects in *Drosophila* by oxidative-mediated inactivation of mitochondrial aconitase. Frataxin overload might induce new *in vivo* function for this protein, by limiting iron availability conducting to aconitase dysfunction. Frataxin has been shown to form *in vitro* iron-binding multimers as ferritin [46]. Alternatively, an excess of frataxin might saturate interactions with its partners leading to mitochondrial dysfunction. The pathogenic effect of frataxin excess in *Drosophila* has essential repercussion in gene therapy.

Several transient knock-down models of *C. elegans* frataxin homologue gene, *frh-1*, by RNAi have been reported. Some authors have observed increase in lifespan when frataxin was knocking-down by RNAi [47]. However, microinjection in L4 larvae showed lifespan significantly reduced and worms have increased sensitivity to oxidative stress that, in turn, might explain the reduction of worms' longevity [48]. These worms showed a consistent pleiotropic phenotype that includes lethargic behaviour, reduced broodsize, egg laying defects, slow growth, abnormal pharyngeal pumping and altered defecation. In a third study, reduction of life expectancy in knock-down worms was confirmed and lifespan was positively correlated with impaired respiration [49]. This frataxin deficient model in *C. elegans* may be a useful biological tool for drug screening in FRDA and to investigate the role of frataxin in biochemical pathways.

Generation of a representative mouse model of FRDA is considered important for the further understanding of disease pathology and the testing of potential therapeutic strategies. Initially, a homozygous knockout of the murine FRDA gene leads to early embryonic lethality without evidence of iron accumulation [41]. This mouse has shown that frataxin plays an important role during early embryonic development. As an attempt to generate a mouse model of the disease, it was generated a striated muscle frataxin deficient line and a neuron/cardiac muscle frataxin deficient line through a conditional gene targeting approach [50]. These mice reproduce the most important pathophysiological and biochemical features of the human disease. To obtain specific and progressive neurological models for FRDA, Puccio and collegues generated inducible knock-out mouse models [51]. They showed that ablation of frataxin in adult mice led to progressive neurological symptoms resembling FRDA. These mutants have specific damage to the large sensory neuron cell bodies in the dorsal root ganglia, and they have degeneration of the posterior columns of spinal cord that appear translucent, because of demyelination and loss of fibers, and severe lesions of neurons in Clarke's columns. Because FRDA is typically caused by a large GAA repeat expansion within the first intron of the *FXN* gene, several mouse models harbouring this mutation have been created. One of them is a knockin mouse that carries a (GAA)230 repeat expansion in the first intron of the mouse frataxin homolog gene (*Frda*) [52]. GAA repeat knockin mice were crossed with frataxin knockout mice to obtain double heterozygous mice expressing 25–36% of wild-type frataxin levels. These mice were viable and did not develop anomalies of motor coordination, iron metabolism or response to iron loading. In addition, repeats were meiotically and mitotically stable. As well, it was generated two lines of human *FXN* YAC transgenic mice containing GAA repeat expansions derived from FRDA patient DNA [53]. Offspring from crosses between these transgenic mice and heterozygous *Fxn*-knockout mice [54] express comparatively reduced levels of human frataxin and rescue the homozygous Fxn-knockout embryonic lethality. These mice represent the first GAA repeat expansion-based FRDA mouse models that exhibit progressive FRDA-like pathology. Moreover, transgenic mice carrying expanded human *FXN* GAA repeats (190 or 82 triplets) showed tissue-specific and age-dependent somatic instability specifically in the cerebellum and dorsal root ganglia [55].

## 17.4 The Function of Frataxin

Frataxin is a soluble protein with no previously known function. No specific domains related to protein families are represented in its polypeptide sequences. Experiments in cell systems, specially in yeast *Saccharomyces cerevisiae*, have provided relevant information about the possible function of frataxin in mitochondria and its role in the pathogenesis of the disease. At least four hypotheses for the primary mitochondrial function of frataxin have been proposed.

### *17.4.1 Frataxin and the Homeostasis of Mitochondrial Iron*

The first biochemical data came from the experiments in yeast and from clinical observations in FRDA patients. They shown a dysregulation of the iron metabolism, had an injury caused by free radicals and mitochondrial dysfunction [56, 57]. Therefore, frataxin was related with mitochondrial iron homeostasis. Yeast lacking Yfh1p accumulates mitochondrial iron at the expense of cytosolic iron. Excess iron in mitochondria when reacting with oxygen generates toxic reactive oxygen species (ROS) by the Fenton reaction. These ROS causes the oxidation of cellular components: damage to proteins and nucleic acids, and lipid peroxidation [32]. Some of the proteins that are more easily affected by free radicals are those containing Fe-S clusters, such as complex I, II and III of the mitochondrial transport chain and aconitase, showing a decrease of activity in human fibroblasts and yeast [58, 59]. In contrast, data from frataxin Knock-out mouse model has raised some questions about the role of iron in the disease pathogenesis. Iron deposits were not detected suggesting that iron accumulation might not be the primary consequence of frataxin deficiency, but a secondary effect [41].

### *17.4.2 Frataxin as an Iron-Storage Protein*

A second hypothesis suggests that the function of frataxin is to bind iron and keep it in soluble and bioavailable form [46]. Yeast frataxin is a soluble monomer that contains no iron. But, when ferrous iron was added in presence of oxygen, it was induced the formation of frataxin trimers that catalyzed iron oxidation [60, 61]. New results confirm that yeast frataxin is stable as an iron-loaded monomer, and the protein can bind two ferrous iron atoms with micromolar binding affinity [62]. On the other hand, high iron concentration led to the protein assembly reaching to sequestrate more than 2000 iron atoms [63]. The higher molecular weight of this complex reminded us the ferritin structure, which contains a large number of ferric iron atoms. Although human frataxin is not able to form those complexes *in vitro,* at physiologic iron concentrations in the mitochondria, close to 10% of human frataxin are able to form a spherical complex which contained this metal [64]. Differences founded between yfh1 and mammalian frataxin could be due to the existence of a mitochondrial ferritin of higher eukaryotes (MtF) [65]. MtF expression in frataxin deficient yeast could rescue the respiratory deficiency, thus allowing cell growth in a non-fermentable carbon source and preventing mitochondrial iron accumulation, so protecting the activity of the iron-sulphur enzymes as well [66]. These data support the idea that MtF could replace frataxin functions in yeast, suggesting that frataxin would be directly involved in mitochondrial iron-storage and detoxification.

A common characteristic of yfh1 and mammalian frataxin is that iron storage is easily reversible by the addition of chelating agents, indicating that iron could be biologically available. Interpreting all these data, we would say that frataxin could be a mitochondrial iron chaperone that prevents Fenton reaction by hiding iron to reactive oxygen species and keeping it for other biosynthetic pathways. The problem to accept this hypothesis is that changes in intracellular iron levels do not have an effect on frataxin expression, so this is not consistent with the protein role on iron storage [67].

### *17.4.3 Frataxin as a Chaperone for Iron*

Ferrochelatase, the final enzyme in the heme biosynthetic pathway that incorporate iron into protoporphyrin IX molecule, is one of the protein that frataxin could acts as a Fe(II) chaperone . Multiple studies have demonstrated a physical interaction between frataxin and ferrochelatase [61, 68, 69]. However, decreases in ferrochelatase activity have not been observed in frataxin deficient HeLa cells or patient lymphoblasts [70, 71]. The only other Fe-S enzyme besides ferrochelatase known to participate in the heme biosynthetic pathway is adrenodoxin which carries out the first step of the conversion of heme O to heme *a*, that is required for cytochrome oxidase activity. Frataxin deficient human oligodendroglial cell line has decreased adrenodoxin activity, decreased heme a and heme c levels and decreased cytochrome oxidase activity [72].

There is evidence in yeast and in mammals that frataxin deletion leads to severe alteration of Fe-S enzymes activities [50, 58, 73]. Fe-S clusters are essential components of respiratory electron transfer complexes as well as the tricarboxylic acid cycle enzymes, aconitase and succinate dehydrogenase. Frataxin is required, but not essential, for Fe-S cluster (ISC) biosynthesis [73, 74]. In eukaryotes, initial ISC biogenesis is mitochondrial [75]. In yeast, [2Fe-2S] cluster assembly is catalyzed by the redundant scaffold proteins Isu1 and Isu2 [76]. Yfh1p has been suggested to be the iron donor in this process [77]. In addition, frataxin interacts with Isu1p in yeast [78, 79]. Sulfur from cysteine is supplied by the cysteine desulfurase Nfs1p in association with essential protein partner Isd11 [76]. New data have shown that human frataxin binds the iron-sulfur biogenesis NFS1/ISCU complex through ISD11 and this interaction is nickel dependent [80]. Recently, studies characterising biophysical and metal binding properties and the abilities to transfer Fe(II) of the fly frataxin, support a frataxin's role as an iron binding protein that delivers Fe(II) to the Fe-S cluster biosynthesis machinery via Isu in an iron dependent manner

[81]. Moreover, Drosophila frataxin enhances the stability of DNA and protect it from oxidative stress because frataxin has the ability to regulate Fenton chemistry, function correlated directly or indirectly to Fe-S cluster formation [81].

Frataxin interacts with aconitase in presence of citrate. Citrate prevents aconitase cluster disassembly and is required for enzyme reactivation. Thus, the citrate requirement for interaction of frataxin and aconitase supports the contention that frataxin can stabilize the $[4Fe\text{-}4S]^{2+}$ cluster and facilitate enzyme reactivation [82].

## *17.4.4 Frataxin as a Regulatory of Oxidative Phosphorylation*

Deletion of the frataxin yeast homologue, *YFH1*, results in mutant strains that show a growth defect on fermentable carbon sources and a reduction in mitochondrial respiration [32, 34], because this mutant has affected the oxidative phosphorylation [58]. In skeletal muscle of FRDA patients, using phosphorus magnetic resonance spectroscopy, it has been demonstrated a deficit in a mitochondrial ATP production [83]. Ristow et al. hypothesized that frataxin might primarily affect oxidative phosphorylation acting as an activator [84]. They demonstrated that overexpression of frataxin in mammalian cells causes an increased mitochondrial membrane potential and results in an elevated cellular ATP content. In *Arabidopsis thaliana* it was observed a higher transcript level of the gene *AtFH* (homologue of frataxin) in flowers, a high energy demanding tissues in plants [85]. Moreover, *AtFH* expression was further increased in flowers from *A. thaliana* lines showing a mitochondrial dysfunction induced by the expression of the unedited version of ATP synthase subunit 9. Thus, the induction of *AtFH* expression in these lines could also be interpreted as a nuclear compensation in response to a decrease in mitochondrial respiration.

Physical interactions between the yeast frataxin Yfh1p and succinate dehydrogenase complex subunits Sdh1p and Sdh2p, which are components of the mitochondrial complex II, have been observed [86]. This result has been also confirmed by functional relationships between Yfh1p and Sdh1p and Sdh2p carrying out synthetic genetic interaction studies of mutated genes. Interaction among human frataxin and the human succinate dehydrogenase subunits SDHA and SDHB have been confirmed [86]. These findings suggest that frataxin is working in the mitochondrial electron transfer chain, perhaps by regulating the delivery of electrons via complex II towards the ubiquinone and also suggest that respiratory chain may have a primary role in the pathogenesis of FRDA. Genetic interaction with a complex II subunit of the respiratory chain, has been observed by synthetic genetic analysis in *C. elegans* [48], confirming similar findings previously observed in *S. cerevisiae* [86]. Finally, co-immunoprecipitation results from mitochondria of lymphoblast and COS-cells confirm the study of frataxin's partners in yeast and *C. elegans*, in which frataxin was observed to interact with succinate dehydrogenase [80].

## 17.5 Therapies for Friedreich Ataxia

Pathophysiology of Friedreich ataxia based on cell and animal models and human studies indicates several possible therapeutic approaches: iron-chelation, antioxidant therapy, therapy with metalloporphyrins and molecules to increases frataxin expression.

### *17.5.1 Antioxidant Therapy*

Coenzyme $Q_{10}$ (CoQ) and vitamin E are important mitochondrial antioxidants. Whereas vitamin E is obtained solely from diet, CoQ is obtained in the diet and also synthesised in every cell via a pathway that shares some of the steps with cholesterol biosynthesis. It is a component of the electron transport chain. Lodi et al. have shown that combined CoQ (400 mg/d) and vitamin E (2100 IU/d) therapy resulted in a significant improvement in cardiac and skeletal muscle energy metabolism in 10 patients with FRDA after 3 months of treatment [87]. A long-term longitudinal follow-up study of these patients with FRDA caused a prolonged improvement in cardiac and skeletal muscle bioenergetics clearly demonstrating its biochemical efficacy [88].

Idebenone is a short-chain analogue of CoQ. It has been demonstrated that idebenone treatment inhibits lipid peroxidation, stimulates mitochondrial functions, and improves the myocardial energy state in cardiac hypertrophy [89]. Most trials demonstrated a positive effect on cardiac hypertrophy [90–92]. The neurological function is in general not modified in adult patients, but a dose-dependent effect was demonstrated in young Friedreich ataxia patients [93]. The treatment with idebenone (5–20 mg/kg/day) for 3–5 years in paediatric and adults patients, prevented progression of cardiomyopathy in both patients, whereas its stabilizing effect on neurological dysfunction was present only in the paediatric population. This suggests that the age at which idebenone treatment is initiated may be an important factor in the effectiveness of the therapy [94].

Mitoquinone (MitoQ) is an antioxidant selectively targeted to mitochondria. This mitochondrially localized antioxidant protects against H2O2-induced apoptosis in cell culture studies, in contrast to untargeted ubiquinone analogues [95]. MitoQ was shown to be 800-fold more potent than idebenone in protecting FRDA fibroblasts from death due to endogenous oxidative stress generated by inhibition of glutathione synthesis [96]. Until now no clinical trials have been reported.

### *17.5.2 Iron Chelators Therapy*

Since excess of mitochondrial iron is likely to play a role in the pathology of FRDA, iron chelators have considered in the treatment of FRDA.

Deferiprone is a chelator specifically targeting mitochondrial iron. This chelator is orally active, membrane permeant, and capable of scavenging iron from mitochondria of specific areas of the brain. Treatment to FRDA patients with deferiprone caused no apparent hematologic or neurologic side effects while reducing neuropathy and ataxic gait in the youngest patients [97]. This was the first clinical demonstration of chelation removing labile iron accumulated in a specific brain area implicated in a neurodegenerative disease. In cell HEK-293 deficient in frataxin, deferiprone restored of impaired mitochondrial membrane, increased ATP production and oxygen consumption and attenuation of mitochondrial DNA damage [98]. However, in skin fibroblasts of FRDA patients and in frataxin depleted neuroblastoma derived cells, a direct consequence of chelation mitochondrial free iron is a concentration and time dependent loss of aconitase activity [99].

A group of iron chelator, 2-pyridylcarboxaldehyde isonicotinoyl hydrazone (PCIH) ligands, has been designed to specifically target mitochondrial iron pools, being the 2-pyridylcarboxaldehyde 2-thiophenecarboxyl hydrazone (PCTH) as the most promising compound. PCTH is highly effective at preventing $H_2O_2$-induced cytotoxicity and at preventing oxidative stress. It also increased FRDA fibroblast cell viability by up to 70% [100].

### *17.5.3 Therapy with Metalloporphyrins*

Hemin is an enzyme inhibitor derived from processed red blood cells. It is an iron containing metalloporphyrin. Exogenous heme administration with hemin increases activity of the iron-sulfur cluster enzymes adrenoxin and aconitase in frataxin deficient human oligodendroglial cell line [72].

### *17.5.4 Therapy to Increase Frataxin Expression*

Recombinant human erythropoietin (rhoEPO) is a glycoprotein hormone that controls erythropoiesis, or red blood cell production. rhoEPO has broad neuroprotective and cardioprotective capabilities [101, 102]. In various cell types: lymphocytes from FRDA patients, primary human cardiac cells and in a neuronal cell line, rhoEPO increases frataxin expression significantly [103].

Heber and colleagues described a screening of small molecule DNA ligands for their effects on EGFP expression in the reporter cell lines [104]. While some of the tested compounds were cytotoxic, some compounds such as Hoeschst 33258, pentamidine, DAPI, distamycin and diminazene, were found to increase GFP expression. Pentamidine was also shown to increase frataxin expression in primary FRDA lymphocytes.

Polyamides possess the ability to bind duplex DNA. These molecules are unique in that they can be programmed to bind predetermined DNA sequences [105]. Polyamide FA1 increased *FXN* transcription approximately 2–3 fold [106].

Butyric acid has been identified to possess the ability to increase frataxin expression or transcription through GAA repeats [107].

Molecules that alter chromatin structure by changing the postsynthetic modification states of the histone proteins, namely histone deacetylase inhibitors, have been screened for their effects on *FXN* gene expression. HDAC inchibitors have been tested in FRDA patient lymphocytes and lymphoblasts. One compound, BML-210 and related analogues (in particular component 4b), showed a significant increase in frataxin message levels by approximately two-fold [108]. Compound 106, a derivative of 4b, increases histone acetylation in the brain at a dose that causes no apparent toxicity in wild type C57Bl/6 mouse or in KIKI mice (a knock-in mouse carrying a $(GAA)_{230}$ repeat in the first intron of the endogenous frataxin gene [52]). This compound was able to restore normal frataxin levels in the central nervous system and heart of KIKI mice, tissues that are relevant targets as they are involved in FRDA pathology [109].

**Acknowledgements** This work is supported by the Spanish Ministry of Science and Innovation and the Fondo de Investigación Sanitaria. The CIBER de Enfermedades Raras is an initiative of the Instituto de Salud Carlos III.

## References

1. Friedreich N. Über degenerative Atrophie der spinalen Hinterstränge. *Virchows Arch Pathol Anat* 1863;**27**:1–26.
2. Friedreich N. Über degenerative Atrophie der spinalen Hinterstränge. *Virchows Arch Pathol Anat* 1863;**26**(433–459).
3. Lopez-Arlandis JM, Vilchez JJ, Palau F, et al. Friedreich's ataxia: an epidemiological study in Valencia, Spain, based on consanguinity analysis. *Neuroepidemiology* 1995;**14**(1):14–19.
4. Polo JM, Calleja J, Combarros O, et al. Hereditary ataxias and paraplegias in Cantabria, Spain. An epidemiological and clinical study. *Brain* 1991; **114 ( Pt 2)**:855–866.
5. Durr A, Cossee M, Agid Y, et al. Clinical and genetic abnormalities in patients with Friedreich's ataxia. *N Engl J Med* 1996;**335** (16):1169–1175.
6. Campuzano V, Montermini L, Molto MD, et al. Friedreich's ataxia: autosomal recessive disease caused by an intronic GAA triplet repeat expansion. *Science* 1996;**271** (5254):1423–1427.
7. Chamberlain S, Shaw J, Rowland A, et al. Mapping of mutation causing Friedreich's ataxia to human chromosome 9. *Nature* 1988;**334**(6179):248–250.
8. Cossee M, Schmitt M, Campuzano V, et al. Evolution of the Friedreich's ataxia trinucleotide repeat expansion: founder effect and premutations. *Proc Natl Acad Sci U S A* 1997;**94**(14):7452–7457.
9. Montermini L, Andermann E, Labuda M, et al. The Friedreich ataxia GAA triplet repeat: premutation and normal alleles. *Hum Mol Genet* 1997;**6**(8):1261–1266.
10. Bidichandani SI, Ashizawa T, Patel PI. Atypical Friedreich ataxia caused by compound heterozygosity for a novel missense mutation and the GAA triplet-repeat expansion. *Am J Hum Genet* 1997;**60**(5):1251–1256.
11. Campuzano V, Montermini L, Lutz Y, et al. Frataxin is reduced in Friedreich ataxia patients and is associated with mitochondrial membranes. *Hum Mol Genet* 1997;**6**(11):1771–1780.
12. Grabczyk E, Mancuso M, Sammarco MC. A persistent RNA.DNA hybrid formed by transcription of the Friedreich ataxia triplet repeat in live bacteria, and by T7 RNAP in vitro. *Nucleic Acids Res* 2007;**35**(16):5351–5359.
13. Grabczyk E, Usdin K. The GAA*TTC triplet repeat expanded in Friedreich's ataxia impedes transcription elongation by T7 RNA polymerase in a length and supercoil dependent manner. *Nucleic Acids Res* 2000;**28**(14):2815–2822.

14. Soragni E, Herman D, Dent SY, et al. Long intronic GAA*TTC repeats induce epigenetic changes and reporter gene silencing in a molecular model of Friedreich ataxia. *Nucleic Acids Res* 2008;**36**(19):6056–6065.
15. Al-Mahdawi S, Pook M, Chamberlain S. A novel missense mutation (L198R) in the Friedreich's ataxia gene. *Hum Mutat* 2000;**16**(1):95.
16. Bartolo C, Mendell JR, Prior TW. Identification of a missense mutation in a Friedreich's ataxia patient: implications for diagnosis and carrier studies. *Am J Med Genet* 1998;**79**(5):396–399.
17. Cossee M, Durr A, Schmitt M, et al. Friedreich's ataxia: point mutations and clinical presentation of compound heterozygotes. *Ann Neurol* 1999;**45**(2):200–206.
18. De Castro M, Garcia-Planells J, Monros E, et al. Genotype and phenotype analysis of Friedreich's ataxia compound heterozygous patients. *Hum Genet* 2000;**106**(1):86–92.
19. De Michele G, Filla A, Cavalcanti F, et al. Atypical Friedreich ataxia phenotype associated with a novel missense mutation in the X25 gene. *Neurology* 2000;**54**(2):496–499.
20. Doudney K PM, Al-Mahdawi S, Carvajal J, Hillerman R, Chamberlain S. A novel site mutation (384+1G-A) in the Friedreich's ataxia gene. *Hum Mutat* 1997;**11**:415.
21. Forrest SM, Knight M, Delatycki MB, et al. The correlation of clinical phenotype in Friedreich ataxia with the site of point mutations in the FRDA gene. *Neurogenetics* 1998;**1**(4):253–257.
22. Labuda M PJaPM. A missense mutation (W155R) in an American patients with Friedreich ataxia. *Hum Mutat* 1999;**13**:506–507.
23. McCormack ML, Guttmann RP, Schumann M, et al. Frataxin point mutations in two patients with Friedreich's ataxia and unusual clinical features. *J Neurol Neurosurg Psychiatry* 2000;**68**(5):661–664.
24. Pook MA, Al-Mahdawi SA, Thomas NH, et al. Identification of three novel frameshift mutations in patients with Friedreich's ataxia. *J Med Genet* 2000;**37**(11):E38.
25. Zhu D, Burke C, Leslie A, et al. Friedreich's ataxia with chorea and myoclonus caused by a compound heterozygosity for a novel deletion and the trinucleotide GAA expansion. *Mov Disord* 2002;**17**(3):585–589.
26. Zuhlke C, Laccone F, Cossee M, et al. Mutation of the start codon in the FRDA1 gene: linkage analysis of three pedigrees with the ATG to ATT transversion points to a unique common ancestor. *Hum Genet* 1998;**103**(1):102–105.
27. Branda SS, Cavadini P, Adamec J, et al. Yeast and human frataxin are processed to mature form in two sequential steps by the mitochondrial processing peptidase. *J Biol Chem* 1999;**274**(32):22763–22769.
28. Koutnikova H, Campuzano V, Koenig M. Maturation of wild-type and mutated frataxin by the mitochondrial processing peptidase. *Hum Mol Genet* 1998;**7**(9):1485–1489.
29. Schmucker S, Argentini M, Carelle-Calmels N, et al. The in vivo mitochondrial two-step maturation of human frataxin. *Hum Mol Genet* 2008;**17**(22):3521–35231.
30. Dhe-Paganon S, Shigeta R, Chi YI, et al. Crystal structure of human frataxin. *J Biol Chem* 2000;**275**(40):30753–30756.
31. Gibson TJ, Koonin EV, Musco G, et al. Friedreich's ataxia protein: phylogenetic evidence for mitochondrial dysfunction. *Trends Neurosci* 1996;**19**(11):465–468.
32. Babcock M, de Silva D, Oaks R, et al. Regulation of mitochondrial iron accumulation by Yfh1p, a putative homolog of frataxin. *Science* 1997;**276**(5319):1709–1712.
33. Foury F, Cazzalini O. Deletion of the yeast homologue of the human gene associated with Friedreich's ataxia elicits iron accumulation in mitochondria. *FEBS Lett* 1997;**411**(2–3):373–377.
34. Koutnikova H, Campuzano V, Foury F, et al. Studies of human, mouse and yeast homologues indicate a mitochondrial function for frataxin. *Nat Genet* 1997;**16**(4):345–351.
35. Radisky DC, Babcock MC, Kaplan J. The yeast frataxin homologue mediates mitochondrial iron efflux. Evidence for a mitochondrial iron cycle. *J Biol Chem* 1999;**274**(8): 4497–4499.

36. Wilson RB, Roof DM. Respiratory deficiency due to loss of mitochondrial DNA in yeast lacking the frataxin homologue. *Nat Genet* 1997;**16**(4):352–357.
37. Cavadini P, Gellera C, Patel PI, et al. Human frataxin maintains mitochondrial iron homeostasis in Saccharomyces cerevisiae. *Hum Mol Genet* 2000;**9**(17):2523–2530.
38. Canizares J, Blanca JM, Navarro JA, et al. dfh is a Drosophila homolog of the Friedreich's ataxia disease gene. *Gene* 2000;**256**(1–2):35–42.
39. Llorens JV, Navarro JA, Martinez-Sebastian MJ, et al. Causative role of oxidative stress in a Drosophila model of Friedreich ataxia. *Faseb J* 2007;**21**(2):333–344.
40. Anderson PR, Kirby K, Hilliker AJ, et al. RNAi-mediated suppression of the mitochondrial iron chaperone, frataxin, in Drosophila. *Hum Mol Genet* 2005;**14**(22):3397–3405.
41. Cossee M, Puccio H, Gansmuller A, et al. Inactivation of the Friedreich ataxia mouse gene leads to early embryonic lethality without iron accumulation. *Hum Mol Genet* 2000;**9**(8):1219–1226.
42. Das N, Levine RL, Orr WC, et al. Selectivity of protein oxidative damage during aging in Drosophila melanogaster. *Biochem J* 2001;**360**(Pt 1):209–216.
43. Prabhu HR, Krishnamurthy S. Ascorbate-dependent formation of hydroxyl radicals in the presence of iron chelates. *Indian J Biochem Biophys* 1993;**30**(5):289–292.
44. Anderson PR, Kirby K, Orr WC, et al. Hydrogen peroxide scavenging rescues frataxin deficiency in a Drosophila model of Friedreich's ataxia. *Proc Natl Acad Sci U S A* 2008;**105**(2):611–616.
45. Runko AP, Griswold AJ, Min KT. Overexpression of frataxin in the mitochondria increases resistance to oxidative stress and extends lifespan in Drosophila. *FEBS Lett* 2008;**582**(5):715–719.
46. Adamec J, Rusnak F, Owen WG, et al. Iron-dependent self-assembly of recombinant yeast frataxin: implications for Friedreich ataxia. *Am J Hum Genet* 2000;**67**(3):549–562.
47. Ventura N, Rea S, Henderson ST, et al. Reduced expression of frataxin extends the lifespan of Caenorhabditis elegans. *Aging Cell* 2005;**4**(2):109–112.
48. Vazquez-Manrique RP, Gonzalez-Cabo P, Ros S, et al. Reduction of Caenorhabditis elegans frataxin increases sensitivity to oxidative stress, reduces lifespan, and causes lethality in a mitochondrial complex II mutant. *Faseb J* 2006;**20**(1):172–174.
49. Zarse K, Schulz TJ, Birringer M, et al. Impaired respiration is positively correlated with decreased life span in Caenorhabditis elegans models of Friedreich Ataxia. *Faseb J* 2007;**21**(4):1271–1275.
50. Puccio H, Simon D, Cossee M, et al. Mouse models for Friedreich ataxia exhibit cardiomyopathy, sensory nerve defect and Fe-S enzyme deficiency followed by intramitochondrial iron deposits. *Nat Genet* 2001;**27**(2):181–186.
51. Simon D, Seznec H, Gansmuller A, et al. Friedreich ataxia mouse models with progressive cerebellar and sensory ataxia reveal autophagic neurodegeneration in dorsal root ganglia. *J Neurosci* 2004;**24**(8):1987–1995.
52. Miranda CJ, Santos MM, Ohshima K, et al. Frataxin knockin mouse. *FEBS Lett* 2002;**512**(1–3):291–297.
53. Al-Mahdawi S, Pinto RM, Ruddle P, et al. GAA repeat instability in Friedreich ataxia YAC transgenic mice. *Genomics* 2004;**84**(2):301–310.
54. Al-Mahdawi S, Pinto RM, Varshney D, et al. GAA repeat expansion mutation mouse models of Friedreich ataxia exhibit oxidative stress leading to progressive neuronal and cardiac pathology. *Genomics* 2006;**88**(5):580–590.
55. Clark RM, De Biase I, Malykhina AP, et al. The GAA triplet-repeat is unstable in the context of the human FXN locus and displays age-dependent expansions in cerebellum and DRG in a transgenic mouse model. *Hum Genet* 2007;**120**(5):633–640.
56. Waldvogel D, van Gelderen P, Hallett M. Increased iron in the dentate nucleus of patients with Friedrich's ataxia. *Ann Neurol* 1999;**46**(1):123–125.
57. Lamarche JB, Cote M, Lemieux B. The cardiomyopathy of Friedreich's ataxia morphological observations in 3 cases. *Can J Neurol Sci* 1980;**7**(4):389–396.

58. Rotig A, de Lonlay P, Chretien D, et al. Aconitase and mitochondrial iron-sulphur protein deficiency in Friedreich ataxia. *Nat Genet* 1997;**17**(2):215–217.
59. Foury F. Low iron concentration and aconitase deficiency in a yeast frataxin homologue deficient strain. *FEBS Lett* 1999;**456**(2):281–284.
60. Park S, Gakh O, Mooney SM, et al. The ferroxidase activity of yeast frataxin. *J Biol Chem* 2002;**277**(41):38589–38595.
61. Park S, Gakh O, O'Neill HA, et al. Yeast frataxin sequentially chaperones and stores iron by coupling protein assembly with iron oxidation. *J Biol Chem* 2003;**278**(33):31340–31351.
62. Cook JD, Bencze KZ, Jankovic AD, et al. Monomeric yeast frataxin is an iron-binding protein. *Biochemistry* 2006;**45**(25):7767–7777.
63. Nichol H, Gakh O, O'Neill HA, et al. Structure of frataxin iron cores: an X-ray absorption spectroscopic study. *Biochemistry* 2003;**42**(20):5971–5976.
64. Cavadini P, O'Neill HA, Benada O, et al. Assembly and iron-binding properties of human frataxin, the protein deficient in Friedreich ataxia. *Hum Mol Genet* 2002;**11**(3): 217–227.
65. Levi S, Corsi B, Bosisio M, et al. A human mitochondrial ferritin encoded by an intronless gene. *J Biol Chem* 2001;**276**(27):24437–24440.
66. Campanella A, Isaya G, O'Neill HA, et al. The expression of human mitochondrial ferritin rescues respiratory function infrataxin-deficient yeast. *Hum Mol Genet* 2004;**13**(19): 2279–2288.
67. Becker EM, Greer JM, Ponka P, et al. Erythroid differentiation and protoporphyrin IX down-regulate frataxin expression in Friend cells: characterization of frataxin expression compared to molecules involved in iron metabolism and hemoglobinization. *Blood* 2002;**99**(10): 3813–3822.
68. Yoon T, Cowan JA. Frataxin-mediated iron delivery to ferrochelatase in the final step of heme biosynthesis. *J Biol Chem* 2004;**279**(25):25943–25946.
69. Lesuisse E, Santos R, Matzanke BF, et al. Iron use for haeme synthesis is under control of the yeast frataxin homologue (Yfh1). *Hum Mol Genet* 2003;**12**(8):879–889.
70. Lange H, Muhlenhoff U, Denzel M, et al. The heme synthesis defect of mutants impaired in mitochondrial iron-sulfur protein biogenesis is caused by reversible inhibition of ferrochelatase. *J Biol Chem* 2004;**279**(28):29101–29108.
71. Napoli E, Taroni F, Cortopassi GA. Frataxin, iron-sulfur clusters, heme, ROS, and aging. *Antioxid Redox Signal* 2006;**8**(3–4):506–516.
72. Napoli E, Morin D, Bernhardt R, et al. Hemin rescues adrenodoxin, heme a and cytochrome oxidase activity in frataxin-deficient oligodendroglioma cells. *Biochim Biophys Acta* 2007;**1772**(7):773–780.
73. Muhlenhoff U, Richhardt N, Ristow M, et al. The yeast frataxin homolog Yfh1p plays a specific role in the maturation of cellular Fe/S proteins. *Hum Mol Genet* 2002;**11**(17): 2025–2036.
74. Duby G, Foury F, Ramazzotti A, et al. A non-essential function for yeast frataxin in iron-sulfur cluster assembly. *Hum Mol Genet* 2002;**11**(21):2635–2643.
75. Lill R, Diekert K, Kaut A, et al. The essential role of mitochondria in the biogenesis of cellular iron-sulfur proteins. *Biol Chem* 1999;**380**(10):1157–1166.
76. Lill R, Muhlenhoff U. Iron-sulfur protein biogenesis in eukaryotes: components and mechanisms. *Annu Rev Cell Dev Biol* 2006;**22**:457–486.
77. Yoon T, Cowan JA. Iron-sulfur cluster biosynthesis. Characterization of frataxin as an iron donor for assembly of [2Fe-2S] clusters in ISU-type proteins. *J Am Chem Soc* 2003;**125**(20):6078–6084.
78. Gerber J, Muhlenhoff U, Lill R. An interaction between frataxin and Isu1/Nfs1 that is crucial for Fe/S cluster synthesis on Isu1. *EMBO Rep* 2003;**4**(9):906–911.
79. Ramazzotti A, Vanmansart V, Foury F. Mitochondrial functional interactions between frataxin and Isu1p, the iron-sulfur cluster scaffold protein, in Saccharomyces cerevisiae. *FEBS Lett* 2004;**557**(1–3):215–220.

80. Shan Y, Napoli E, Cortopassi G. Mitochondrial frataxin interacts with ISD11 of the NFS1/ISCU complex and multiple mitochondrial chaperones. *Hum Mol Genet* 2007;**16**(8):929–941.
81. Kondapalli KC, Kok NM, Dancis A, et al. Drosophila frataxin: an iron chaperone during cellular Fe-S cluster bioassembly. *Biochemistry* 2008;**47**(26):6917–6927.
82. Bulteau AL, O'Neill HA, Kennedy MC, et al. Frataxin acts as an iron chaperone protein to modulate mitochondrial aconitase activity. *Science* 2004;**305**(5681):242–245.
83. Lodi R, Cooper JM, Bradley JL, et al. Deficit of in vivo mitochondrial ATP production in patients with Friedreich ataxia. *Proc Natl Acad Sci U S A* 1999;**96**(20):11492–11495.
84. Ristow M, Pfister MF, Yee AJ, et al. Frataxin activates mitochondrial energy conversion and oxidative phosphorylation. *Proc Natl Acad Sci U S A* 2000;**97**(22):12239–12243.
85. Busi MV, Zabaleta EJ, Araya A, et al. Functional and molecular characterization of the frataxin homolog from Arabidopsis thaliana. *FEBS Lett* 2004;**576**(1–2):141–144.
86. Gonzalez-Cabo P, Vazquez-Manrique RP, Garcia-Gimeno MA, et al. Frataxin interacts functionally with mitochondrial electron transport chain proteins. *Hum Mol Genet* 2005;**14**(15):2091–2098.
87. Lodi R, Hart PE, Rajagopalan B, et al. Antioxidant treatment improves in vivo cardiac and skeletal muscle bioenergetics in patients with Friedreich's ataxia. *Ann Neurol* 2001;**49**(5):590–596.
88. Hart PE, Lodi R, Rajagopalan B, et al. Antioxidant treatment of patients with Friedreich ataxia: four-year follow-up. *Arch Neurol* 2005;**62**(4):621–626.
89. Geromel V, Darin N, Chretien D, et al. Coenzyme Q(10) and idebenone in the therapy of respiratory chain diseases: rationale and comparative benefits. *Mol Genet Metab* 2002;**77**(1–2):21–30.
90. Mariotti C, Solari A, Torta D, et al. Idebenone treatment in Friedreich patients: one-year-long randomized placebo-controlled trial. *Neurology* 2003;**60**(10):1676–1679.
91. Ribai P, Pousset F, Tanguy ML, et al. Neurological, cardiological, and oculomotor progression in 104 patients with Friedreich ataxia during long-term follow-up. *Arch Neurol* 2007;**64**(4):558–564.
92. Rustin P, von Kleist-Retzow JC, Chantrel-Groussard K, et al. Effect of idebenone on cardiomyopathy in Friedreich's ataxia: a preliminary study. *Lancet* 1999;**354**(9177):477–479.
93. Di Prospero NA, Baker A, Jeffries N, et al. Neurological effects of high-dose idebenone in patients with Friedreich's ataxia: a randomised, placebo-controlled trial. *Lancet Neurol* 2007;**6**(10):878–886.
94. Pineda M, Arpa J, Montero R, et al. Idebenone treatment in paediatric and adult patients with Friedreich ataxia: long-term follow-up. *Eur J Paediatr Neurol* 2008;**12**(6):470–475.
95. Kelso GF, Porteous CM, Coulter CV, et al. Selective targeting of a redox-active ubiquinone to mitochondria within cells: antioxidant and antiapoptotic properties. *J Biol Chem* 2001;**276**(7):4588–4596.
96. Jauslin ML, Meier T, Smith RA, et al. Mitochondria-targeted antioxidants protect Friedreich Ataxia fibroblasts from endogenous oxidative stress more effectively than untargeted antioxidants. *Faseb J* 2003;**17**(13):1972–1974.
97. Boddaert N, Le Quan Sang KH, Rotig A, et al. Selective iron chelation in Friedreich ataxia: biologic and clinical implications. *Blood* 2007;**110**(1):401–408.
98. Kakhlon O, Manning H, Breuer W, et al. Cell functions impaired by frataxin deficiency are restored by drug-mediated iron relocation. *Blood* 2008;**112**(13):5219–5227.
99. Goncalves S, Paupe V, Dassa EP, et al. Deferiprone targets aconitase: implication for Friedreich's ataxia treatment. *BMC Neurol* 2008;**8**:20.
100. Lim CK, Kalinowski DS, Richardson DR. Protection against hydrogen peroxide-mediated cytotoxicity in Friedreich's ataxia fibroblasts using novel iron chelators of the 2-pyridylcarboxaldehyde isonicotinoyl hydrazone class. *Mol Pharmacol* 2008;**74**(1):225–235.
101. Siren AL, Ehrenreich H. Erythropoietin–a novel concept for neuroprotection. *Eur Arch Psychiatry Clin Neurosci* 2001;**251**(4):179–184.

102. Smith KJ, Bleyer AJ, Little WC, et al. The cardiovascular effects of erythropoietin. *Cardiovasc Res* 2003;**59**(3):538–548.
103. Sturm B, Stupphann D, Kaun C, et al. Recombinant human erythropoietin: effects on frataxin expression in vitro. *Eur J Clin Invest* 2005;**35**(11):711–717.
104. Grant L, Sun J, Xu H, et al. Rational selection of small molecules that increase transcription through the GAA repeats found in Friedreich's ataxia. *FEBS Lett* 2006;**580**(22):5399–53405.
105. Dervan PB, Edelson BS. Recognition of the DNA minor groove by pyrrole-imidazole polyamides. *Curr Opin Struct Biol* 2003;**13**(3):284–299.
106. Burnett R, Melander C, Puckett JW, et al. DNA sequence-specific polyamides alleviate transcription inhibition associated with long GAA.TTC repeats in Friedreich's ataxia. *Proc Natl Acad Sci U S A* 2006;**103**(31):11497–11502.
107. Sarsero JP, Li L, Wardan H, et al. Upregulation of expression from the FRDA genomic locus for the therapy of Friedreich ataxia. *J Gene Med* 2003;**5**(1):72–81.
108. Herman D, Jenssen K, Burnett R, et al. Histone deacetylase inhibitors reverse gene silencing in Friedreich's ataxia. *Nat Chem Biol* 2006;**2**(10):551–8.
109. Rai M, Soragni E, Jenssen K, et al. HDAC inhibitors correct frataxin deficiency in a Friedreich ataxia mouse model. *PLoS ONE* 2008;**3**(4):e1958.

# Chapter 18
# Genetics and Pathogenesis of Inherited Ataxias and Spastic Paraplegias

**Carmen Espinós and Francesc Palau**

**Abstract** Inherited ataxias and hereditary spastic paraplegias are two heterogeneous groups of neurodegenerative disorders with a wide spectrum of clinical symptoms and also, with a remarkable number of involved loci/genes. Inherited ataxias are clinically characterized by progressive degeneration of cerebellum and spinocerebellar tracts of the spinal cord associated with a variable combination of signs of central and peripheral nervous system. Hereditary spastic paraplegias (HSPs) are characterized by slowly progressive spasticity and weakness of lower limbs, due to pyramidal tract dysfunction. The classification of these diseases is extremely difficult because of overlapping symptoms among different clinical forms. For this reason, the genetic classification for both inherited ataxias and HSP forms, based on the causative loci/genes has reached general acceptance. The aim of this review is to summarize the genetics and the pathogenic mechanisms involved in these two groups of neurodegenerative spinocerebellar disorders.

**Keywords:** Inherited ataxia · Autosomal recessive cerebellar ataxia (ARCA) · Autosomal dominant cerebellar ataxia (ADCA) · Spinocerebellar ataxia (SCA) · Hereditary spastic paraplegia (HSP)

## 18.1 Inherited Ataxias

Inherited ataxias are a heterogeneous group of neurodegenerative disorders in which progressive degeneration of cerebellum and spinocerebellar tracts of the spinal cord are associated with a variable combination of signs of central and peripheral nervous system. Inherited ataxic disorders can be divided into two main groups, depending on whether onset of symptoms is before or after the age of 20 years. Most of the early onset disorders are autosomal recessive (Autosomal

---

C. Espinós (✉)
Laboratory of Genetics and Molecular Medicine, Instituto de Biomedicina de Valencia, CSIC, and CIBER de Enfermedades Raras (CIBERER), Valencia, Spain
e-mail: cespinos@ibv.csic.es

C. Espinós et al. (eds.), *Inherited Neuromuscular Diseases*, Advances in Experimental Medicine and Biology 652, DOI 10.1007/978-90-481-2813-6_18, 

Recessive Cerebellar Ataxias, ARCAs), and the later onset ones autosomal dominant (Autosomal Dominant Cerebellar Ataxias, ADCAs). Finally, a minor and less known third group that includes the X-linked forms has also been described.

## *18.1.1 Autosomal Recessive Cerebellar Ataxias*

Autosomal recessive cerebellar ataxias (ARCAs) are a clinically and genetically heterogeneous group of rare neurological diseases involving both central and peripheral nervous system, and in some case other systems and organs, and characterized by degeneration or abnormal development of cerebellum and spinal cord and, in most cases, early onset occurring before the age of 20 years. This group encompasses a large number of rare diseases, the most frequent in Caucasian population being Friedreich ataxia (estimated prevalence 2–4/100,000), ataxia-telangiectasia (1–2.5/100,000) and early onset cerebellar ataxia with retained tendon reflexes (1/100,000). Other ARCA forms are much less common.

A classification of ARCAs into four categories have been proposed: (1) congenital or developmental ataxias; (2) metabolic ataxias, including ataxias caused by enzymatic defects; (3) ataxia due to DNA repair defects; (4) degenerative and progressive ataxic disorders that include ataxias with known cause and pathogenesis, and ataxias of unknown etiology [1] (Table 16.1).

### 18.1.1.1 Congenital Ataxias

Some rare development anomalies, such as dysgenesis or agenesis of the vermis, cerebellar hemispheres or parts of the brainstem, may give rise to congenital ataxias. Among these ataxias, the Joubert syndrome (JBTS) represents one of the most common forms.

#### Joubert Syndrome

This is an autosomal recessive disorder characterized by congenital cerebellar ataxia, hypotonia, psychomotor delay and variable occurrence of oculomotor apraxia and neonatal breathing abnormalities. The neuroradiological hallmark of JBTS is an abnormal configuration of the superior cerebellar penduncles that connect the cerebellum to the midbrain and thalamus known as the "molar tooth sign" (MTS). The MTS has subsequently been reported in a group of syndromes termed Joubert syndrome and related disorders (JSRD) [2]. To date, two loci and six genes have been associated with JS (Table 18.1) [2–8]. Nearly all these loci/genes encode for proteins expressed in the primary cilium or in the centrosome, and some of them are also causative of overlapping diseases such as isolated nephronophthisis (NPH), Senior-Loken syndrome (SLS) and Meckel syndrome (MKS). Thus, only two JBTS loci/genes (JBTS1 and JBTS3) are exclusively associated with Joubert syndrome. The remaining JBTS loci/genes are implicated in Joubert syndrome and other ciliopathies. For instance, mutations in the *NPHP1* gene that encodes for nephrocystin

**Table 18.1** Genes and loci causative of ARCAs

| | Protein (GENE or LOCUS) | Location | MIM |
|---|---|---|---|
| ***Congenital ataxias*** | | | |
| Joubert syndrome | (JBTS1) | 9q34 | 213300 |
| JBTS1 | (JBTS2) | 11p12-q13 | 608091 |
| JBTS2 | Jouberin (*AHI1*) | 6q23 | 608629 |
| JBTS3 | Nephrocystin (*NPHP1*) | 2q13 | 609583 |
| JBTS4 | CEP290 (*CEP290*) | 12q21 | 610188 |
| JBTS5 | Meckelin (*TMEM67*) | 8q22 | 610688 |
| JBTS6 | RPGRIP1L (*RPGRIP1L*) | 16q12 | 611560 |
| JBTS7 | | | |
| Cayman ataxia | Cayataxin (*ATCAY*) | 19p13.3 | 601238 |
| ***Metabolic ataxias*** | | | |
| Ataxia with isolated vitamin E deficiency (AVED) | Alpha-tocopherol transfer protein ($\alpha$-*TTP*) | 8q13 | 277460 |
| Abetalipoproteinemia | Microsomal trygliceride transfer protein (*MTP*) | 4q22-q24 | 200100 |
| Cerebrotendinous xanthomatosis | Sterol 27-hydroxylase (*CYP27*) | 2q33-qter | 213700 |
| Refsum disease | Phytanoyl-CoA hydroxylase (*PhyH*) | 10pter-p11.2 | 266500 |
| | Peroxisomal biogenesis factor-7 (*PEX7*) | 6q22-q24 | |
| ***DNA repair defects*** | | | |
| Ataxia telangiectasia | (*ATM*) | 11q22.3 | 208900 |
| Ataxia with oculomotor apraxia 1 (AOA1) | Aprataxin (*APTX*) | 9p13 | 208920 |
| Ataxia with oculomotor apraxia 2 (AOA2) | Senataxin (*SETX*) | 9q34 | 606002 |
| Ataxia with oculomotor apraxia 3 (AOA3) | ? | ? | |

**Table 18.1** (continued)

| | Protein (GENE or LOCUS) | Location | MIM |
|---|---|---|---|
| Ataxia-telangiectasia-like disorder (ATLD) | *MRE11A* | 11q21 | 604391 |
| Spinocerebellar ataxia with axonal neuropathy (SCAN1) | Tyrosyl-DNA phosphodiesterase 1 (*TDP1*) | 14q31 | 607250 |
| Xeroderma Pigmentosum (XP) | XPA (*XPA*) | 9q22.3 | 278700 |
| XP of complementation group A | XPB/ERCC3·(*XPB/ERCC3*) | 2q21 | 133510 |
| XP of complementation group B | XPC (*XPC*) | 3p25 | 278720 |
| XP of complementation group C | XPD/ERCC2 (*XPD/ERCC2*) | 19q13.2-q13.3 | 278730 |
| XP of complementation group D | XPE (*DDB2*) | 11p12-p11 | 278740 |
| XP of complementation group E | XPF/ERCC4 (*XPF/ERCC4*) | 16p13.3-p13.3 | 278760 |
| XP of complementation group F | XPG/ERCC5 (*XPG/ERCC5*) | 13q32-q33 | 133530 |
| XP of complementation group G | POLH (*POLH*) | 6p21.1-p12 | 278750 |
| XP variant (XPV) or XP with normal DNA repair rates | | | |
| ***Degenerative ataxias*** | | | |
| Friedreich ataxia | Frataxin (*FXN*) | 9q13 | 229300 |
| Mitochondrial recessive ataxic syndrome (MIRAS) | Polymerase $\gamma$ (*POLG*) | 15q26.1 | 174763 |
| Charlevoix-Saguenay spastic ataxia | Sacsin (*SACS*) | 13q12 | 270550 |
| Early onset cerebellar ataxia with retained tendon reflexes (EOCARR) | | 13q11-12 | 212895 |
| Infantile onset spinocerebellar ataxia (IOSCA) | Twinkle (*C10orf2*) | 10q22.3-q24.1 | 271245 |
| Marinesco-Sjögren Syndrome (MSS): | SIL1 (*SIL1*) | 5q32 | 248800 |
| Classical MSS | ? | 18qter | |
| MSS with myoglobinuria | | | |
| Primary coenzyme $Q_{10}$ deficiency with cerebellar ataxia | (*COQ8/CABC1/ADCK3*) | 1q42.2 | 607426 |
| Posterior column ataxia and retinitis pigmentosa (PCARP) | (*AXPC1*) | 1q31 | 609033 |

are responsible for JBTS4 and also for NPHP1 and SLSN1, or mutations in the *CEP290* gene that encodes for a centrosomal protein can cause JBTS5 and also NPHP6, SLSN6 and MKS4.

#### 18.1.1.2 Metabolic Ataxias

Metabolic ataxias include progressive ataxias, disorders associated with intermittent ataxia (e.g. syndromes with hyperammonemias, disorders of pyruvate and lactate metabolism), and metabolic diseases in which ataxia occurs as a minor feature (e.g. leukodystrophy). Some of the most relevant ataxias that belong to this group are ataxia with isolated vitamin E deficiency and abetalipoproteinemia.

##### Ataxia with Isolated Vitamin E Deficiency (AVED)

AVED is a rare autosomal recessive neurodegenerative disease, mostly detected in the Mediterranean basin, caused by mutations in the $\alpha$-*TTP* (α-tocopherol transfer protein) gene [9]. In North-African and South-Italian populations, the most frequent mutation responsible for the disease is the c.744delA mutation and is distributed as a result of a founder effect [9, 10]. In AVED families of North European origin the c.513insTT mutation has been often characterised and in the Japanese population the p.H101Q mutation is the most frequent change [9, 11, 12]. Since the $\alpha$-*TTP* gene was first identified as the defective gene in this disease near 18 different mutations in the $\alpha$-*TTP* gene have been reported and they are placed along the entire gene [13]. AVED is clinically characterized by age at onset before 20 years gait and limb ataxia, dysarthria, lower limb areflexia, loss of vibration and positional sense, and bilateral extensor plantar reflexes. The deficit reduces the capacity to incorporate $\alpha$ -tocopherol (the most biologically active form of vitamin E) into very low density lipoproteins secreted by the liver and therefore into plasma and tissues. In fact, diagnosis is based on the finding of low serum vitamin E values (<2.5 mg/L; reference values 6–15 mg/L), in absence of intestinal fat malabsorption and/or abetalipoproteinemia [14]. The vitamin E supplementation stops the progression of neurologic signs and symptoms in most of the patients [13, 15] what suggests that a prompt genetic characterisation of AVED may promote an early effective treatment of the disease.

##### Abetalipoproteinemia

Abetalipoproteinemia or Bassen-Kornzweig syndrome is an autosomal recessive inherited inborn error of lipoprotein metabolism caused by molecular abnormalities in the gene coding for the largest subunit of the microsomal trygliceride transfer protein (MTP) [16–19]. The MTP catalyzes the transport of triglyceride, cholesteryl ester and phospholipid between phospholipid surfaces. Thus, the assembly or secretion of plasma lipoproteins that contain apolipoprotein B is thought to be the basic pathogenetic defect. The absence of plasma lipoproteins containing apoB causes fat malabsorption, acanthocytosis, retinopathy, and progressive ataxic neuropathy. Patients become deficient especially in vitamin E. Total cholesterol is low (<70 mg/dL) and triglycerides are almost undetectable. Low-density lipoproteins (LDL) and VLDL are practically absent. Symptoms observed in some patients with abetalipoproteinemia are indistinguishable from those of patients with homozygous hypobetalipoproteinemia (HBLP), a codominant genetic disorder characterized by decreased or absent plasma levels of apolipoprotein (apo) B, due to mutations in the *apo B* gene [20, 21].

#### 18.1.1.3 DNA Repair Defects

Ataxia telangiectasia (AT), ataxia telangiectasia-like disorder (ATLD), ataxia with oculomotor apraxia type 1 (AOA1), type 2 (AOA2) and type 3 (AOA3), and spinocerebellar ataxia with axonal neuropathy (SCAN1) belong to a group of recessively inherited disorders characterized by progressive ataxia and because the known causative genes for these disorders are associated with the DNA/RNA quality control system. The impairment of DNA/RNA integrity results in selective neuronal loss in these ARCAs.

##### Ataxia Telangiectasia (AT)

This is a multisystem disease characterized by early onset progressive cerebellar ataxia, oculomotor apraxia, oculocutaneous teleangiectasia, coreoathetosis, susceptibility to bronchopulmonary disease, a variable immunodeficiency state with involvement of cellular and humoral immunity, high risk of malignancy especially leukaemia and lymphoma, and unusually sensitivity to ionizing radioactivity. AT has an incidence of about 1 in 40,000–100,000 births and an estimated carrier frequency of approximately 1%. AT is due to mutations located on the *ATM* gene [22, 23]. Mutations are distributed in every part of the gene with no hot spots (http://www.vmresearch.org/atm.htm). Most of patients are compound heterozygous for mutations that give truncated proteins (85% of mutations) and less than 15% are missense mutations or short in-frame deletions or insertions [24, 25, 27]. Some AT patients have a milder phenotype. These patients used to be compound heterozygous for a null mutation with a milder mutation that makes possible the expression of a small amount of protein [28–30]. ATM protein is almost exclusively nuclear and is expressed in most of the tissues. The protein is a member of the phosphoinositol-3 kinase (PI-3 K)-like serine/threonine kinases involved in cell cycle checkpoint control and DNA repair such us the DNA-dependent protein kinase and the ATR kinase [31, 32]. Checkpoints occur to introduce a pause in proliferation, in order to address cellular stress. For this reason, it is believed that proteins that influence checkpoints are required to prevent cancer, and factors involved in DNA damage response are often linked to the activation of checkpoints. ATM has multiple substrates; one of them is p53. The stabilization and activation of p53 is defective in AT cells, and these cells are also characterized, by a defective G1-S checkpoint, in which p53 has a central role [33, 34].

### Ataxia-Telangiectasia-Like Disorder (ATLD)

To date only near 17 ATLD cases have been reported. This ARCA is a very rare syndrome that shares some similarities with AT including progressive cerebellar ataxia, dysarthria, and oculomotor apraxia. In contrast to AT, ATLD patients present a later age of onset, slower progression and milder neurologic symptoms. The causative gene, *MRE11A* , maps very close to ATM and encodes the Mre11 protein [30]. A protein complex consisting of Mre11-Rad50-Nbs1 (MRN complex) is involved in the cellular response to DNA double-strand breaks. It appears to act by sensing double-strand breaks and as an effector in regulating cell-cycle checkpoints and downstream effects [35].

### Ataxia with Oculomotor Apraxia (AOA)

Three forms of AOA have been described. AOA type 1 (AOA1) is characterized by variable onset (1–16 years), cerebellar atrophy and ataxia (uncoordinated movement gait), late axonal peripheral neuropathy, and oculomotor apraxia. Other features appear in AOA1 including dystonia, masked facies or mental retardation. Clinical laboratory findings include normal AFP ($\alpha$-fetoprotein) level, hypoalbuminemia and hypercholesterolemia [36, 37]. AOA1 is a frequent disease in Portugal, where it represents the second most frequent form of recessive ataxia (~21%), behind Friedreich's ataxia (~38%) [38]. The causative gene for AOA1 is *APTX* that encodes aprataxin [38]. Aprataxin is an expressed nuclear protein composed of three domains that share homology with the amino-terminal domain of polynucleotide kinase 3'-phosphatase (PNKP), with histidine-triad (HIT) proteins and with DNA-binding C2H2 zinc-finger proteins, respectively. PNKP is involved in DNA single-strand break repair (SSBR) machinery following exposure to ionizing radiation and reactive oxygen species [39, 40]. Aprataxin catalyzes the hydrolytic release of DNA end-blocking lesions, serving as a proofreader for the cellular single-strand DNA repair.

Ataxia with oculomotor apraxia type 2 (AOA2) is also referred to as non-Friedreich spinocerebellar ataxia type 1 (SCAR1). It is an autosomal recessive disorder and its relative frequency represents approximately 8% of non-Friedreich ARCA [41]. It is characterized by spinocerebellar ataxia with onset between 11 and 22 years, choreoathetosis and dystonic posturing with walking. Oculomotor apraxia has been reported in 50% of patients. Creatin kinase (CK) and gamma immunoglobulins are elevated [41]. Cerebellar atrophy is observed in some patients. Electrophysiology studies show absence of sensory potentials. The causative gene is *SETX* and encodes senataxin, a protein that contains at its C terminus a classic 7-motif domain found in the superfamily 1 of helicases [42]. Senataxin is a nuclear protein whose function remains unclear. Senataxin may act in the DNA repair pathway and also may be a nuclear RNA helicase with a role in the splicing machinery. Response to oxidative stress-induced double-strand DNA damage but not to irradiation-induced double-strand DNA or single-strand DNA damage has been described [43]. Mutations in *SETX* gene have also been identified in an autosomal dominant form of amyotrophic lateral sclerosis known as ALS4 [44].

A third type of AOA, AOA3, has been described in a single patient who developed ataxia and oculomotor apraxia at 8 years of age [45]. He was dysarthric and had a mild immunodeficiency. He did not have ocular telangiectasia and had normal levels of AFP, albumin and cholesterol. His brain MRI showed cerebellar atrophy. A mutational analysis of candidate genes (*ATM* , *APTX* , and *SETX*) did not yield any positive result and therefore, the causative gene still has not identified.

#### Spinocerebellar Ataxia with Axonal Neuropathy (SCAN1)

SCAN1 shares with AOA1 the cerebellar atrophy and axonal sensorimotor neuropathy, but not ocular apraxia. Mild hypercholesterolemia and hypoalbuminemia are similarly present. To date, SCAN1 is restricted to a consanguineous family from Saudi Arabian that made possible to map the disease to chromosome 14q31 [46]. Patients were carriers of a homozygous mutation, p.H493R, in the gene encoding topoisomerase I-dependent DNA damage repair enzyme (TDP1) [44]. TDP1 is an enzyme that repairs stalled covalently bound topoisomerase I-DNA complexes, which lead to single-strand breaks during DNA unwiding. Maybe, AOA1 and SCAN1 pathology proceeds from the same biochemical pathway in which aprataxin might contribute to SSBR system with its hydrolase activity [47].

### 18.1.1.4 Degenerative Ataxias

In the last 10 years causative genes and pathogenic mechanisms have been described for a number of degenerative and progressive ARCAs. This is especially relevant for Friedreich ataxia, the most common inherited ataxia in the Caucasian population (see specific chapter for this disease in this book). However, there are ataxic syndromes caused by defective mitochondrial proteins encoded by the nuclear genome, as the infantile onset spinocerebellar ataxia (IOSCA) and the mitochondrial recessive ataxia syndrome (MIRAS).

#### Mitochondrial Recessive Ataxia Syndrome (MIRAS)

Mutations in the *POLG* gene that encodes for the only DNA polymerase found in mitochondria have emerged as one of the most common causes of inherited mitochondrial disease in children and adults [48]. The *POLG* mutation database can be found at http://tools.niehs.nih.gov/polg/. Several neurodegenerative disorders, including progressive external ophthalmoplegia (PEO), Alpers syndrome and SANDO (sensory ataxic neuropathy, dysarthria, and ophtalmoparesis), are associated with mutations of *POLG* . Several *POLG* mutations also produce ataxic syndromes in an autosomal recessive manner. Mitochondrial recessive ataxia syndrome (MIRAS) is the most prevalent late-onset ataxia in Finland [49]. Clinical picture is characterized by progressive gait unsteadiness, dysarthria, decreased or absent deep-tendon reflexes in the lower limbs, decreased vibration or joint position sense, nystagmus and other eye-movement abnormalities. Epilepsy may be the presenting symptom. Some patients may show mild to moderate cognitive impairment, involuntary movements may be present. The clinical picture may be heterogeneous. Most of patients are homozygous for the p.W148S change associated in *cis* with the p.E143G polymorphism [49]. A more recent study of primarily Norwegian patients who were homozygous for p.A467T, or homozygous for p.W148S (p.E143G) or compound heterozygous for p.A467T/p.W148S (p.E143G), showed syndromes clinically identical [50]. How these *POLG* changes affect the normal function of the protein and contribute to the precise pathogenesis of the disorder is unknown.

Infantile Onset Spinocerebellar Ataxia (IOSCA)

IOSCA is a severe neurological disorder with onset before 2 years belonging to the group of autosomal recessive cerebellar ataxias. Clinical picture includes cerebellar ataxia, hypotonia, sensory neuropathy with areflexia, optic atrophy, ophthalmoplegia, hearing loss, involuntary movements and epilepsy. This disorder has been described in some counties of Finland [51, 52]. This rare ataxia is caused by mutations in the *PEO1* gene (chromosome 10q22.3-q24.1) encoding Twinkle, a mitochondrial DNA helicase involved in DNA replication, and a rarer splicing variant Twinky with unclear function [53]. Similar to *POLG* , mutations in *PEO* have also been reported in individuals showing autosomal dominant progressive external ophtalmoplegia (PEO) [52].

Hakonen et al. [54] performed a quantitative and qualitative analyses of mtDNA as well as the respiratory chain proteins from the tissues of patients with MIRAS or IOSCA. These authors provided evidence of tissue-specific depletion of mtDNA in IOSCA and neuronal complex I (CI)-defect in both of these mitochondrial ataxias.

### *18.1.2 Autosomal Dominant Cerebellar Ataxias*

Autosomal dominant cerebellar ataxias (ADCAs) are a clinically and genetically heterogeneous group of disorders characterized by a slowly progressive cerebellar syndrome presenting as ataxia of gait, stance and limbs, dysarthria and/or oculomotor disorder due to cerebellar degeneration in the absence of coexisting diseases. The degenerative process can be limited to the cerebellum or can spread further to retina, optic nerve, ponto-medullary systems, basal ganglia, cerebral cortex, spinal tracts or peripheral nerves. Harding proposed a classification of the ADCAs on the basis of the clinical features, and differentiated this group of disorders into three main groups [55]. She distinguished between dominant ataxias with a complex phenotype in which ataxia is accompanied by varying non-ataxia symptoms, that she named autosomal dominant cerebellar ataxia type I (ADCA-I), a rare phenotype in which ataxia is accompanied by progressive visual loss, ADCA-II, and a less frequent phenotype with a purely cerebellar syndrome, ADCA-III. In the molecular classification ADCAs are referred to as spinocerebellar ataxias (SCAs). The majority of SCA disorders including the most frequent forms, SCA1, SCA2 and SCA3, fall into the ADCA-I category. The ADCA-II phenotype is exclusively found in SCA7. SCA6 is the prototype of the ADCA-III category, but there are other purely cerebellar disorders such as SCA5 or SCA11 [56–58] (Table 18.2). The Harding classification is still useful for clinical purposes. However, the genetic classification, as a result of the reported ataxia loci/genes, is the first classification that reached general acceptance. Since the chromosomal mapping of the first SCA (SCA1) in 1977 [59] the list has rapidly grown (Table 18.2).

Epidemiological studies conducted in different European regions found prevalence rates of SCAs ranging from 0.9 to 3.0:100,000 [60]. Data suggest that SCA3 is the commonest subtype worldwide and together with SCA1, SCA2, SCA6, and SCA7 are the most frequent forms and account for 50–80% of ADCA families. The remaining forms are thought to be <1% and, at least 20% of ADCA families do not carry one of the yet identified SCA genes [61, 62]. Frequencies of the different types of ataxias may vary among regions and ethnic groups. For example, SCA2 is common in Korea, and SCA3 is much more common in Japan and Germany than in the United Kingdom [58, 63–66].

At least 28 SCA loci have been identified and in 14 of these disorders the involved genes are known (Table 18.2). The most common SCAs (SCA1, SCA2, SCA3, SCA6, SCA7 and SCA17) are due to CAG repeat expansions that encode a pure repeat of the amino acid glutamine in the disease protein. These diseases are polyglutamine (polyQ) expansion disorders as Huntington's disease, spinobulbar muscular atrophy (SBMA), and dentatorubropallidoluysian atrophy (DRPLA). Together, the six polyglutamine SCAs accounts for well over the 50% of affected families in nearly all regions of the world.

A second group of SCAs comprises those due to repeat expansions falling outside of the protein-coding region of the respective disease genes and therefore, the pathogenic expansion does not encode any amino acid. SCAs belonging to this group are SCA8, SCA10 and SCA12, although the pathogenic effect of the CTG expansion associated with SCA8 is still debated [67]. SCA10 is due to a massive expansion of an ATTCT pentanucleotide in intron 9 of a gene with unknown function [68]. SCA12 is caused by CAG expansions with more than 66 repeats placed at the 5' region of the *PPP2R2B* gene [69]. In all three diseases, it remains unclear how a noncoding repeat causes neurodegeneration.

Finally, at least five SCAs are caused by conventional mutations (substitutions, duplications, deletions, insertions): SCA5, SCA13, SCA14, SCA27 and 16q22-linked SCA. The genes responsible for these SCAs belong to different biological pathways what suggests that cerebellar and brainstem degeneration can be the biological consequence of alterations in any one of many distinct cellular pathways.

#### 18.1.2.1 Common Features of Spinocerebellar Ataxias

SCAs are a clinically heterogeneous group [60] (Table 18.2). The most notable unifying characteristic is the pattern of neurodegeneration: all the SCAs are associated with the typical clinical features reflecting damage to the cerebellum. Indeed, many SCAs have extensive cerebellar atrophy involving molecular, Purkinje cell and

**Table 18.2.** Classification of spinocerebellar ataxias(SCAs).

| SCA subtype | MIM | Locus | Gene | Mutation type | Protein | ADCA type | Clinical features |
|---|---|---|---|---|---|---|---|
| SCA1 | 164400 | 6p23 | *ATXN1* | polyQ disease(CAG>39) | Ataxin-1 | I | Ataxia, pyramidal signs, neuropathy, dysphagia, restless legs syndrome |
| SCA2 | 183090 | 12q24.1 | *ATXN2* | polyQ disease(CAG>32) | Ataxin-2 | I | Ataxia, slow saccades, neuropathy, restless legs syndrome. |
| SCA3/MJD | 109150 | 14q32.1 | *ATXN3* | polyQ disease(CAG>54) | Ataxin-3 | I | Ataxia, pyramidal signs, ophthalmoplegia, neuropathy, dystonia, restless legs syndrome. |
| SCA4 | 600223 | 16q22.1 | ¿? | ¿? | ¿? | I | Ataxia, sensory neuropathy. |
| SCA5 | 600224 | 11 | *SPTBN2* | Point mutation | Beta-III spectrin | III | Almost purely cerebellar ataxia. |
| SCA6 | 183086 | 19q13 | *CACNA1A* | polyQ disease(CAG>19) | Calcium channel subunit | III | Almost purely cerebellar ataxia. |
| SCA7 | 164500 | 3p14 | *ATXN7* | polyQ disease(CAG>37) | Ataxin-7 | II | Ataxia, ophthalmoplegia, visual loss. |
| SCA8 | 603680 | 13q21 | *ATXN8OS* | 3' untranslated repeat (CTG>110-250?) | Ataxin-8 | I | Ataxia, sensory neuropathy, spasticity. |
| SCA10 | 603516 | 22q13 | *ATXN10* | Intronic repeat expansion (ATTCT>550) | Ataxin-10 | I | Ataxia, epilepsy. |
| SCA11 | 604432 | 15q14-q21.3 | ¿? | ¿? | ¿? | III | Almost purely cerebellar ataxia. |
| SCA12 | 604326 | 5q31-33 | *PPP2R2B* | 5' untranslated repeat (CAG>66) | Phosphatase subunit | I | Ataxia, tremor. |
| SCA13 | 605259 | 19q13.3-q13.4 | *KCNC3* | Point mutation | Potassium channel | I | Ataxia, mental retardation. |
| SCA14 | 605361 | 19q13.4-qter | *PRKCG* | Point mutation | Protein kinase Cγ | III | Ataxia, myoclonus, dystonia, sensory loss. |
| SCA15 | 606658 | 3p24.2-3pter | ¿? | ¿? | ¿? | III | Almost purely cerebellar ataxia. |

**Table 18.2.** (continue)

| SCA subtype | MIM | Locus | Gene | Mutation type | Protein | ADCA type | Clinical features |
|---|---|---|---|---|---|---|---|
| SCA16 | 606364 | 8q22.1-q24.1 | ¿? | ¿? | ¿? | III | Almost purely cerebellar ataxia. |
| SCA17 | 607136 | 6q27 | *TBP* | polyQ disease(CAG>44) | TATA binding protein | I | Ataxia, dystonia, chorea, dementia, psychiatric abnormalities. |
| SCA18 | 607458 | 7q22-q32 | ¿? | ¿? | ¿? | I | Ataxia, sensory neuropathy, neurogenic muscle atrophy. |
| SCA19/SCA22 | 607346 | 1p21-q21 | ¿? | ¿? | ¿? | I | Ataxia, myoclonus, cognitive impairment. |
| SCA20 | 608687 | 11 | ¿? | ¿? | ¿? | I | Ataxia, dysphonia. |
| SCA21 | 607454 | 7p21.3-p15.1 | ¿? | ¿? | ¿? | I | Ataxia, parkinsonim. |
| SCA22/SCA19 | 607346 | 1p21-q23 | ¿? | ¿? | ¿? | I | Pure cerebellar ataxia with dysarthria, nystagmus. |
| SCA23 | 610245 | 20p13-12.3 | ¿? | ¿? | ¿? | I | Ataxia, sensory neuropathy, pyramidal signs. |
| SCA24 | 607317 | 1p36 | ¿? | ¿? | ¿? | I | Ataxia, myoclonus, sensory neuropathy, dysarthria |
| SCA25 | 608703 | 2p21-p13 | ¿? | ¿? | ¿? | I | Ataxia, sensory neuropathy. |
| SCA26 | 609306 | 19p13.3 | ¿? | ¿? | ¿? | III | Almost purely cerebellar ataxia. |
| SCA27 | 609307 | 13q34 | *FGF14* | Point mutation | Fibroblast growth factor 14 | I | Ataxia, tremor, mental retardation. |
| SCA28 | 610246 | 18p11.22-q11.2 | ¿? | ¿? | ¿? | I | Ataxia, opthalmoparesis, pyramidal signs |
| SCA29 | 117360 | 3p26 | ¿? | ¿? | ¿? | I | Early onset, nonprogressive ataxia, vermian hypoplasia |
| SCA 16q22-linked | 117210 | 16q22.1 | *PLEKHG4* | Point mutation | Puratrophin | III | Almost purely cerebellar ataxia. |

*Abbreviations included in Table 2: MJD= Machado-Joseph disease; ADCA= Autosomal dominant cerebellar ataxia; polyQ= polyglutamine.*

granule cell layers as well as the deep cerebellar nuclei. SCA6 is typically a pure cerebellar ataxia in which the degeneration is largely confined to Purkinje neurons without major involvement of other parts of the brain. However, the extracerebellar features usually define the clinical forms. Thus, SCA7 is characterized by visual loss and SCA10 by epilepsy. SCAs also shared their relentless progression. In most SCAs an inexorably progressive degenerative process leads to death over 15–30 years. SCA6 is an exception: SCA6 patients show an adult onset ataxia.

Most SCAs are due to dynamic mutations. These are caused by an expansion of trinucleotide sequences in or adjacent to a protein-coding gene [70, 71] and are characterised by the intergenerational instability of trinucleotide repeats and by an increasing bias during transmission. In this way, the existence of intermediate-sized nonpathologic alleles near the upper limit of the normal-sized range is thought to act as a reservoir from which *de novo* mutations arise in several generations, given that the larger the expanded alleles are, the more instable the expansion becomes [72–77]. Dynamic mutations could also explain the clinical variability indeed within a family: as the repeat length may increase from generation to generation, there is often a tendency to earlier disease onset in subsequent generations. This phenomenon is named anticipation and this implies that disease has a tendency to worsen from generation to generation within a family. Anticipation is mainly severe in SCA7, in which severe infantile forms can be caused by large expansion (>250 repeats) [78]. SCA8 by contrast, is characterized by dramatic repeat instability and a high degree of reduced penetrance: extremely large repeats (800 repeats) may be associated with an absence of clinical symptoms [79].

#### 18.1.2.2 Insights into Pathogenesis

Polyglutamine Spinocerebellar Ataxias

Except for SCA6, which encodes the α1A-subunit of a P/Q-type calcium channel [80], and SCA17, which encodes TATA-box binding protein (TBP) [81, 82], the function of the mutated proteins with anomalous polyglutamine expansions is unknown. These proteins have not common sequences or domains and that is why, the pathogenesis must be directly linked to the expanded polyglutamine tracts [83].

The discovery that human polyglutamine disease brain contains intracellular inclusions of the disease protein suggested that the expansion promotes misfolding of the disease protein, resulting in aggregation [84]. It is assumed that the common toxic gain-of-function mechanism for the polyglutamine-containing protein is aggregation and deposition of misfolded proteins leading to neuronal dysfunction and eventually cell death. These inclusions contain cellular components such as ubiquitin, the proteasome, HSP70 and transcription factors [85–87]. Whether the toxicity is a direct result of the aggregate or results from intermediary structures formed during the process of aggregation remains to be determined. Expanded CAG repeats engineered to be expressed at the mRNA level, but not at the protein level, do not display toxicity when introduced into cells or animals what supports the toxic protein model.

It is crucial to discriminate the pathogenic significance of large macromolecular inclusions *versus* small aggregates or oligomers and that is why, researches have also focused in earlier steps in the aggregation pathway. Multiple studies have now dissociated large inclusions from toxicity *in vivo* [88, 89] and *in vitro* [90, 91]. However, polyglutamine proteins *in vitro* clearly form small aggregates or oligomers [92–94]. The cellular chaperonin protein, TRiC, suppressed polyglutamine toxicity while promoting its assembly into larger, non-toxic 500 kDa oligomers [95]. These larger oligomers proved to be conformationally distinct from smaller 200 kD complexes that are associated with toxicity. The importance of protein oligomerization in pathogenesis is increasing in polyglutamine diseases and also in other neurodegenerative diseases, including Alzheimer's disease and Parkinson's disease [84, 96]. Thus, misfolded protein, small soluble oligomers, may directly interfere with critical cellular events, challenge the cell's ability to prevent more widespread misfolding, and compromise its ability to keep up with protein degradation.

The relationship between SCA neurodegeneration and the ubiquitin-dependent proteasome system (UPS), the main cellular machinery to degrade aberrantly folded proteins, is certain since the findings of ubiquitin-positive protein aggregates in neuropathological researches [84]. A few ataxins, such as ataxin-1, ataxin-2 and ataxin-7, are susceptible and targeted by the proteasome for degradation and clearance [97–99]. Protein misfolding exerted by the expanded polyglutamine might lead to difficulties in the recognition and degradation process by the proteasome and, hence, in subsequent impaired clearance of mutant proteins. In SCA1 the polyglutamine expansions in ataxin-1 may interfere with its modulation of the transcriptional repressor capicua in a regulatory complex [100]. Duplication of an ataxin-1-like gene competes mutant ataxin-1 away from the capicua complex and suppresses SCA1 disease in mice [101]. The normal ataxin-3, an ubiquitin-specific cysteine protease that associates with the proteasome [99], suppresses neurodegeneration caused by mutant ataxin-3, and this suppression depends on its ubiquitin-binding activity and protease activities [102]. In both, SCA1 and SCA3, the expanded polyglutamine within the proteins might alter its normal function and produce functional disruptions of the UPS pathway.

Interactions of expanded polyglutamine proteins with specific transcription factors may perturb gene expression and thus, initiate neurodegeneration. Such interactions could involve sequestration of a target protein by polyglutamine protein monomers, or recruitment into aggregates. CBP (CREB-binding protein) has been found in nuclear inclusions formed by several polyglutamine expanded proteins including ataxin-1 [103]. Proteasomes and CBP remained highly dynamic components of inclusions, indicating that although enriched with ataxin-1, they are not irreversibly trapped. Ataxin-1 also forms a complex with retinoid-related orphan receptor α (RORα), a transcription factor important for cerebellar development. Expression of mutant ataxin-1 leads to depletion of this critical transcription factor, which likely contributes to pathogenesis [104].

SCA7 is the only SCA associated with retinal degeneration and transcriptional repression might explain cell specificity. Ataxin-7 is an integral component of SAGA-like complexes (SAGA in yeast contains the Gcn5 acetylase), the TATA-binding protein-free TAF-containing complex (TFTC) and the SPT3-TAF9-GCN5 acetyltransferase complex (STAGA) [105, 106], and interacts with the photoreceptor-specific transcriptional activator CRX [107]. Ataxin-7 recruits TFTC/STAGA to promoters of retina-specific genes. Polyglutamine expanded ataxin-7 suppresses the activities of both CRX and the acetyltransferase component of the TFTC/STAGA complex, and thus inhibits the expression of genes vital for retinal function [106, 107].

### Noncoding Repeat Spinocerebellar Ataxias

Thus far, three SCAs are supposed to be caused by repeat expansions that are not translated: SCA8, SCA10 and SCA12. The pathogenic mechanisms remain uncertain.

SCA8 is characterized by dramatic repeat instability and a high degree of reduced penetrance. The SCA8 locus was mapped to chromosome 13q21 in a large ADCA family in which, repeats of 110 to 250 CTG in the 3'UTR of the *ATXN8OS* gene are associated with the disease, whereas smaller repeats (71-110 CTG) as well as larger alleles (250-800 CTG) are postulated to show reduced penetrance [108]. The finding of SCA8 expansions in healthy controls and in patients with variable diseases, such as schizophrenia, bipolar affective psychosis and Lafora disease, as well as in SCA6 families have raised questions regarding the disease-causing character of the CTG expansion at the SCA8 locus [109]. An interesting pathogenic mechanism for SCA8 has been described. Two genes spanning the SCA8 repeat are expressed in opposite directions: the novel gene *ataxin-8* (*ATXN8*) encodes a nearly pure polyglutamine expansion protein in the CAG direction, and *ataxin-8 opposite strand* (*ATXN8OS*) expresses a noncoding CUG expansion RNA [110]. Both, the CUG and the CAG expansions, are known to be toxic in other diseases and therefore, SCA8 may involve both RNA and protein gain of function mechanisms [111, 112].

SCA10 is characterized by cerebellar atrophy, ataxia and seizures and has been found exclusively in Mexicans and Brazilians [113, 114]. SCA10 mutant alleles contain a huge expansion (800–4500 repeats) of an ATTCT pentanucleotide in the intron 9 of the *ATXN10* gene [68, 115]. This gene encodes a cytoplasmic protein, ataxin-10, with unknown function but strongly expressed in brain. The ataxin-10 contains two armadillo (arm) repeat domains and interacts with the heterotrimeric GTP-binding protein [116]. The arm repeats mediate protein-protein interactions to modulate a myriad of cellular processes, including intracellular signalling, cytoskeletal regulation, nuclear transport and regulation of gene expression [117]. It has been suggested that the ataxin-10 could regulate important cellular processes through the mediation of G-protein intracellular signalling. Thus, the G-protein β2 subunit (Gβ2) has been defined as an ataxin 10-interacting protein. Constitutive expression of ataxin-10 induces neuritogenesis in neural cells by activating the Ras-MAP kinase-Elk-1 cascade and enhances dramatically neuronal differentiation induced by coexpressing Gβ2 [118]. In the *Sca10* (mouse ataxin 10 homolog)-null mice, the mutant *ATXN10* allele is transcribed at the normal level and in patient-derived cells, the pre-mRNA with an expanded allele is processed normally. *Sca10* -null mice exhibit embryonic lethality whereas heterozygous mutants do not present with SCA10 phenotype [119]. This suggests that a simple gain/loss of function of *ATXN10* is unlikely to be the major pathogenic mechanism.

SCA12 is associated with an expansion of a CAG repeat in the 5' region of the gene *PPP2R2B* (mutant alleles with 55–78 triplets and normal alleles with 9–28 triplets) [69]. The *PPP2R2B* encodes a brain-specific regulatory subunit of the protein phosphatase PP2A, an enzyme implicated in multiple cellular functions, including cell cycle regulation, tau phosphorylation, and apoptosis [120]. It has been proposed that the SCA12 repeat expansion may alter the levels of expression of one splice variant of *PPP2R2B* by influencing the efficiency of the promoter driving expression, or have an effect on *PPP2R2B* splicing, or modify gene expression in some other way. To date, little is known about how this anomalous expansion causes SCA12, although PP2 is described to mediate in neurodegenerative process in SCAs 1 and 14 [121, 122], and also in Alzheimer's disease [123].

### Spinocerebellar Ataxias Caused by Conventional Mutations

To date, five SCAs are described not caused by dynamic mutations: SCA5, SCA13, SCA14, SCA27 and SCA 16q22-linked. The genes responsible for these SCA forms have very different functions. This fact highlights that very different biological mechanisms could produce a cerebellar degeneration.

In frame deletions in the β-III spectrin (*SPTBN2*) gene lead to SCA5 [124]. It has been proposed that these mutations modify the levels, distribution and stability of the β-III spectrin-associating protein and Purkinje cell-specific glutamate transporter EAAT4. Downregulation of both β-III spectrin and EAAT4 transcripts found by microarray analysis in two mouse ataxias models, SCA transgenic and staggered mice [125] suggests the convergence of pathogenic mechanisms triggered by distinct mutations. The involvement of a well-known cytoskeletal protein suggests that destabilization of membrane proteins, glutamate signalling and vesicle trafficking deficits could play a role in causing neurodegeneration in SCA5 and in other neurodegenerative diseases, such as SCA1, Alzheimer and Huntington's diseases and amyotrophic lateral sclerosis.

SCA13 is caused by mutations in the *KCNC3* gene that encodes a voltage-gated $K^+$ channel (Kv3.3) highly enriched in cerebellum [126]. Kv3.3 is a fast-rectifying voltage-gated Shaw subtype potassium channel abundantly expressed in the cerebellum. Two missense mutations have been identified in this gene and both mutations seem to alter *KCNC3* function in a *Xenopous laevis* oocyte expression system: mutations appear to shift the activation curve in the negative direction and slowed channel losing. These mutations appear to have a dominant effect on electrophysiological properties of the multisubunit $K^+$ channel and therefore, might change the output characteristics of fast-skipping cerebellar neurons, in which KCNC channels confer capacity for high-frequency firing [127].

The disease symptoms of SCA14 are attributable to various mutations in the *PRKCG* , also known as *PKCγ* (protein kinase Cγ) gene which result in altering a highly conserved residue in the cystein rich region of the respective protein [121,

128]. This member of the family of serine/threonine kinases is highly expressed in brain and spinal cord, with particularly high expression in Purkinje cells of the cerebellar cortex during dendritic development, where it seems to act a negative regulator of dendritic growth and branching [129]. The molecular mechanisms by which *PKCγ* controls Purkinje cell development and neuronal connectivity are still poorly understood. Most disease-causing mutations are placed in the PKCγC1B regulatory subdomain and recent studies have shown that mutations located in the C1B domain affect accessibility and kinase activity leading to aberrant mitogen-activated protein kinase (MAPK) signalling [130].

The fibroblast growth factor 14 (*FGF14*) gene has been identified as the disease-causing gene for the SCA27 [131]. FGF14 is a member of a subclass of fibroblast growth factors that is expressed in the developing and adult central nervous system. Symptoms of the *Fgf14* knockout mice and SCA27 patients are very similar what has favoured the understanding of this neurodegenerative disorder. The *Fgf14* knockout mouse has impaired the synaptic transmission at hippocampal Schaffer collateral-CA1 synapses and short and long-term potentiation, and exhibits spatial memory deficits in the Morris water maze [132, 133]. These findings suggest a role for FGF14 in regulating synaptic plasticity by controlling the mobilization, trafficking or docking of synaptic vesicles to presynaptic active zones and also, in spatial learning and synaptic plasticity.

Mutations in the *PLEKHG4* gene underlie the SCA 16q22-linked, a spinocerebellar ataxia subtype characterized by pure cerebellar atrophy and sensorineural hearing impairment. This gene encodes for puratrophin-1, also known as pleckstrin homology domain containing family G protein 4, a protein implicated in intracellular signalling and actin dynamics at the Golgi apparatus. The SCA 16q22-linked has been exclusively described in Japan [134, 135] where is one of the most common forms of SCAs [135, 136]. Most of patients carry the same mutation, a c.-16C>T change in the *PLEKHG4* gene, suggesting a founder event [137, 138]. Interestingly, the *PLEKHG4* gene is placed on the same chromosomal region where the SCA4 locus is linked [139]. To date two SCA4 families have been reported: one with Scandinavian origin [139] and another from Germany [140]. The *PLEKHG4* gene has been analyzed in the patients from the German family and no mutation has been found [141]. These findings would indicate that SCA4 and 16q22-linked Japanese ataxia are not allelic.

### 18.1.3 X-Linked Cerebellar Ataxias

X-linked spinocerebellar ataxia (SCAX) is a clinically and genetically heterogeneous disorder. Exhaustive studies with detailed clinical and genetic descriptions are not available. Only a few cases of SCAX families are known and most of these clinical reports are not recent studies, Clinical heterogeneity, time-dependent evolution of symptoms, and overlapping phenotypes make difficult to achieve a definitive diagnosis. No locus or gene associated with any of the described SCAX forms has been characterised. Only two chromosomal locations have been identified and

**Table 18.3** Classification of X-linked spinocerebellar ataxias (SCAXs)

| SCAX type | MIM | Location | Clinical Features |
|---|---|---|---|
| SCAX1 | 302500 | Xp11.21-q21.3 | Olivopontocerebellar ataxia Gait and limb ataxia, intention tremor, dysmetria, dysdiadochokinesia, dysarthria, and nystagmus. |
| SCAX2 | 302600 | – | Cerebellar ataxia with extrapyramidal involvement; early onset |
| SCAX3 | 301790 | – | Hypotonia, ataxia, sensorineural deafness, developmental delay, esotropia, and optic atrophy; death in childhood. |
| SCAX4 | 301840 | – | Ataxia, pyramidal tract signs and adult-onset dementia; early onset. |
| SCAX5 | 300703 | Xq25-q27.1 | Neonatal hypotonia, delayed motor development, nonprogressive ataxia, nystagmus, and dysarthria; early onset. |

therefore, molecular diagnosis is not possible. To date five SCAX forms have been described (Table 18.3):

- The first X-linked SCA type was described in three unrelated families [142]. This disorder is characterized by heterogeneous clinical aspects. Illarioshkin et al. [143] mapped a locus for X-linked recessive congenital ataxia in a Russian family to a large genetic interval (54 cM) on Xp11.21-q24. This interval was narrowed (24 cM) by Bertini et al. [144].
- SCAX2 was described in 1958 by Malamud and Cohen [145] in a male infant with developed ataxia at age 10 months after normal early development. His family history revealed multiple other affected males related through females, consistent with X-linked recessive inheritance. Mapping studies to identify the locus/gene responsible for this disease has not been performed.
- The three type of an X-linked SCA was reported in a large family and no mapping studies are known [146].
- Only one large family has been described as SCAX4 [147]. Preliminary linkage studies using RFLPs suggested that Xq26-qter and much of the short arm could be excluded as sites for the gene.
- Finally, a large American family was recently reported with spinocerebellar ataxia inherited in an X-linked recessive pattern linked to a new locus, Xq25-q27.1 [148].

## 18.2 Hereditary Spastic Paraplegias: Definition and Classification

Hereditary spastic paraplegias (HSPs), also known as familial spastic paraparesis or Strümpell-Lorrain disease, are a clinically and genetically heterogeneous group of disorders characterized by slowly progressive spasticity and weakness of lower limbs, due to pyramidal tract dysfunction. Epidemiological studies conducted in different populations found prevalence rates of HSPs ranging from 2.0 to 9.6:100,000 [149–152]. Such variation is probably due to a combination of differing diagnostic criteria, variable epidemiological methodology, and geographical factors. Several classifications have been proposed based on the mode of inheritance, the age of onset of symptoms, and the presence of additional clinical features.

Clinically, HSP have been classified as pure and complicated or complex forms, according to a classification suggested by Harding [153]. Pure HSP refers to forms with spasticity in the lower limbs alone. HSP is classified as complex when associated with other neurological signs, including ataxia, mental retardation, dementia, extrapyramidal signs, visual dysfunction or epilepsy, among others, or with extra-neurological symptoms [154]. Further subdivision of pure HSP based on the age of onset of the disease has also been proposed: type I with onset before 35 years and type II with onset after 35 years. Type I patients have a slow and variable course compared with the more rapidly evolving type II, in which muscle weakness, urinary symptoms, and sensory loss were more marked. Neither of these classifications is ideal, with many families not easily fitting the criteria.

As in other neurodegenerative disorders, the molecular classification seems to have reached general acceptance. All modes of inheritance have been described associated with HSPs. Autosomal dominant (AD) is the main mode of inheritance (Table 18.4), accounting for 70–80% of all HSP forms in Western countries [155]. Autosomal recessive (AR) HSP forms are common in inbred populations [156, 157] and they may comprise a significant proportion of apparently sporadic cases. Each HSP form is associated with multiple loci/genes. All genetically defined HSPs are assigned the symbol SPG (spastic gait) followed by a number. Thirty-two HSP loci and sixteen genes have been identified. Autosomal dominant (AD) HSPs are almost invariably pure in clinical terms, whereas autosomal recessive (AR) HSPs appear to be complex with an earlier age of onset.

### *18.2.1 Pure HSP Forms*

In pure HSP disease progression, extent of disability and age of symptoms onset are variable. The disease usually progresses slowly over the years, without remissions. Patients experience progressive difficulty in walking and they could end up wheelchair-bound. Urinary symptoms are frequent and range from urinary urgency to incontinence. The main clinical difference however, is the age of onset of symptoms. Disease can have a childhood or adolescence onset (as in SPG3, SPG10 and SPG12) or have an adult onset (as in SPG19). Other forms such as SPG4, PG8 and SPG13, are characterized by a much wider range of ages of onset, spanning several decades.

SPG4 encodes spastin and is the major gene responsible for AD HSP, with an overall frequency close to 40% [158]. All types of mutations (missense, nonsense, splicing site, deletions) have been detected in spastin. This suggests

**Table 18.4** Autosomal dominant forms of hereditary spastic paraplegias (HSPs)

| SPG type | MIM | Locus | Gene/protein | Putative roles of the protein | Associated signs |
|---|---|---|---|---|---|
| *Pure forms* | | | | | |
| SPG3A | 182600 | 14q12-q21 | *SPG3A/* Atlastin | GTPase, ER to Golgi transfer, spastin partner | |
| SPG4 | 182601 | 2p22 | *SPAST* /Spastin | Microtubule-severing activity, early secretory pathway | |
| SPG6 | 600363 | 15q11.2-q12 | *NIPA1* | $Mg^{2+}$transporter, endosomal trafficking | |
| SPG8 | 603563 | 8q24 | *KIAA0196/* Strumpellin | Spectrin domain protein | |
| SPG10 | 604187 | 12q13 | *KIF5A* | Kinesin heavy chain motor protein | |
| SPG12 | 604805 | 19q13 | ? | ? | |
| SPG13 | 605280 | 2q24-q34 | *HSP60* | Mitochondrial chaperone | |
| SPG19 | 607152 | 9q33-q34 | ? | ? | |
| SPG31 | 610250 | 2p12 | *REEP1* | Endosomal trafficking, mitochondrial chaperone | |
| SPG33 | 610244 | 10q24.2 | *ZFYVE27/* Protrudin | Endosomal trafficking, spastin partner | |
| SPG37 | 611945 | 8p21.1-q13.3 | ? | ? | |
| *Complex forms* | | | | | |
| SPG9 | 601162 | 10q23.3-q24.2 | ? | ? | Cataracts, gastroesophageal reflux, motor neuropathy, skeletal abnormalities. |
| SPG17 | 270685 | 11q12-q14 | *BSCL2* /Seipin | ER integral protein | Slow disease progression, motor neuron involvement, abnormal vibration sense, pes cavus and other foot deformities. Allelic to CMT4D. |
| SPG29 | 609727 | 1p31-p21 | ? | ? | Hearing impairment, hiatal hernia. |

that SPG4 is caused by a loss-function pathogenic mechanism implying that a threshold level of spastin protein expression is critical for axonal preservation. The second most frequent form is SPG3A, caused by mutations in the atlastin, accounting for approximately 10% of cases. The predominant mutations in SPG3A are missense type, which argues a mechanism involving gain of function. Near 7% of patients who are negative for both SPG3A and SPG4, have mutations in the *SPG31* gene that encodes for a mitochondrial protein, REEP1 [159]. To date only missense mutations have been characterized in the *SPG31* gene. The remaining HSP AD pure forms have been described in a few families: two SPG6 families [160, 161], three SPG8 families [162], four SPG10 families [163–166], one SPG13 family [167] and one SPG33 family [168]. Regarding to the remaining HSP (SPG12, SPG19 and SPG37) involved genes still have not been identified.

HSP AR pure forms are far less common that the AD forms (Table 18.5). SPG7 was initially described as a HSP AR pure form, but it has been shown to be either pure or associated with cerebellar atrophy and variable degrees of cerebellar dysfunction and mental deficit. To date four loci have been described as HSP AR pure form: SPG5, SPG24, SPG28 and SPG30. Recently, in five families, mutations in the *CYP7B1* gene have been related to SPG5 [169]. For the remaining HSP AR pure forms, no gene has yet been identified for these forms.

### *18.2.2 Complex HSP Forms*

Complicated or complex HSP forms have additional neurologic or extraneurologic signs or symptoms, including mental retardation, peripheral neuropathy, cerebellar ataxia, epilepsy, optic atrophy, retinitis pigmentosa, deafness, and cataracts. Tables 18.4, 18.5, and 18.6 shows the main clinical features associated to each complex HSP form.

Only three loci (SPG9, SPG17 and SPG29) have been described in rare complex forms of AD HSP (Table 18.4). Moreover, mutations in genes involved in pure HSP forms, have also been found in patients presenting with complex HSP forms: SPG10 mutations have been detected in an important proportion (10%) of complicated patients [170].

Eleven complex HSP forms are AR. The most common form is *SPG11/KIAA1840* gene, accounting for 21% of AR HSP cases [171, 172], followed by SPG15, thought to account for 15% of AR HSP [173]. The remaining AR HSP whose responsible gene is known, SPG20, SPG21 (reported only in Amish populations) and SPG7 (approximately 4%) are not so frequent. SPG7 is particular because it contains a large number of polymorphisms and they often are in a heterozygous state [174, 175]. The possibility of some of these polymorphisms could act as a genetic modifier has been postulated. In fact, the p.A510V change has been found to alter the function of the SPG7 protein, in agreement with its frequent association with heterozygous mutations [174, 175].

**Table 18.5** Autosomal recessive forms of hereditary spastic paraplegias (HSPs)

| SPG type | MIM | Locus | Gene/protein | Putative roles of the protein | Associated signs |
|---|---|---|---|---|---|
| *Pure forms* | | | | | |
| SPG5 | 270800 | 8p | *CYP7B1* | Cholesterol and neurosteroid metabolism | |
| SPG24 | 607584 | 13q | ? | ? | |
| SPG28 | 609340 | 14q | ? | ? | |
| SPG30 | 610357 | 2q | ? | ? | |
| *Complex forms* | | | | | |
| SPG7 | 602783 | 16q | *SPG7* /Paraplegin | Mitochondrial ATPase | Cerebellar signs. Polyneuropathy. Additional features: pes cavus, optic atrophy. |
| SPG14 | 605229 | 3q | ? | ? | Mental retardation, istal motor neuropathy, pes cavus. |
| SPG27 | 609041 | 10q | ? | ? | Cerebellar ataxia, neuropathy, mental retardation, facial and skeletal dysmorphia |
| SPG11 (AR-HSP-TCC) | 610844 | 15q | *KIAA1840* /Spatacsin | ? | Cognitive impairment, thin corpus callosum variably, upper extremity weakness, dysarthria, and nistagmus. |
| SPG15 (Kjelli syndrome) | 270700 | 14q | *ZFYVE26* /Spastizin | Endosomal trafficking | Pigmented maculopathy, distal amyotrophy, dysarthria, mental retardation, and further intellectual deterioration. |
| SPG20 (Troyer syndrome) | 275900 | 13q | *KIAA0610* /Spartin | Microtubule interaction, endosomal trafficking | Spastic tetraparesis, dysarthria, distal amyotrophy, short stature, learning difficulties. Mild cerebellar signs. |
| SPG21 (Mast syndrome) | 248900 | 15q | *ACP33* /Maspardin | Endosomal, trans-Golgi trafficking | Dementia, cerebellar and extrapyramidal signs, thin corpus callosum, and white matter abnormalities. Cognitive impairment in childhood. Gait disturbance in adolescence. |
| SPG23 (Lison syndrome) | 270750 | 1q | ? | ? | Skin pigment abnormality. |
| SPG25 | 608220 | 5q | ? | ? | Spinal disc herniations. |
| SPG26 | 609105 | 12cen | ? | ? | Progressive spastic paraparesis, dysarthria, distal amyotrophy, intellectual impairment. |
| SPG32 | 611251 | 14q | ? | ? | Ataxia, sensory neuropathy. |

**Table 18.6** X-linked forms of hereditary spastic paraplegias (HSPs)

| SPG type | MIM | Locus | Protein | Putative roles of the protein | Clinical features |
|---|---|---|---|---|---|
| *Complex forms* | | | | | |
| SPG1 | 312900 | Xq28 | L1CAM | Cell adhesion, neurite outgrowth, myelination | Congenital. Spastic paraplegia plus mental retardation and adducted thumbs. Allelic to other conditions: X-linked hydrocephalus, MASA syndrome, CRASH syndrome. |
| SPG2 | 312920 | Xq22 | PLP1 | Primary constituent of myelin | Patients may exhibit either severe infantile dysmyelination (Pelizeaus-Merzbacher syndrome) or slowly progressive spastic paraplegia (SPG2). Nystagmus, hypotonia, cognitive impairment, sevre spasticity and ataxia with onset in early childhood and shortened life span. |
| SPG16 | 300266 | Xq11.2 | ? | ? | Onset in early childhood. Facial hypotonia, strabismus and reduced vision, bowel dysfunction, skeletal abnormalities, mental retardation, aphasia, restlessness. |

All X-linked HSP forms (SPG1, SPG2 and SPG16) are complex. Table 18.6 shows the main clinical features of these three HSP forms. To date only two genes are known: *SPG1* that encodes L1CAM and *SPG2* that encodes the proteolipid protein 1 (PLP1). SPG1 and MASA (mental retardation, aphasia, shuffling gait, and adducted thumbs) syndrome are allelic disorders. SPG1 also presents with spasticity. Both conditions have altered the L1CAM protein. SPG2 and Pelizaeus-Merzbacher disease (PMD) are allelic disorders and both diseases are caused by mutations in the *PLP1* gene, which encodes one of the major components of myelin. Different types of mutations (point mutations, duplications, deletions) have been identified in the *PLP1* gene, which suggests that different molecular pathogenic mechanisms underlie these disorders. The number of reported families presenting with SPG1 or SPG2 is high (near 100 cases), although the number of families with spastic paraplegia is much more less [176]. Finally, the gene responsible for SPG16 remains unknown. Only one family has been reported with this phenotype [177].

### *18.2.3 Pathogenic Mechanisms*

The molecular mechanisms leading to axonal degeneration are probably as divers and complex as the genetics of HSPs. The identification of causative genes and insight into the functions of the proteins they encode has suggested that aberrant intracellular-trafficking dynamics by alterations of the Golgi apparatus, endosomes, or axonal transport seems to be a common process for the specific pattern of neurodegeneration related to HSPs [178]. Other processes have also been observed impaired in HSPs: mitochondrial dysfunction, myelination, cholesterol/neurosteroid metabolism, protein folding and axon guidance.

To date, 16 HSP genes have been identified and 9 of which encode for proteins belonging to pathways involved in intracellular trafficking (Tables 18.4 and 18.5). Spastin (SPG4), an adenosine triphosphate (ATPase), is involved in various cellular activities according to its three main domains: a microtubule interacting and endosomal trafficking domain (MIT), a second microtubule interacting domain, and the ATPase AAA domain [179]. Spastin interacts with microtubules and probably is implicated in microtubules dynamics since its overexpression results in a microtubule-disassembly phenotype. Impairment of its function leads to the abnormal accumulation of cellular organelles and cytoskeletal components causing axonal swellings in mice that are deficient in spastin [180]. Spastin interacts with two other HSP proteins, atlastin-1 (SPG3A) and protrudin (ZFYVE27/SPG33), suggesting that defects in any of these proteins might initiate the same pathophysiological process [181]. Protrudin is expressed in endoplasmic reticulum (ER) and endosomes, probably using its FYVE zinc-finger domain, which is known to bind phosphatidylinositol 3-phosphate from membranes during endosomal trafficking [182]. Atlastin (SPG3A) a dynamin-like large GTPase, is expressed in the ER, the Golgi, neuritis and growth cones, and affects the neurite outgrowth. Mutations in the GTPase domain interfered with the maturation of Golgi complexes by preventing the budding of vesicles from the ER, whereas mutations in other regions of the protein disrupted fission of endoplasmic reticulum-derived vesicles or their migration to their Golgi target [183].

Functions of several HSP proteins are not so clear, but their involvement in intracellular transport is suspected. Spartin (SPG20) like spastin, is an AAA protein with an MIT domain and has been implicated in the endocytosis and transport of the epidermal growth factor receptor [184]. Preliminary functional studies suggest that Maspardin (SPG21) has a role in vesicle-mediated trafficking and protein sorting within the cytoplasm [185]. NIPA1 (SPG6) is a neuron-specific transmembrane protein principally localized in the early endosomal compartment and on the plasma membrane, where it is thought to be a magnesium transporter [186]. The homologue in yeast of protein encode by *REEP1* (SPG31) interacts with Rab proteins and it is involved in the tubular morphology in the ER [159]. ZFYVE26 (SPG15), which encodes a zinc-finger protein with a FYVE domain that we named spastizin, colocalized partially with markers of endoplasmic reticulum and endosomes in cultured cells, suggesting a role in intracellular trafficking [173]. Finally, KIF5A (kinesin-1 motor protein) responsible for SPG10 is part of a hetero tetrameric motor protein complex involved in the transport of cargoes along microtubules in an anterograde direction. A mutated KIF5A protein has a lower affinity for microtubules and/or reduced gliding velocity of microtubule-dependent anterograde axonal transport [187].

Mytochondrial dysfunction is the second more frequent process that causes HSPs (Tables 18.4 and 18.5). REEP1 (SPG31) besides of being involved in intracellular trafficking has been found in mitochondria in a variety of cell types [159]. HSP60 (SPG13) was initially described as an important mitochondrial protein for folding key proteins after import into the mitochondria. However, HSP60 can also be cytosolic and binds to Bax suggesting a key regulatory role in apoptosis [188]. Paraplegin (SPG7) is highly homologous to members of the mitochondrial AAA protease family [189] and coassembles with homologous AFG3L2 in the mitochondrial inner membrane [190]. In paraplegin-deficient mice axonal swellings caused by massive accumulation of organelles and neurofilaments, similar to those observe in spastin-deficient mice [191]. Additionally, spartin (SPG20) also involved in intracellular trafficking, when mutated loss its interaction with mitochondria [192]. In conclusion, altered axonal trafficking and mitochondrial deficiencies may be intimately related.

Finally, two HSP proteins have particular functions. Seipin (SPG17), an ER resident membrane protein, is an N-glycosylated protein that is proteolytically cleaved into N- and C-terminal fragments and is polyubiquitinated. In culture cells, expression of mutant forms activates the unfolded protein response (UPR) pathway and induces ER stress-mediated cell death [193]. CYP7B1 (SPG5) is a member of the cytochrome P450 superfamily of monooxygenases, involved in the metabolism of cholesterol, neurosteroids, and other lipids [169]. The findings indicate a primary metabolic route for the modification of neurosteroids in the brain and a pivotal role of altered cholesterol metabolism in the pathogenesis of motor-neuron degenerative disease.

Regarding the X-linked HSP forms, two genes have been characterised, *L1CAM* (SPG1) and *PLP1* (SPG2) (Table 18.6). Defects on these genes lead to an abnormal development. L1CAM is a glycoprotein expressed during development on the surface of long axons and growth cones, including those of the corticospinal tract. L1CAM is critical for neuronal migration and differentiation. L1CAM knockout mice exhibit phenotypes resulting from impairment of axonal guidance [194, 195]. *PLP1* gene encodes one of the major components of myelin and mutations on it would alter myelination. Oligodendrocytes in PLP1 knockout mice differentiate normally and produce compacted myelin sheaths but develop axonal swelling followed by degeneration of long axons [196, 197].

**Acknowledgements** This work is supported by the Spanish Ministry of Science and Innovation and the Fondo de Investigación Sanitaria. The CIBER de Enfermedades Raras is an initiative of the Instituto de Salud Carlos III.

## References

1. Palau F, Espinós C. Autosomal recessive cerebellar ataxias. Orphanet J Rare Dis 2006;1:47.
2. Valente EM, Brancati F, Dallapiccola B. Genotypes and phenotypes of Joubert syndrome and related disorders. Eur J Med Genet 2008;51(1):1–23.
3. Saar K, Al-Gazali L, Sztriha L, et al. Homozygosity mapping in families with Joubert syndrome identifies a locus on chromosome 9q34.3 and evidence for genetic heterogeneity. Am J Hum Genet 1999;65(6):1666–1671.

4. Keeler LC, Marsh SE, Leeflang EP, et al. Linkage analysis in families with Joubert syndrome plus oculo-renal involvement identifies the CORS2 locus on chromosome 11p12-q13.3. Am J Hum Genet 2003;73(3):656–662.
5. Valente EM, Salpietro DC, Brancati F, et al. Description, nomenclature, and mapping of a novel cerebello-renal syndrome with the molar tooth malformation. Am J Hum Genet 2003;73(3):663–670.
6. Lagier-Tourenne C, Boltshauser E, Breivik N, et al. Homozygosity mapping of a third Joubert syndrome locus to 6q23. J Med Genet 2004;41(4):273–277.
7. Valente EM, Silhavy JL, Brancati F, et al. Mutations in CEP290, which encodes a centrosomal protein, cause pleiotropic forms of Joubert syndrome. Nat Genet 2006;38(6):623–625.
8. Cantagrel V, Silhavy JL, Bielas SL, et al. Mutations in the cilia gene ARL13B lead to the classical form of Joubert syndrome. Am J Hum Genet 2008;83(2):170–179.
9. Ouahchi K, Arita M, Kayden H, et al. Ataxia with isolated vitamin E deficiency is caused by mutations in the alpha-tocopherol transfer protein. Nat Genet 1995;9(2):141–145.
10. Cavalier L, Ouahchi K, Kayden HJ, et al. Ataxia with isolated vitamin E deficiency: heterogeneity of mutations and phenotypic variability in a large number of families. Am J Hum Genet 1998;62(2):301–310.
11. Gotoda T, Arita M, Arai H, et al. Adult-onset spinocerebellar dysfunction caused by a mutation in the gene for the alpha-tocopherol-transfer protein. N Engl J Med 1995;333(20): 1313–1318.
12. Yokota T, Shiojiri T, Gotoda T, et al. Friedreich-like ataxia with retinitis pigmentosa caused by the His101Gln mutation of the alpha-tocopherol transfer protein gene. Ann Neurol 1997;41(6):826–832.
13. Fernandez-Burriel M, Martinez-Rubio D, Lupo V, et al. A novel delins mutation in the alpha-TTP gene in a family segregating ataxia with isolated vitamin E deficiency. Pediatr Res 2008;64(3):262–264.
14. Koenig M. Ataxia with isolated vitamin E deficiency. In: Klockgether T , ed. Handbook of Ataxia Disorders. New York: Marcel Dekker, Inc. 2000:223–234.
15. Mariotti C, Gellera C, Rimoldi M, et al. Ataxia with isolated vitamin E deficiency: neurological phenotype, clinical follow-up and novel mutations in TTPA gene in Italian families. Neurol Sci 2004;25(3):130–137.
16. Wetterau JR, Aggerbeck LP, Bouma ME, et al. Absence of microsomal triglyceride transfer protein in individuals with abetalipoproteinemia. Science 1992;258(5084):999–1001.
17. Sharp D, Blinderman L, Combs KA, et al. Cloning and gene defects in microsomal triglyceride transfer protein associated with abetalipoproteinaemia. Nature 1993;3656441):65–69.
18. Shoulders CC, Brett DJ, Bayliss JD, et al. Abetalipoproteinemia is caused by defects of the gene encoding the 97 kDa subunit of a microsomal triglyceride transfer protein. Hum Mol Genet 1993;2(12):2109–2116.
19. Narcisi TM, Shoulders CC, Chester SA, et al. Mutations of the microsomal triglyceride-transfer-protein gene in abetalipoproteinemia. Am J Hum Genet 1995;57(6): 1298–1310.
20. Linton MF, Farese RV, Jr., Young SG. Familial hypobetalipoproteinemia. J Lipid Res 1993;34(4):521–541.
21. Ohashi K, Ishibashi S, Yamamoto M, et al. A truncated species of apolipoprotein B (B-38.7) in a patient with homozygous hypobetalipoproteinemia associated with diabetes mellitus. Arterioscler Thromb Vasc Biol 1998;18(8):1330–1334.
22. Gatti RA, Berkel I, Boder E, et al. Localization of an ataxia-telangiectasia gene to chromosome 11q22-23. Nature 1988;336(6199):577–580.
23. Gatti RA, Lange E, Rotman G, et al. Genetic haplotyping of ataxia-telangiectasia families localizes the major gene to an approximately 850 kb region on chromosome 11q23.1. Int J Radiat Biol 1994;66(6 Suppl):S57–S62.
24. Wright J, Teraoka S, Onengut S, et al. A high frequency of distinct ATM gene mutations in ataxia-telangiectasia. Am J Hum Genet 1996;59(4):839–846.

25. Becker-Catania SG, Chen G, Hwang MJ, et al. Ataxia-telangiectasia: phenotype/genotype studies of ATM protein expression, mutations, and radiosensitivity. Mol Genet Metab 2000;70(2):122–133.
26. Li A, Swift M. Mutations at the ataxia-telangiectasia locus and clinical phenotypes of A-T patients. Am J Med Genet 2000;92(3):170–177.
27. Lavin MF, Gueven N, Bottle S, et al. Current and potential therapeutic strategies for the treatment of ataxia-telangiectasia. Br Med Bull 2007;81–82:129–147.
28. Willems PJ, Van Roy BC, Kleijer WJ, et al. Atypical clinical presentation of ataxia telangiectasia. Am J Med Genet 1993;45(6):777–782.
29. Gilad S, Chessa L, Khosravi R, et al. Genotype-phenotype relationships in ataxia-telangiectasia and variants. Am J Hum Genet 1998;62(3):551–561.
30. Stewart GS, Maser RS, Stankovic T, et al. The DNA double-strand break repair gene hMRE11 is mutated in individuals with an ataxia-telangiectasia-like disorder. Cell 1999;99(6):577–587.
31. Savitsky K, Bar-Shira A, Gilad S, et al. A single ataxia telangiectasia gene with a product similar to PI-3 kinase. Science 1995;268(5218):1749–1753.
32. Chen G, Lee E. The product of the ATM gene is a 370-kDa nuclear phosphoprotein. J Biol Chem 1996;271(52):33693–33697.
33. Kastan MB, Zhan Q, el-Deiry WS, et al. A mammalian cell cycle checkpoint pathway utilizing p53 and GADD45 is defective in ataxia-telangiectasia. Cell 1992;71(4): 587–597.
34. Khanna KK, Lavin MF. Ionizing radiation and UV induction of p53 protein by different pathways in ataxia-telangiectasia cells. Oncogene 1993;8(12):3307–3312.
35. Williams RS, Williams JS, Tainer JA. Mre11-Rad50-Nbs1 is a keystone complex connecting DNA repair machinery, double-strand break signaling, and the chromatin template. Biochem Cell Biol 2007;85(4):509–520.
36. Klein C, Wenning GK, Quinn NP, et al. Ataxia without telangiectasia masquerading as benign hereditary chorea. Mov Disord 1996;11:217–220.
37. Tranchant C, Fleury M, Moreira MC, et al. Phenotypic variability of aprataxin gene mutations. Neurology 2003;60(5):868–870.
38. Barbot C, Coutinho P, Chorao R, et al. Recessive ataxia with ocular apraxia: review of 22 Portuguese patients. Arch Neurol 2001;58(2):201–205.
39. Moreira MC, Barbot C, Tachi N, et al. Homozygosity mapping of Portuguese and Japanese forms of ataxia-oculomotor apraxia to 9p13, and evidence for genetic heterogeneity. Am J Hum Genet 2001;68(2):501–508.
40. Whitehouse CJ, Taylor RM, Thistlethwaite A, et al. XRCC1 stimulates human polynucleotide kinase activity at damaged DNA termini and accelerates DNA single-strand break repair. Cell 2001;104(1):107–117.
41. Le Ber I, Bouslam N, Rivaud-Pechoux S, et al. Frequency and phenotypic spectrum of ataxia with oculomotor apraxia 2: a clinical and genetic study in 18 patients. Brain 2004;127(Pt 4):759–767.
42. Moreira MC, Klur S, Watanabe M, et al. Senataxin, the ortholog of a yeast RNA helicase, is mutant in ataxia-ocular apraxia 2. Nat Genet 2004;36(3):225–227.
43. Suraweera A, Becherel OJ, Chen P, et al. Senataxin, defective in ataxia oculomotor apraxia type 2, is involved in the defense against oxidative DNA damage. J Cell Biol 2007;177(6):969–979.
44. Chen YZ, Bennett CL, Huynh HM, et al. DNA/RNA helicase gene mutations in a form of juvenile amyotrophic lateral sclerosis (ALS4). Am J Hum Genet 2004;74(6):1128–1135.
45. Gueven N, Chen P, Nakamura J, et al. A subgroup of spinocerebellar ataxias defective in DNA damage responses. Neuroscience 2007;145(4):1418–1425.
46. Takashima H, Boerkoel CF, John J, et al. Mutation of TDP1, encoding a topoisomerase I-dependent DNA damage repair enzyme, in spinocerebellar ataxia with axonal neuropathy. Nat Genet 2002;32(2):267–272.

47. Caldecott KW. DNA single-strand break repair and spinocerebellar ataxia. Cell 2003; 112:7–10.
48. Wong LJ, Naviaux RK, Brunetti-Pierri N, et al. Molecular and clinical genetics of mitochondrial diseases due to POLG mutations. Hum Mutat 2008;29(9):E150–E72.
49. Hakonen AH, Heiskanen S, Juvonen V, et al. Mitochondrial DNA polymerase W748S mutation: a common cause of autosomal recessive ataxia with ancient European origin. Am J Hum Genet 2005;77(3):430–441.
50. Tzoulis C, Engelsen BA, Telstad W, et al. The spectrum of clinical disease caused by the A467T and W748S POLG mutations: a study of 26 cases. Brain 2006;129(Pt 7):1685–1692.
51. Lonnqvist T, Paetau A, Nikali K, et al. Infantile onset spinocerebellar ataxia with sensory neuropathy (IOSCA): neuropathological features. J Neurol Sci 1998;161(1):57–65.
52. Nikali K, Suomalainen A, Saharinen J, et al. Infantile onset spinocerebellar ataxia is caused by recessive mutations in mitochondrial proteins Twinkle and Twinky. Hum Mol Genet 2005;14(20):2981–2990.
53. Naviaux RK, Nguyen KV. POLG mutations associated with Alpers syndrome and mitochondrial DNA depletion. Ann Neurol 2005;58(3):491.
54. Hakonen AH, Goffart S, Marjavaara S, et al. Infantile-onset spinocerebellar ataxia and mitochondrial recessive ataxia syndrome are associated with neuronal complex I defect and mtDNA depletion. Hum Mol Genet 2008;17(23):3822–3835.
55. Harding AE. The clinical features and classification of the late onset autosomal dominant cerebellar ataxias. A study of 11 families, including descendants of the 'the Drew family of Walworth'. Brain 1982;105(Pt 1):1–28.
56. Dubourg O, Durr A, Cancel G, et al. Analysis of the SCA1 CAG repeat in a large number of families with dominant ataxia: clinical and molecular correlations. Ann Neurol 1995;37(2):176–180.
57. Burk K, Abele M, Fetter M, et al. Autosomal dominant cerebellar ataxia type I clinical features and MRI in families with SCA1, SCA2 and SCA3. Brain 1996;119 ( Pt 5): 1497–1505.
58. Schols L, Amoiridis G, Buttner T, et al. Autosomal dominant cerebellar ataxia: phenotypic differences in genetically defined subtypes? Ann Neurol 1997;42(6):924–932.
59. Jackson JF, Currier RD, Terasaki PI, et al. Spinocerebellar ataxia and HLA linkage: risk prediction by HLA typing. N Engl J Med 1977;296(20):1138–1141.
60. Klockgether T. The clinical diagnosis of autosomal dominant spinocerebellar ataxias. Cerebellum 2008; 7(2):101–105.
61. van de Warrenburg BP, Sinke RJ, Verschuuren-Bemelmans CC, et al. Spinocerebellar ataxias in the Netherlands: prevalence and age at onset variance analysis. Neurology 2002;58(5):702–708.
62. Schols L. Autosomal dominant spinocerebellar ataxias. Orphanet Encyclopedia, http://www.orpha.net/data/patho/GB/uk-SCA.pdf 2003.
63. Leggo J, Dalton A, Morrison PJ, et al. Analysis of spinocerebellar ataxia types 1, 2, 3, and 6, dentatorubral-pallidoluysian atrophy, and Friedreich's ataxia genes in spinocerebellar ataxia patients in the UK. J Med Genet 1997;34(12):982–985.
64. Watanabe H, Tanaka F, Matsumoto M, et al. Frequency analysis of autosomal dominant cerebellar ataxias in Japanese patients and clinical characterization of spinocerebellar ataxia type 6. Clin Genet 1998;53(1):13–19.
65. Kim JY, Park SS, Joo SI, et al. Molecular analysis of Spinocerebellar ataxias in Koreans: frequencies and reference ranges of SCA1, SCA2, SCA3, SCA6, and SCA7. Mol Cells 2001;12(3):336–341.
66. Silveira I, Miranda C, Guimaraes L, et al. Trinucleotide repeats in 202 families with ataxia: a small expanded (CAG)n allele at the SCA17 locus. Arch Neurol 2002;59(4): 623–629.
67. Ito H, Kawakami H, Wate R, et al. Clinicopathologic investigation of a family with expanded SCA8 CTA/CTG repeats. Neurology 2006;67(8):1479–1481.

68. Matsuura T, Yamagata T, Burgess DL, et al. Large expansion of the ATTCT pentanucleotide repeat in spinocerebellar ataxia type 10. Nat Genet 2000;26(2):191–194.
69. Holmes SE, O'Hearn EE, McInnis MG, et al. Expansion of a novel CAG trinucleotide repeat in the 5' region of PPP2R2B is associated with SCA12. Nat Genet 1999;23(4):391–392.
70. Ashley CT, Jr., Warren ST. Trinucleotide repeat expansion and human disease. Annu Rev Genet 1995;29:703–728.
71. Cummings CJ, Zoghbi HY. Fourteen and counting: unraveling trinucleotide repeat diseases. Hum Mol Genet 2000;9(6):909–916.
72. Richards RI, Holman K, Friend K, et al. Evidence of founder chromosomes in fragile X syndrome. Nat Genet 1992;1(4):257–260.
73. Goldberg YP, Kremer B, Andrew SE, et al. Molecular analysis of new mutations for Huntington's disease: intermediate alleles and sex of origin effects. Nat Genet 1993;5(2):174–179.
74. Imbert G, Kretz C, Johnson K, et al. Origin of the expansion mutation in myotonic dystrophy. Nat Genet 1993;4(1):72–76.
75. Myers RH, MacDonald ME, Koroshetz WJ, et al. De novo expansion of a (CAG)n repeat in sporadic Huntington's disease. Nat Genet 1993;5(2):168–173.
76. Kunst CB, Warren ST. Cryptic and polar variation of the fragile X repeat could result in predisposing normal alleles. Cell 1994;77(6):853–861.
77. Paulson HL, Fischbeck KH. Trinucleotide repeats in neurogenetic disorders. Annu Rev Neurosci 1996;19:79–107.
78. Benton CS, de Silva R, Rutledge SL, et al. Molecular and clinical studies in SCA-7 define a broad clinical spectrum and the infantile phenotype. Neurology 1998;51(4):1081–1086.
79. Ranum LP, Moseley ML, Leppet MF. Massive CTG expansions and deletions may reduce penetrance of spinocerebellar ataxia type 8. Am J Hum Genet 1999;S65:A466(2648).
80. Zhuchenko O, Bailey J, Bonnen P, et al. Autosomal dominant cerebellar ataxia (SCA6) associated with small polyglutamine expansions in the alpha 1A-voltage-dependent calcium channel. Nat Genet 1997;15(1):62–69.
81. Koide R, Kobayashi S, Shimohata T, et al. A neurological disease caused by an expanded CAG trinucleotide repeat in the TATA-binding protein gene: a new polyglutamine disease? Hum Mol Genet 1999;8(11):2047–2053.
82. Nakamura K, Jeong SY, Uchihara T, et al. SCA17, a novel autosomal dominant cerebellar ataxia caused by an expanded polyglutamine in TATA-binding protein. Hum Mol Genet 2001;10(14):1441–1448.
83. Zoghbi HY, Orr HT. Glutamine repeats and neurodegeneration. Annu Rev Neurosci 2000;23:217–247.
84. Ross CA, Poirier MA. Protein aggregation and neurodegenerative disease. Nat Med 2004;10 Suppl:10–17.
85. Cummings CJ, Mancini MA, Antalffy B, et al. Chaperone suppression of aggregation and altered subcellular proteasome localization imply protein misfolding in SCA1. Nat Genet 1998;19(2):148–154.
86. McCampbell A, Taylor JP, Taye AA, et al. CREB-binding protein sequestration by expanded polyglutamine. Hum Mol Genet 2000;9 (14):2197–2202.
87. Schmidt T, Lindenberg KS, Krebs A, et al. Protein surveillance machinery in brains with spinocerebellar ataxia type 3: redistribution and differential recruitment of 26S proteasome subunits and chaperones to neuronal intranuclear inclusions. Ann Neurol 2002;51(3): 302–310.
88. Klement IA, Skinner PJ, Kaytor MD, et al. Ataxin-1 nuclear localization and aggregation: role in polyglutamine-induced disease in SCA1 transgenic mice. Cell 1998;95(1): 41–53.
89. Slow EJ, Graham RK, Osmand AP, et al. Absence of behavioral abnormalities and neurodegeneration in vivo despite widespread neuronal huntingtin inclusions. Proc Natl Acad Sci U S A 2005;102(32):11402–11407.

90. Saudou F, Finkbeiner S, Devys D, et al. Huntingtin acts in the nucleus to induce apoptosis but death does not correlate with the formation of intranuclear inclusions. Cell 1998;95(1): 55–66.
91. Arrasate M, Mitra S, Schweitzer ES, et al. Inclusion body formation reduces levels of mutant huntingtin and the risk of neuronal death. Nature 2004;431(7010):805–810.
92. Kayed R, Head E, Thompson JL, et al. Common structure of soluble amyloid oligomers implies common mechanism of pathogenesis. Science 2003;300 (5618):486–489.
93. Schaffar G, Breuer P, Boteva R, et al. Cellular toxicity of polyglutamine expansion proteins: mechanism of transcription factor deactivation. Mol Cell 2004;15(1):95–105.
94. Wacker JL, Zareie MH, Fong H, et al. Hsp70 and Hsp40 attenuate formation of spherical and annular polyglutamine oligomers by partitioning monomer. Nat Struct Mol Biol 2004;11(12):1215–1222.
95. Behrends C, Langer CA, Boteva R, et al. Chaperonin TRiC promotes the assembly of polyQ expansion proteins into nontoxic oligomers. Mol Cell 2006;23(6):887–897.
96. Haass C, Selkoe DJ. Soluble protein oligomers in neurodegeneration: lessons from the Alzheimer's amyloid beta-peptide. Nat Rev Mol Cell Biol 2007;8(2):101–112.
97. Cummings CJ, Reinstein E, Sun Y, et al. Mutation of the E6-AP ubiquitin ligase reduces nuclear inclusion frequency while accelerating polyglutamine-induced pathology in SCA1 mice. Neuron 1999;24(4):879–892.
98. Matilla A, Gorbea C, Einum DD, et al. Association of ataxin-7 with the proteasome subunit S4 of the 19S regulatory complex. Hum Mol Genet 2001;10(24):2821–2831.
99. Chai Y, Berke SS, Cohen RE, et al. Poly-ubiquitin binding by the polyglutamine disease protein ataxin-3 links its normal function to protein surveillance pathways. J Biol Chem 2004;279(5):3605–3611.
100. Lam YC, Bowman AB, Jafar-Nejad P, et al. ATAXIN-1 interacts with the repressor Capicua in its native complex to cause SCA1 neuropathology. Cell 2006;127(7):1335–1347.
101. Bowman AB, Lam YC, Jafar-Nejad P, et al. Duplication of Atxnll suppresses SCA1 neuropathology by decreasing incorporation of polyglutamine-expanded ataxin-1 into native complexes. Nat Genet 2007;39(3):373–379.
102. Warrick JM, Morabito LM, Bilen J, et al. Ataxin-3 suppresses polyglutamine neurodegeneration in Drosophila by a ubiquitin-associated mechanism. Mol Cell 2005;18(1):37–48.
103. Stenoien DL, Mielke M, Mancini MA. Intranuclear ataxin1 inclusions contain both fast- and slow-exchanging components. Nat Cell Biol 2002;4(10):806–810.
104. Serra HG, Duvick L, Zu T, et al. RORalpha-mediated Purkinje cell development determines disease severity in adult SCA1 mice. Cell 2006;127(4):697–708.
105. Helmlinger D, Hardy S, Sasorith S, et al. Ataxin-7 is a subunit of GCN5 histone acetyltransferase-containing complexes. Hum Mol Genet 2004;13(12):1257–1265.
106. Palhan VB, Chen S, Peng GH, et al. Polyglutamine-expanded ataxin-7 inhibits STAGA histone acetyltransferase activity to produce retinal degeneration. Proc Natl Acad Sci U S A 2005;102(24):8472–8477.
107. La Spada AR, Fu YH, Sopher BL, et al. Polyglutamine-expanded ataxin-7 antagonizes CRX function and induces cone-rod dystrophy in a mouse model of SCA7. Neuron 2001;31(6):913–927.
108. Koob MD, Moseley ML, Schut LJ, et al. An untranslated CTG expansion causes a novel form of spinocerebellar ataxia (SCA8). Nat Genet 1999;21(4):379–384.
109. Schols L, Bauer I, Zuhlke C, et al. Do CTG expansions at the SCA8 locus cause ataxia? Ann Neurol 2003;54(1):110–115.
110. Moseley ML, Zu T, Ikeda Y, et al. Bidirectional expression of CUG and CAG expansion transcripts and intranuclear polyglutamine inclusions in spinocerebellar ataxia type 8. Nat Genet 2006;38(7):758–769.
111. Ikeda Y, Dalton JC, Moseley ML, et al. Spinocerebellar ataxia type 8: molecular genetic comparisons and haplotype analysis of 37 families with ataxia. Am J Hum Genet 2004;75(1):3–16.

112. Ikeda Y, Daughters RS, Ranum LP. Bidirectional expression of the SCA8 expansion mutation: One mutation, two genes. Cerebellum 2008;7(2):150–158.
113. Rasmussen A, Matsuura T, Ruano L, et al. Clinical and genetic analysis of four Mexican families with spinocerebellar ataxia type 10. Ann Neurol 2001;50(2):234–239.
114. Teive HA, Roa BB, Raskin S, et al. Clinical phenotype of Brazilian families with spinocerebellar ataxia 10. Neurology 2004;63(8):1509–1512.
115. Zu L, Figueroa KP, Grewal R, et al. Mapping of a new autosomal dominant spinocerebellar ataxia to chromosome 22. Am J Hum Genet 1999;64(2):594–599.
116. Lin X, Ashizawa T. Recent progress in spinocerebellar ataxia type-10 (SCA10). Cerebellum 2005;4(1):37–42.
117. Coates JC. Armadillo repeat proteins: beyond the animal kingdom. Trends Cell Biol 2003;13(9):463–471.
118. Waragai M, Nagamitsu S, Xu W, et al. Ataxin 10 induces neuritogenesis via interaction with G-protein beta2 subunit. J Neurosci Res 2006;83(7):1170–1178.
119. Wakamiya M, Matsuura T, Liu Y, et al. The role of ataxin 10 in the pathogenesis of spinocerebellar ataxia type 10. Neurology 2006;67(4):607–613.
120. Sontag E. Protein phosphatase 2A: the Trojan Horse of cellular signaling. Cell Signal 2001;13(1):7–16.
121. Chen DH, Brkanac Z, Verlinde CL, et al. Missense mutations in the regulatory domain of PKC gamma: a new mechanism for dominant nonepisodic cerebellar ataxia. Am J Hum Genet 2003;72(4):839–849.
122. Chen HK, Fernandez-Funez P, Acevedo SF, et al. Interaction of Akt-phosphorylated ataxin-1 with 14-3-3 mediates neurodegeneration in spinocerebellar ataxia type 1. Cell 2003;113(4):457–468.
123. Lim J, Lu KP. Pinning down phosphorylated tau and tauopathies. Biochim Biophys Acta 2005;1739(2-3):311–322.
124. Ikeda Y, Dick KA, Weatherspoon MR, et al. Spectrin mutations cause spinocerebellar ataxia type 5. Nat Genet 2006;38(2):184–190.
125. Gold DA, Baek SH, Schork NJ, et al. RORalpha coordinates reciprocal signaling in cerebellar development through sonic hedgehog and calcium-dependent pathways. Neuron 2003;40(6):1119–1131.
126. Waters MF, Minassian NA, Stevanin G, et al. Mutations in voltage-gated potassium channel KCNC3 cause degenerative and developmental central nervous system phenotypes. Nat Genet 2006;38(4):447–451.
127. Waters MF, Pulst SM. Sca13. Cerebellum 2008;7(2):165–169.
128. Yabe I, Sasaki H, Chen DH, et al. Spinocerebellar ataxia type 14 caused by a mutation in protein kinase C gamma. Arch Neurol 2003;60(12):1749–1751.
129. Schrenk K, Kapfhammer JP, Metzger F. Altered dendritic development of cerebellar Purkinje cells in slice cultures from protein kinase Cgamma-deficient mice. Neuroscience 2002;110(4):675–689.
130. Verbeek DS, Goedhart J, Bruinsma L, et al. PKC gamma mutations in spinocerebellar ataxia type 14 affect C1 domain accessibility and kinase activity leading to aberrant MAPK signaling. J Cell Sci 2008;121(Pt 14):2339–2349.
131. van Swieten JC, Brusse E, de Graaf BM, et al. A mutation in the fibroblast growth factor 14 gene is associated with autosomal dominant cerebellar ataxia [corrected]. Am J Hum Genet 2003;72(1):191–199.
132. Xiao M, Xu L, Laezza F, et al. Impaired hippocampal synaptic transmission and plasticity in mice lacking fibroblast growth factor 14. Mol Cell Neurosci 2007;34(3): 366–377.
133. Wozniak DF, Xiao M, Xu L, et al. Impaired spatial learning and defective theta burst induced LTP in mice lacking fibroblast growth factor 14. Neurobiol Dis 2007;26(1):14–26.
134. Owada K, Ishikawa K, Toru S, et al. A clinical, genetic, and neuropathologic study in a family with 16q-linked ADCA type III. Neurology 2005;65(4):629–632.

135. Ouyang Y, Sakoe K, Shimazaki H, et al. 16q-linked autosomal dominant cerebellar ataxia: a clinical and genetic study. J Neurol Sci 2006;247(2):180–186.
136. Basri R, Yabe I, Soma H, et al. Spectrum and prevalence of autosomal dominant spinocerebellar ataxia in Hokkaido, the northern island of Japan: a study of 113 Japanese families. J Hum Genet 2007;52(10):848–855.
137. Ohata T, Yoshida K, Sakai H, et al. A -16C>T substitution in the 5' UTR of the puratrophin-1 gene is prevalent in autosomal dominant cerebellar ataxia in Nagano. J Hum Genet 2006;51(5):461–466.
138. Amino T, Ishikawa K, Toru S, et al. Redefining the disease locus of 16q22.1-linked autosomal dominant cerebellar ataxia. J Hum Genet 2007;52(8):643–649.
139. Flanigan K, Gardner K, Alderson K, et al. Autosomal dominant spinocerebellar ataxia with sensory axonal neuropathy (SCA4): clinical description and genetic localization to chromosome 16q22.1. Am J Hum Genet 1996;59(2):392–399.
140. Hellenbroich Y, Bubel S, Pawlack H, et al. Refinement of the spinocerebellar ataxia type 4 locus in a large German family and exclusion of CAG repeat expansions in this region. J Neurol 2003;250(6):668–671.
141. Hellenbroich Y, Bernard V, Zuhlke C. Spinocerebellar ataxia type 4 and 16q22.1-linked Japanese ataxia are not allelic. J Neurol 2008;255(4):612–613.
142. Shokeir MHK. X-linked cerebellar ataxia. Clin Genet 1970;1:225–231.
143. Illarioshkin SN, Tanaka H, Markova ED, et al. X-linked nonprogressive congenital cerebellar hypoplasia: clinical description and mapping to chromosome Xq. Ann Neurol 1996;40(1):75–83.
144. Bertini E, des Portes V, Zanni G, et al. X-linked congenital ataxia: a clinical and genetic study. Am J Med Genet 2000;92(1):53–56.
145. Malamud N, Cohen P. Unusual form of cerebellar ataxia with sex-linked inheritance. Neurology 1958;8(4):261–266.
146. Schmidley JW, Levinsohn MW, Manetto V. Infantile X-linked ataxia and deafness: a new clinicopathologic entity? Neurology 1987;37(8):1344–1349.
147. Farlow MR, DeMyer W, Dlouhy SR, et al. X-linked recessive inheritance of ataxia and adult-onset dementia: clinical features and preliminary linkage analysis. Neurology 1987;37(4):602–607.
148. Zanni G, Bertini E, Bellcross C, et al. X-linked congenital ataxia: a new locus maps to Xq25-q27.1. Am J Med Genet A 2008;146A(5):593–600.
149. Polo JM, Calleja J, Combarros O, et al. Hereditary ataxias and paraplegias in Cantabria, Spain. An epidemiological and clinical study. Brain 1991;114(Pt 2):855–866.
150. Filla A, De Michele G, Marconi R, et al. Prevalence of hereditary ataxias and spastic paraplegias in Molise, a region of Italy. J Neurol 1992;239(6):351–353.
151. Leone M, Bottacchi E, D'Alessandro G, et al. Hereditary ataxias and paraplegias in Valle d'Aosta, Italy: a study of prevalence and disability. Acta Neurol Scand 1995;91(3): 183–187.
152. Silva MC, Coutinho P, Pinheiro CD, et al. Hereditary ataxias and spastic paraplegias: methodological aspects of a prevalence study in Portugal. J Clin Epidemiol 1997;50(12):1377–1384.
153. Harding AE. Hereditary "pure" spastic paraplegia: a clinical and genetic study of 22 families. J Neurol Neurosurg Psychiatry 1981;44(10):871–883.
154. Harding AE. Classification of the hereditary ataxias and paraplegias. Lancet 1983;1(8334):1151–1155.
155. McMonagle P, Webb S, Hutchinson M. The prevalence of "pure" autosomal dominant hereditary spastic paraparesis in the island of Ireland. J Neurol Neurosurg Psychiatry 2002;72(1):43–46.
156. Coutinho P, Barros J, Zemmouri R, et al. Clinical heterogeneity of autosomal recessive spastic paraplegias: analysis of 106 patients in 46 families. Arch Neurol 1999;56(8): 943–949.

157. Boukhris A, Stevanin G, Feki I, et al. Hereditary spastic paraplegia with mental impairment and thin corpus callosum in Tunisia: SPG11, SPG15, and further genetic heterogeneity. Arch Neurol 2008;65(3):393–402.
158. Depienne C, Fedirko E, Forlani S, et al. Exon deletions of SPG4 are a frequent cause of hereditary spastic paraplegia. J Med Genet 2007;44(4):281–284.
159. Zuchner S, Wang G, Tran-Viet KN, et al. Mutations in the novel mitochondrial protein REEP1 cause hereditary spastic paraplegia type 31. Am J Hum Genet 2006;79(2):365–369.
160. Rainier S, Chai JH, Tokarz D, et al. NIPA1 gene mutations cause autosomal dominant hereditary spastic paraplegia (SPG6). Am J Hum Genet 2003;73(4):967–971.
161. Reed JA, Wilkinson PA, Patel H, et al. A novel NIPA1 mutation associated with a pure form of autosomal dominant hereditary spastic paraplegia. Neurogenetics 2005;6(2):79–84.
162. Valdmanis PN, Meijer IA, Reynolds A, et al. Mutations in the KIAA0196 gene at the SPG8 locus cause hereditary spastic paraplegia. Am J Hum Genet 2007;80(1):152–161.
163. Reid E, Kloos M, Ashley-Koch A, et al. A kinesin heavy chain (KIF5A) mutation in hereditary spastic paraplegia (SPG10). Am J Hum Genet 2002;71(5):1189–1194.
164. Fichera M, Lo Giudice M, Falco M, et al. Evidence of kinesin heavy chain (KIF5A) involvement in pure hereditary spastic paraplegia. Neurology 2004;63(6):1108–1110.
165. Blair MA, Ma S, Hedera P. Mutation in KIF5A can also cause adult-onset hereditary spastic paraplegia. Neurogenetics 2006;7(1):47–50.
166. Lo Giudice MN, Falco M, Sturnio M, Calzolari E, Di Benedetto D, Fichera M. A missense mutation in the coiled-coil domain of the KIF5A gene and late-onset hereditary spastic paraplegia. Arch Neurol 2006;63:284–287.
167. Hansen JJ, Durr A, Cournu-Rebeix I, et al. Hereditary spastic paraplegia SPG13 is associated with a mutation in the gene encoding the mitochondrial chaperonin Hsp60. Am J Hum Genet 2002;70(5):1328–1332.
168. Klebe S, Durr A, Bouslam N, et al. Spastic paraplegia 5: Locus refinement, candidate gene analysis and clinical description. Am J Med Genet B Neuropsychiatr Genet 2007;144B(7):854–861.
169. Tsaousidou MK, Ouahchi K, Warner TT, et al. Sequence alterations within CYP7B1 implicate defective cholesterol homeostasis in motor-neuron degeneration. Am J Hum Genet 2008;82(2):510–515.
170. Goizet C, Boukhris A, Mundwiller E, et al. Complicated forms of autosomal dominant hereditary spastic paraplegia are frequent in SPG10. Hum Mutat 2009;30(2):376–385.
171. Stevanin G, Santorelli FM, Azzedine H, et al. Mutations in SPG11, encoding spatacsin, are a major cause of spastic paraplegia with thin corpus callosum. Nat Genet 2007;39(3):366–372.
172. Denora PS, Schlesinger D, Casali C, et al. Screening of ARHSP-TCC patients expands the spectrum of SPG11 mutations and includes a large scale gene deletion. Hum Mutat 2009; 30(3):500–519.
173. Hanein S, Martin E, Boukhris A, et al. Identification of the SPG15 gene, encoding spastizin, as a frequent cause of complicated autosomal-recessive spastic paraplegia, including Kjellin syndrome. Am J Hum Genet 2008;82(4):992–1002.
174. Elleuch N, Depienne C, Benomar A, et al. Mutation analysis of the paraplegin gene (SPG7) in patients with hereditary spastic paraplegia. Neurology 2006;66(5):654–659.
175. Arnoldi A, Tonelli A, Crippa F, et al. A clinical, genetic, and biochemical characterization of SPG7 mutations in a large cohort of patients with hereditary spastic paraplegia. Hum Mutat 2008;29(4):522–531.
176. Depienne C, Stevanin G, Brice A, et al. Hereditary spastic paraplegias: an update. Curr Opin Neurol 2007;20(6):674–680.
177. Tamagaki A, Shima M, Tomita R, et al. Segregation of a pure form of spastic paraplegia and NOR insertion into Xq11.2. Am J Med Genet 2000;94(1):5–8.
178. Crosby AH, Proukakis C. Is the transportation highway the right road for hereditary spastic paraplegia? Am J Hum Genet 2002;71(5):1009–1016.

179. White SR, Evans KJ, Lary J, et al. Recognition of C-terminal amino acids in tubulin by pore loops in Spastin is important for microtubule severing. J Cell Biol 2007;176(7):995–1005.
180. Tarrade A, Fassier C, Courageot S, et al. A mutation of spastin is responsible for swellings and impairment of transport in a region of axon characterized by changes in microtubule composition. Hum Mol Genet 2006;15(24):3544–3558.
181. Evans K, Keller C, Pavur K, et al. Interaction of two hereditary spastic paraplegia gene products, spastin and atlastin, suggests a common pathway for axonal maintenance. Proc Natl Acad Sci U S A 2006;103(28):10666–10671.
182. Mannan AU, Krawen P, Sauter SM, et al. ZFYVE27 (SPG33), a novel spastin-binding protein, is mutated in hereditary spastic paraplegia. Am J Hum Genet 2006;79(2):351–357.
183. Namekawa M, Muriel MP, Janer A, et al. Mutations in the SPG3A gene encoding the GTPase atlastin interfere with vesicle trafficking in the ER/Golgi interface and Golgi morphogenesis. Mol Cell Neurosci 2007;35(1):1–13.
184. Bakowska JC, Jupille H, Fatheddin P, et al. Troyer syndrome protein spartin is mono-ubiquitinated and functions in EGF receptor trafficking. Mol Biol Cell 2007;18(5): 1683–1692.
185. Simpson MA, Cross H, Proukakis C, et al. Maspardin is mutated in mast syndrome, a complicated form of hereditary spastic paraplegia associated with dementia. Am J Hum Genet 2003;73(5):1147–1156.
186. Goytain A, Hines RM, El-Husseini A, et al. NIPA1(SPG6), the basis for autosomal dominant form of hereditary spastic paraplegia, encodes a functional Mg2+ transporter. J Biol Chem 2007;282(11):8060–8068.
187. Ebbing B, Mann K, Starosta A, et al. Effect of spastic paraplegia mutations in KIF5A kinesin on transport activity. Hum Mol Genet 2008;17(9):1245–1252.
188. Gupta S, Knowlton AA. HSP60, Bax, apoptosis and the heart. J Cell Mol Med 2005;9(1): 51–58.
189. Casari G, De Fusco M, Ciarmatori S, et al. Spastic paraplegia and OXPHOS impairment caused by mutations in paraplegin, a nuclear-encoded mitochondrial metalloprotease. Cell 1998;93(6):973–983.
190. Atorino L, Silvestri L, Koppen M, et al. Loss of m-AAA protease in mitochondria causes complex I deficiency and increased sensitivity to oxidative stress in hereditary spastic paraplegia. J Cell Biol 2003;163(4):777–787.
191. Ferreirinha F, Quattrini A, Pirozzi M, et al. Axonal degeneration in paraplegin-deficient mice is associated with abnormal mitochondria and impairment of axonal transport. J Clin Invest 2004;113(2):231–242.
192. Lu J, Rashid F, Byrne PC. The hereditary spastic paraplegia protein spartin localises to mitochondria. J Neurochem 2006;98(6):1908–1919.
193. Ito D, Suzuki N. Seipinopathy: a novel endoplasmic reticulum stress-associated disease. Brain 2009;132:8–15.
194. Rosenthal A, Jouet M, Kenwrick S. Aberrant splicing of neural cell adhesion molecule L1 mRNA in a family with X-linked hydrocephalus. Nat Genet 1992;2(2):107–112.
195. Dahme M, Bartsch U, Martini R, et al. Disruption of the mouse L1 gene leads to malformations of the nervous system. Nat Genet 1997;17 (3):346–349.
196. Boison D, Stoffel W. Disruption of the compacted myelin sheath of axons of the central nervous system in proteolipid protein-deficient mice. Proc Natl Acad Sci U S A 1994;91(24):11709–11713.
197. Klugmann M, Schwab MH, Puhlhofer A, et al. Assembly of CNS myelin in the absence of proteolipid protein. Neuron 1997;18(1):59–70.

# Index



C. Espinós et al. (eds.), *Inherited Neuromuscular Diseases*, Advances in Experimental Medicine and Biology 652, DOI 10.1007/978-90-481-2813-6_BM2,

## C



## D